T0295749

Internet of Multimedia Things (IoMT)

Techniques and Applications

Internet of Multimedia Things (IoMT)

Techniques and Applications

Edited by

Shailendra Shukla
MNNIT Allahabad, Allahabad, India

Amit Kumar Singh
NIT Patna, Patna, India

Gautam Srivastava
Brandon University, Brandon, MB, Canada

Series Editor Fatos Xhafa
Universitat Politècnica de Catalunya, Spain

Academic Press is an imprint of Elsevier
125 London Wall, London EC2Y 5AS, United Kingdom
525 B Street, Suite 1650, San Diego, CA 92101, United States
50 Hampshire Street, 5th Floor, Cambridge, MA 02139, United States
The Boulevard, Langford Lane, Kidlington, Oxford OX5 1GB, United Kingdom

ISBN: 978-0-323-85845-8

For information on all Academic Press publications
visit our website at https://www.elsevier.com/books-and-journals

Publisher: Mara Conner
Editorial Project Manager: Ivy Dawn Torre
Production Project Manager: Sojan P. Pazhayattil
Designer: Vicky Pearson Esser

Typeset by VTeX

Contents

Contributors

Chady Abou Jaoude
Antonine University, Faculty of Engineering – TICKET Lab, Beirut, Lebanon

Gaurav Bhatnagar
Department of Mathematics, Indian Institute of Technology Jodhpur, India

Zahi Al Chami
Antonine University, Faculty of Engineering – TICKET Lab, Beirut, Lebanon
Universite de Pau et des Pays de l'Adour, E2S UPPA, LIUPPA, Anglet, France

Chiranjoy Chattopadhyay
Department of Computer Science and Engineering, Indian Institute of Technology Jodhpur, India

Richard Chbeir
Universite de Pau et des Pays de l'Adour, E2S UPPA, LIUPPA, Anglet, France

Rupesh Kumar Dewang
Department of Computer Science & Engineering, Motilal Nehru National Institute of Technology Allahabad, Prayagraj, India

Vivek K. Dwivedi
Department of Electronics and Communication Engineering, Jaypee Institute of Information Technology, Noida, India

Shreya Goyal
Department of Computer Science and Engineering, Indian Institute of Technology Jodhpur, India

Puneet Kumar Jain
Department of Computer Science and Engineering, National Institute of Technology Rourkela, Odisha, India

Bailey Janeczko
Department of Math and Computer Science, Brandon University, Brandon, MB, Canada

Niharika Keshari
Department of Computer Science & Engineering, Motilal Nehru National Institute of Technology Allahabad, Prayagraj, India

Prateek Ishwar Khade
Department of Computer Science & Information Systems, Birla Institute of Technology and Science, Pilani, Jhunjhunu, Rajasthan, India

Alok Kumar
Department of Electronics and Communication Engineering, Jaypee University of Information Technology, Solan, India

Om Prakash Mahela
Power System Planning Division, Rajasthan Rajya Vidyut Prasaran Nigam Ltd., Jaipur, India

Ashish Kumar Maurya
Department of Computer Science & Engineering, Motilal Nehru National Institute of Technology Allahabad, Prayagraj, India

Shweta Pandit
Department of Electronics and Communication Engineering, Jaypee University of Information Technology, Solan, India

Amitesh Singh Rajput
Department of Computer Science & Information Systems, Birla Institute of Technology and Science, Pilani, Jhunjhunu, Rajasthan, India

Yasar Abbas Ur Rehman
TCL Corporate Research Hong Kong, Hong Kong

Shailendra Shukla
Computer Science and Engineering Department, MNNIT Allahabad, Allahabad, India

Amit Kumar Singh
Computer Science and Engineering Department, NIT Patna, Patna, India

Dinesh Singh
Department of Computer Science & Engineering, Motilal Nehru National Institute of Technology Allahabad, Prayagraj, India

Ghanshyam Singh
Centre for Smart Information and Communication Systems, Department of Electrical and Electronics Engineering Science, University of Johannesburg, Johannesburg, South Africa

Shyam Pratap Singh
Department of Electronics and Communication Engineering, Galgotia College of Engineering and Technology, Greater Noida, Uttar Pradesh, India

Gautam Srivastava

Department of Math and Computer Science, Brandon University, Brandon, MB, Canada

Muhammad Tariq

National University of Computer and Emerging Sciences (NUCES), Peshawar, Pakistan

Prabhat Thakur

Department of Electrical and Electronics Engineering Science, University of Johannesburg, Johannesburg, South Africa

Asheesh Kumar Mani Tripathi

Information and Cyber Security Services, HCL, Noida, India

CHAPTER

A review on Internet of Multimedia Things (IoMT) routing protocols and quality of service

1

Dinesh Singh, Ashish Kumar Maurya, Rupesh Kumar Dewang, and Niharika Keshari

Department of Computer Science & Engineering, Motilal Nehru National Institute of Technology Allahabad, Prayagraj, India

1.1 Introduction

During the last decade, Internet of Things (IoT) has grown very fast and connected numerous devices worldwide. IoT can provide anytime, anyplace connectivity for anyone and anything [1,2]. Presently, 23.8 billion units of devices are connected to the Internet, and this amount is expected to jump 41.2 billion units by 2025 [3]. The IoT devices have limited memory, size, energy, and computing capabilities [4]. Thus, these IoT connected devices highly depend on reliable communication network and efficient routing protocols [5]. Technical developments in IoT operating systems [6], IoT enabling platforms [7], 5G IoT [8,9], wireless sensor networks (WSNs) [10, 11], software defined networks (SDNs) [12], mobile ad hoc networks (MANETs) [13–16], vehicular ad hoc networks (VANETs) [17,18], fog/edge computing [19,20], cloud computing [21], are facilitating IoT to comprehend to connect anything and anywhere [22].

The massive growth in multimedia on-demand traffics such as images, audios, and videos, has evolved the concept of Internet of Multimedia Things (IoMT) from IoT [22]. IoMT devices produce massive amount of multimedia data and are more constrained than IoT devices. They need more processing power, massive memory storage, adequate bandwidth, and consumes more power to support the underlying application efficiently [22,23]. The comparison between key data characteristics of IoT and IoMT [22] has shown in Table 1.1. The IoMT has diverse applications like smart grid, smart cities, industrial IoT, smart homes, health IoT, smart farming and agriculture, traffic monitoring, soil health monitoring, and satellite control systems. Most of these applications are interactive applications that involve reliable and timely delivery of data. Hence, it demands efficient routing protocols, and enforces stringent quality of service (QoS) parameters [23]. Due to transmission of unstructured, bulky and multimedia data over the network, IoMT needs revision of existing routing pro-

Internet of Multimedia Things (IoMT). https://doi.org/10.1016/B978-0-32-385845-8.00006-X

Table 1.1 Comparison between key data characteristics of IoT and IoMT.

S. No.	IoT data	Multimedia data
1	Linear data	Gigantic data
2	Limited bandwidth	Adequate Bandwidth
3	Less memory storage	Massive memory storage
4	Limited processing	Excessive processing
5	Delay tolerant	Delay sensitive
6	Less power consumption	More power consumption

tocols of IoT which focus on energy aware computing, efficient feature extraction, and optimizing routing criteria to minimize delays [23]. The quality of service performance in IoMT is determined by various metrics such as bandwidth, packet loss ratio, delay, throughput, resource management, and energy conservation [23]. The quality of service in technical white paper [24] is defined by Microsoft as: "Network QoS refers to the ability of the network to handle this traffic such that it meets the service needs of certain applications." The frequent packet loss and redundant packet delivery degrade the quality of constrained multimedia applications in IoMT network. The massive amount of data produced in IoMT network from many heterogeneous sources creates the processing overburden on the routing devices, resulting in packet loss.

The multimedia traffic is growing very fast which increases newer challenges for computing, sharing, storing, and transmitting the multimedia data. Computing data in IoMT needs novel methods for fog/edge and cloud computing devices. Similarly, for storing multimedia data, different compression/decompression methods are required in IoMT [22]. For example, utilizing a standard IoT routing protocol called RPL (Routing protocol for low-power and lossy networks) [25] in IoMT deployment scenarios, requires more improvements in terms of fault tolerance, energy-awareness, delay-awareness, and load balancing [22]. The Green-RPL [26], Context-Aware and Load Balancing RPL (CLRPL) [27], and free bandwidth (FreeBW)-RPL [28] are some modified versions of RPL protocol for IoMT. In this chapter, we give a comprehensive review on routing protocols and quality of service in IoMT.

The remainder of the chapter is organized as follows: In section 1.2, we illustrate the working methodology of routing protocols used in the IoMT network. The QoS routing protocols are explained in section 1.3 of this chapter. The conclusion and future directions is presented in section 1.4.

1.2 Routing protocols in IoMT

Wireless Sensor Network (WSN) performs a crucial role in assisting IoMT in diverse application spectrum. The energy constraint of nodes in WSN somehow restricts it to fulfill the expectations of IoMT applications. The efficient routing protocols save the energy of WSN nodes and thus, allow to carry off a massive amount of IoMT data. Recently, cluster-based routing has attracted due to less communication overhead and reduced routing energy consumption. The cluster members of a cluster can save their routing energy expense because of the routing assistance of the cluster head. But, since everything has its pros and cons, cluster-based protocol has its disadvantage.

In case of battery discharge or malfunctioning of the CH, all the cluster members would fail to transmit their collected data to the destination, which would defeat the smooth functioning of the IoMT application. Thus, the fault-tolerant approaches are in use in cooperation with cluster-based routing protocol for IoMT applications. Here, we illustrate the working principle of a fault-tolerant-based routing protocol used for IoMT.

1.2.1 Fault tolerant routing protocol

The fault-tolerant routing protocol [29] uses cluster-based scenario for the processing of IoMT application data. It uses the cluster join method for handling faulty cluster heads (CHs). Whenever a fault is detected in a CH, its cluster members are the first ones to catch it as they fail to receive an acknowledgment from their CH upon receiving transmitted data. So, they broadcast a help message to their neighboring CHs for re-selecting their CH. Out of all the respondents of the help message, the closest one is selected as the best backup CH. But, even this method has some leftover loopholes described as follows:

- If there are many faulty CHs, they will introduce a new problem of help message explosion.
- Each member of the faulty CH cluster may select a different CH as its backup for data routing, and then there would be a problem in handling that cluster's data.

The possible solutions to the above problems are:

- One of the solutions is the re-clustering of the entire network nodes. But, this is a laborious and costly process.
- Another solution is to keep a CH entirely for backing up the failed CHs by not initially assigning a cluster to it.
- The other solution includes identifying overlapping nodes (two nodes having similar coverage area) and putting one of such nodes into sleep mode, then wake it up only when the existing CH fails.

In view of these solutions to overcome the issues, a fault-tolerant routing protocol is illustrated in [29]. The protocol performs the following operations:

1. The protocol initially detects the failure in CHs and record faulty and nonfaulty CHs.
2. It analyzes the energy levels of nonfaulty CHs and attempts to construct the virtual CH. The data is routed through this virtual CH to the destination node.

3. The maximum fault-tolerant capability of nonfaulty CHs is calculated to accommodate the members of faulty CHs.
4. The protocol uses a flow-bipartite graph (FBG) to estimate the energy cost of IoMT data routing. A flow-based pairs of faulty and fault-free CHs are formed to create an FBG.
5. Using the FBG, a flow transmission pattern is identified between the faulty and fault-free CHs. This indicates that which of the faulty CH is tolerated by which of the fault-free CHs.

The virtual CH and FBG construction methods used in this routing protocol are explained below:

Virtual CH

The steps involved for creating of a virtual CH and virtual super frame are:

1. The destination node organizes the available energy of all the failure-free CHs, as shown in Fig. 1.1 [29]. There are three components in that total available energy: the energy taken for receiving the IoMT data from all the failure-free members, the energy is taken to aggregate the IoMT data obtained from the cluster members, and the energy taken to route the aggregated data from the CH to the destination node.
2. The remaining energy of a failure-free CH after using the above three components is used for virtual CH construction. The failure-free CHs receive IoMT data from failure-affected cluster members, aggregate the data and then transmit them to the destination node, as shown in Fig. 1.2 [29]. But, which failure-free CH is tolerated by which of the faulty node remains unknown. Thus, the concept of average transmission energy consumption is used to analyze energy used in fault tolerance.

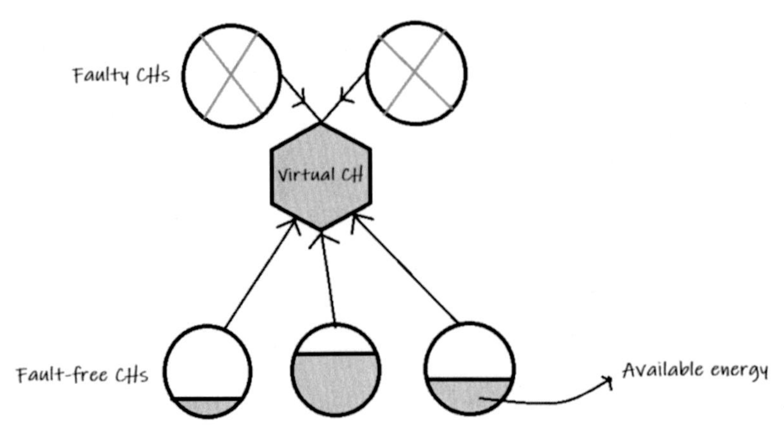

FIGURE 1.1

Organization of the virtual CH with the available energy of all the fault-free CHs.

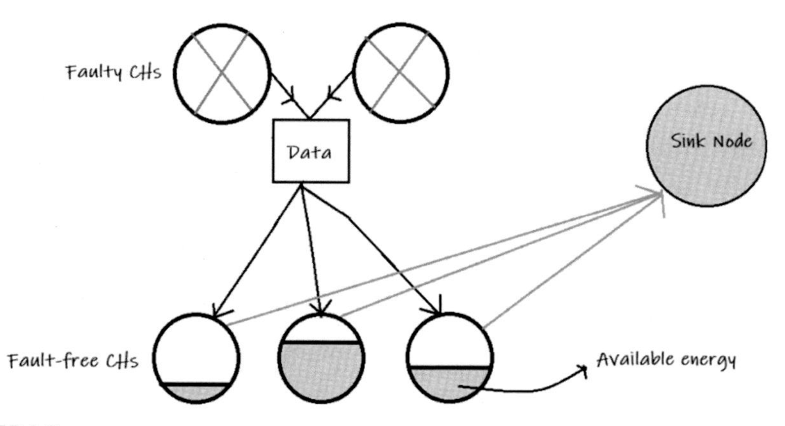

FIGURE 1.2

Data aggregation of faulty CHs by fault-free CHs.

3. A virtual superframe structure is constituted for some failure-free and some failure-affected members. The number of such virtual super frames transmitted by the virtual CH is defined as its transmission capability. Thus, the transmission capability is derived by dividing the sum of the energies of all failure-free CHs by sum of failure free and fault-tolerant energy consumption.
4. Finally, the verification of fault-tolerant capability begins. After confirmation of one or more CH failures, the minimum data of faulty CHs required to be handled by the fault-free CHs is calculated. If that data is greater than the availability of data received in fault-free CHs, then the purpose of IoMT application is not solved.

Flow-Bipartite Graph

After creating virtual CH, flow-based pairs of faulty and fault-free CHs are formed to develop a flow-bipartite graph (FBG). An FBG, as shown in Fig. 1.3 [29], is a graph whose vertices are divided into the source and destination sets such that every edge connects a vertex in source to a vertex in destination and is associated with a transmission cost and an energy cost. Each destination vertex is attached with a capacity cap. To establish an FBG, the following steps must be followed –

1. All faulty CHs act as source, and failure-free CHs as destination node.
2. The amount of input flow from each source vertex is set to the demanded amount of data (i.e., the data to be gathered at failure-free CHs due to failed CHs).
3. The capacity cap of the destination vertex is set to the available energy of the failure-free CHs.
4. The distance between the farthest member of a faulty CH and a fault-free CH is calculated and compared with a node's communication range. If the former is larger, an edge can be created between those source and destination nodes.

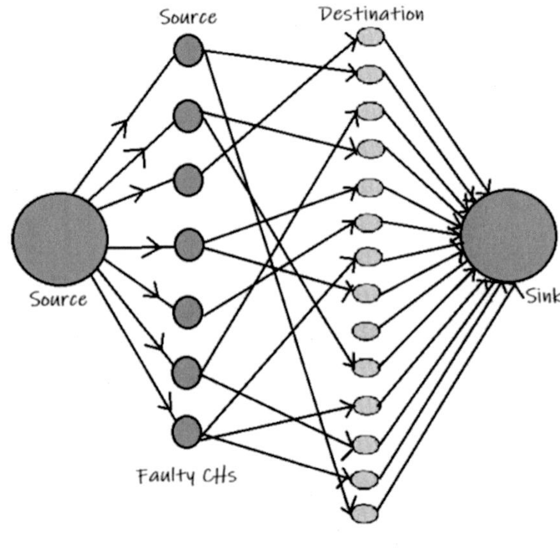

FIGURE 1.3

An example illustration of flow-bipartite graph formed by faulty and fault-free CHs.

5. The transmission cost of an FBG is calculated. It is the sum of cost of transmitting data from failure affected members to the chosen fault-free CH and for transmitting from failure-free CH to the destination node.

   Now, the fault-tolerant energy cost of FBG is calculated.

1. It is the sum of energy taken to assist the transmissions of failure-affected members to the destination node and the energy taken to transmit the IoMT data of the failure-free cluster members to the destination node.
2. With this, fault-tolerant load distribution can be achieved by making two or more fault-free CHs tolerate a single faulty CH.
3. The minimum cost flow (MCF) problem is also solved using the FBG approach such that the total minimum transmission cost can be obtained.

1.2.2 DDSV routing protocol

An optimizing Delay and Delivery Ratio for Multimedia Big Data Collection in Mobile Sensing Vehicles (DDSV) routing protocol [30] uses optimized routing criteria so that the delay involves in data collection and data delivery is minimum. It considers a road scenario that has multiple intersections on a fixed distance. The two types of moving vehicles, Bus and taxi, are considered on the road to receive the IoMT data generated from the source, i.e., traffic light. The vehicles deliver the received IoMT data to the other vehicles or the Data Collection (DCs) centers located at the

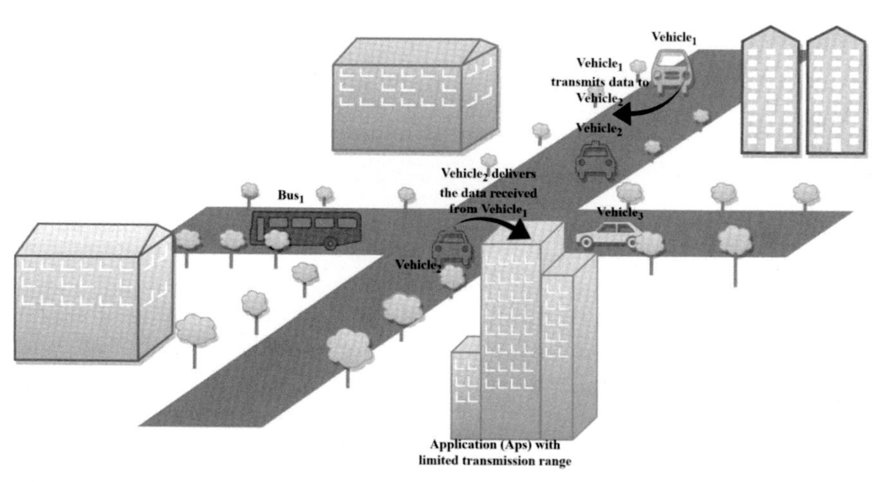

FIGURE 1.4

DDSV routing scenario.

intersection of the road segment based on the optimized routing criteria. The DDSV routing scenario is shown in Fig. 1.4 [30].

The route for IoMT data transfer consists of intersections and road segments. The protocol takes data packet priority as well as vehicular priority for decision-making purposes. The data packet priorities are assigned on the basis of location or area from where the data is generated. The routing scenario is partitioned into many urban and suburban sections.

1. At each intersection i of the road, the routing decisions are taken. The routing decision is based on the minimum of expected data delay (D_i) on the possible routes.
2. The expected data delay (D_i) is computed as the function of two components: movement probability ($P^{\varepsilon}_{i,j}$) of vehicle ε at intersection i to the routing decision θ_i and data forwarding probability ($d^{\varepsilon}_{i,j}$).
3. There are multiple routes between intersection i and intersection j to transfer the IoMT data. The movement probability ($P^{\varepsilon}_{i,j}$) of vehicle ε at intersection i is computed as

$$p^{\varepsilon}_{i,j} = \alpha \times P^{\varepsilon}_{i,j} + (1-\alpha) \times (1 - S_{\varepsilon}) \tag{1.1}$$

   where $P^{\varepsilon}_{i,j}$ is the priority of vehicle ε and S_{ε} is the probability of the vehicle ε to travel in the suburban area. The S_{ε} is defined as the ratio of time spent by a vehicle in suburban area to the time it taken in moving as per the historical trajectory datasets.
4. A vehicle finds the following possibilities, as shown in Fig. 1.5 [30], at each intersection i to route the IoMT data.

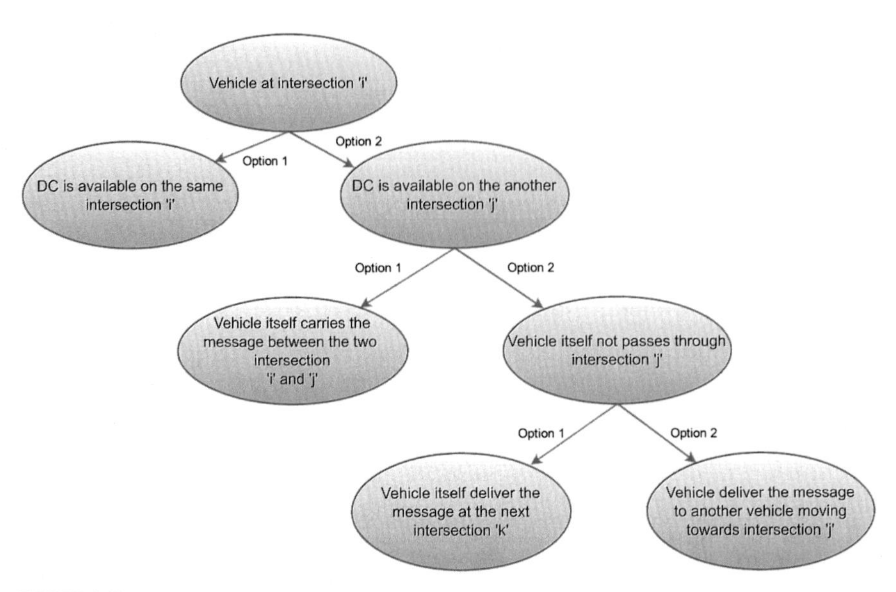

FIGURE 1.5

DDSV outing decision at each intersection.

The movement probability ($P^{\varepsilon}_{i,j}$) of vehicle ε at intersection i in view of the above routing possibilities can be computed as

$$P^{\varepsilon}_{i,j}(\theta_i) = \left[\prod_{edges}\left(1-\varphi^{\varepsilon 1}_{i,k}\right)\right] \times \left[\varphi^{\varepsilon}_{i,j}\left(1-\sum_{edges}\omega^{\varepsilon 1}_{i,k}\right) + \omega^{\varepsilon}_{i,j} - \omega^{\varepsilon}_{i,j}\varphi^{\varepsilon}_{i,j}\right] \quad (1.2)$$

Where $\varphi^{\varepsilon}_{i,j}$ is the probability of the vehicle ε to move on the road segment between intersection i and intersection j and $\omega^{\varepsilon}_{i,j}$ is the probability of the vehicle ε to deliver the data packet to another vehicle moving on the road segment between intersection i and intersection j.

5. The data forwarding probability ($d^{\varepsilon}_{i,j}$) for a vehicle ε depends on the vehicle type, i.e., Bus or taxi.
   a. The Bus vehicle has predefined trajectories and the probability depends on the length of the road segment and average speed of the vehicle only. For Bus type vehicles, it is calculated as:

$$d^{\varepsilon}_{i,j} = \sum_{s_{e,r} \in S_{i,j}} \frac{l_{e,r}}{v^{\varepsilon}_{e,r}} \quad (1.3)$$

   where $l_{e,r}$ is the length of one of the subset of the road segment between intersection i and intersection j and $v_{e,r}$ is the average speed of the Bus on the respective road segment.

b. The taxi type vehicles have no fixed trajectories and thus, the probability considers the transmission range, vehicle speed as well as the wireless transmission delay for computation. It is calculated as:

$$d^{\varepsilon}_{i,j} = \left(1 - e^{-r.o_{i,j}}\right) \times \frac{l_{i,j} \times c}{r} + e^{-r.o_{i,j}} \times \frac{l_{i,j}}{v^{\varepsilon}_{i,j}} \tag{1.4}$$

Where r is the transmission range of vehicle's Wi-Fi channel, c is the wireless delivery delay, $o_{i,j}$ and $v_{i,j}$ is the density degree and vehicle's speed on the road segment between intersection i and intersection j, respectively.

6. For routing purpose, it is assumed that the priority of the vehicle moving in the suburban area is high as compared to vehicles moving in urban area.

1.2.3 Optimal routing for multihop social-based D2D communications

An optimal routing algorithm for multihop communication between Device-to-Device (D2D) is given in [31]. The algorithm is well suited for efficient IoMT data communication and 5G networks. It considers the social behavior of the devices in communication to their neighbors. The algorithm computes the trust probability for D2D communications based on the rank model. Both, random and fixed locations of the base stations are considered for measuring the connection probability (CP) between any pair of devices. The measurement of CP is taken with and without, channel state information (CSI). The network model of the Social-aware multihop D2D routing algorithm is shown in Fig. 1.6 [31]. The transmission of information between D2D transmitter and D2D destination using the number of D2D relays is taking place. Here, the interference due to the cellular communication equipment (CUEs) and position of base stations (BSs) is considered in follow-up. The Nodes are working in half-duplex mode. A single antenna and D2D transmitter are assumed to be at the origin, and D2D destination is at a fixed distance away from the origin, whereas using the Poison Point Process (PPP), BS and CUEs are modeled.

With the help of real data traces from online social networks, trust connectivity of D2D based on rank-based trust model is calculated. The probability that D_{i+1} is trusted by D_i is given as

$$P^{(t)}_{i,i+1} = \frac{1}{GR^{\beta}_{i}\,(i+1)} \tag{1.5}$$

where β is the parameter from the rank-based model and $G = \sum_{n=1}^{N}[\frac{1}{n^{\beta}}]$ is a normalizing factor.

A trusted connection will be established only when two nodes can trust and communicate with each other and thus, the trusted connectivity probability (T-CP) is defined as

$$P^{(c)}_{i,i+1} = P^{(t)}_{i,i+1}P^{(c)}_{i,i+1} \tag{1.6}$$

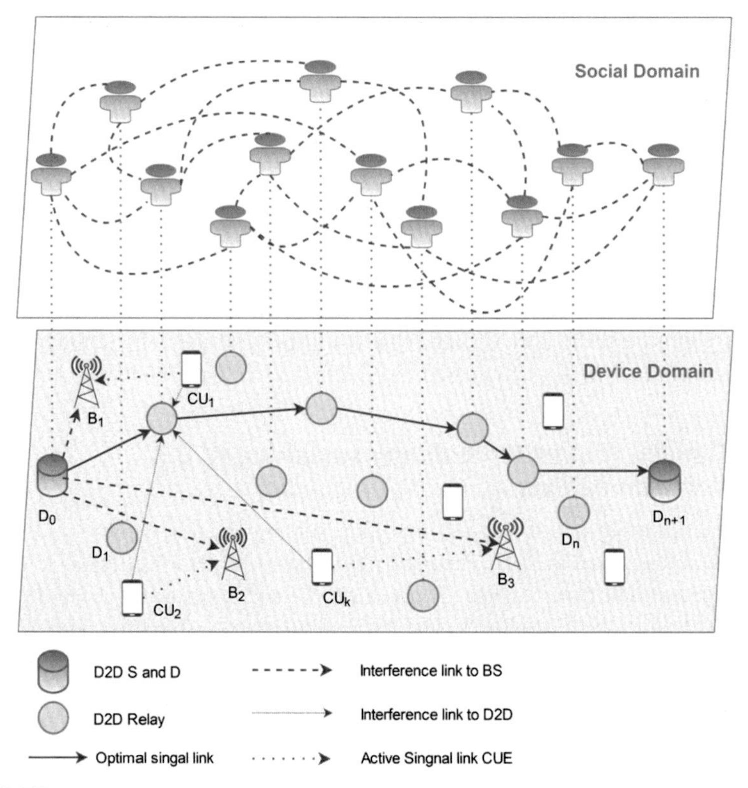

FIGURE 1.6

Network model of optimal routing for multihop social-based D2D Communications.

where the connectivity probability between D_i and D_{i+1} is denoted as $P_{i,i+1}^{(c)}$.

In case of random BS and fixed BS scenario, both CSI aware and not CSI aware situations are actively taken for the computation of the CP. In CSI aware situation, the channel state information between BSs and D2D transmitter is used in CP computation. However, none of the channel state information between BSs and D2D transmitter is used in CP computation. In random BS scenario, the CP between any D2D devices for given $D_{i,i+1}$ depends on the intensity of BS and CUE, path lose exponent, transmit power of each CUE, threshold I_{th}. However, in fixed BS scenario, rest other parameters are similar to one that we use in random scenario except the two. Here, in place of intensity, the location and number of BSs is used in CP computation between any D2D devices for given $D_{i,i+1}$.

The objective of the routing algorithm is to find the optimal path between the D2D transmitter and the receiver. The path selection is based on the maximum value of the T-CP at every intermediate link. The maximum T-CP between D2D transmitter and D2D destination in presence of multiple D2D relays can be obtained by routing algorithm which helps in selecting optimal path. The optimal path satisfies:

$$P_{T-CP}\left(\pi^*\right) = \max_{\pi^* \epsilon S_\pi} \prod_{i \epsilon S_\pi} P_{i,i+1} \tag{1.7}$$

where S_π denotes the set of all potential paths between the D2D transmitter and receiver. The routing algorithm finds the maximum T-CP between source and destination in a weighted graph with the help of standard Dijkstra's algorithm. Initially, at each D2D node, distance between itself and other D2D devices and BSs are calculated and stored in topology information, which contains the neighbor list. The adjacency T-CP matrix is obtained by using updated topology information. Therefore, the D2D transmitter is set up as permanent node and all other nodes as temporary nodes. Here, we calculate the T-CP for all possible neighbors links and find the link with maximum T-CP. The intermediate destination node in selected link with maximum T-CP is set as permanent node and now keeps on finding maximum T-CP unless all the nodes are marked as permanent node. Thus optimal path and maximum TCP of multihop D2D communication are obtained. The Computational complexity of routing algorithm is as same as that of Dijkstra algorithm, i.e., $O(N^2)$.

1.2.4 Green-RPL routing protocol

The Green-RPL [26] is an energy-efficient green routing protocol for IoMT. It is an enhanced version of the RPL (Routing Protocol for Low-Power and Lossy Networks) protocol [25]. RPL is a routing protocol for resource-constrained devices that uses Destination Oriented Directed Acyclic Graph (DODAG) to maintain network topology. This DAG comprises multihop routes from leaf nodes to the root node. RPL optimizes an objective function and chooses the best path by selecting the desired predecessor nodes starting from leaf nodes. The previous RPL protocol implementations are not feasible for IoMT and do not consider the multimedia data.

In contrast, the Green-RPL protocol considers the data generated from multimedia devices. The Green-RPL protocol reduces energy consumption and carbon footprint emissions together with QoS requirements of applications. To guarantee QoS for a particular multimedia application, it determines the delay bound for the application. For example, in VoIP applications, the delay limitation usually is 120 msec. To ensure energy efficiency, the protocol considers the features of all the intermediary links between the leaf nodes and the root node and estimates the energy consumption by the chosen immediate predecessor node to support traffic needs for one more immediate successor node. An optimization model is given for this protocol based on various requirements and constraints.

In Green-RPL routing protocol, the desired immediate predecessor node is selected according to the objective function minimizing the emissions of cumulative path carbon footprints on all the links from the immediate predecessor node to the root node, and satisfying the constraints such as cumulative path link energy, cumulative path delay, idle time, and battery status of the immediate predecessor node. When more than one immediate predecessor node satisfies the given constraints, then the node that provides the most greener path is chosen as the desired immediate predecessor node. In cumulative path link energy constraint, energy consumption by a node is calculated based on the quality of links in the selected route. The cumulative path delay constraint specifies that the predefined delay threshold should not be increased. According to the idle time constraint, an immediate predecessor node should be chosen as the desired node only if its idle time is sufficient to support another child node. In the last constraint battery status, a child node should choose an immediate predecessor node as the desired one if the battery level of its energy resources is greater than the predefined threshold.

1.2.5 Context-aware and load balancing RPL (CLRPL)

A modified version of RPL, known as Context-aware and load balancing RPL under heavy and highly dynamic load for IoT and IoMT applications is discussed in [27]. It addresses the power depletion and packet loss problems associated with RPL. In CLRPL, one objective function called as Context-Aware Objective Function (COAF) and one routing metric called as Context-Aware Routing metric (CARF) has been given based on the contexts of the nodes. Here, context means any information related to nodes in the IoT network that can be used to characterize the nodes.

Through COAF, CLRPL selects the immediate predecessor of nodes by finding their rank, Expected Transmission Count (ETX), and residual power level. Normally, ETX is used to give a realistic estimate of the channel status that is based on link losses. To avoid the loss during the initial phases of congestion and monitor both queue status and link status, ETX is used with other metrics that can give the status of the node queue in the network. The objective functions used in the previous implementation of RPL do not consider the latter states of nodes in the route while finding the rank. The rank value of a node may suffer from queue overflow and power drainage, which may further affect the rank of all downward nodes in the path. Thus, COAF recursively computes the latter states of the sequence of nodes in the route towards the root. COAF circumvents the thundering herd effect in the network, which happened in RPL, by actively considering the network states and gradually shifting toward the real rank value from a high-rank value.

The protocol uses the routing metric, CARF, which considers the network traffic dynamicity index, node rank, and buffer queue utilization to calculate the route throughout heavy traffic conditions. Here, the network traffic dynamicity index shows that how much link has been utilized in the past. CARF reduces the effect of upstream immediate predecessors as it grows a long way in the path toward the root. Instead of just considering only the rank of nodes, this metric considers the other factors related to nodes and networks to select the suitable immediate predecessors in a network with heavy traffic. CLRPL avoids routing loops through the use of a countermeasure while selecting the best immediate predecessors using CARF and other parameters. This protocol also ensures load balancing by taking the number of nodes having an immediate predecessor from the list of candidate immediate predecessors for selection of the immediate predecessor.

Another enhanced version of RPL protocol named free bandwidth (FreeBW)-RPL is given in [28] which suggests a novel objective function called FreeBW that takes the FreeBW computations in the network layer. This protocol calculates the maximum FreeBW using the volume of the bandwidth to select the routing path to provide better performance.

1.2.6 Energy-Harvesting-Aware (EHA) routing protocol

Energy harvesting (EH) is evolving as an essential technology for IoT and field deployable WSNs applications. In these applications, the deployed nodes or devices may recharge their batteries to perform actions by extracting energy from renewable sources such as radio frequency (RF) and solar-based radiation through the use of EH techniques. Energy-Harvesting-Aware (EHA) routing protocol proposed in [32] addresses the issues of Quality of Service and energy efficiency for IoT applications at the network layer. The protocol can efficiently perform according to the harvested-energy at IoT devices, the residual energy, and the dynamic traffic capacity produced by IoT applications. EHA routing protocol is based on the energy prediction process and energy-backoff process that describes cost metrics to choose the best path. In the energy prediction process, an enhanced estimate of existing energy levels from different sources has been determined using the Kalman filter method that considers current statistics and previous time steps. In the energy backoff process, network lifetime has been increased by keeping the nodes having the lowest energy in sleep mode until their energy level is restored to perform operations. The protocol also considers hybrid energy harvesting sources and uses three EH techniques such as solar-based EH technique, moving vehicle-based EH technique, and RF-based EH technique. The EHA routing protocol first selects the node having the maximum level of residual energy and then finds the link with the least cost using the shortest path algorithm given by Dijkstra to send the data packets to the target node from the origin nodes. Further, by using the EH techniques, the protocol aims to determine the sustainable paths to route the data packets.

1.2.7 Optimized 3-D deployment with lifetime constraint (O3DwLC) protocol

The deployment of sensing devices is a very challenging task to satisfy both robust connectivity and the long lifetime of devices subject to cost minimization. The most common solution for handling such a situation is to install the relay nodes in the target environment. The function of relay nodes is to forward the sensed data up to the maximum distance sink node and thus requires additional storage and more power. The energy of sensing nodes can be saved for further collecting forwarding data. But installing the relay nodes is more challenging because of the roughness of the environment and the requirement of 3-D infrastructure. The roughness of the environment is due to natural calamities and nonexpected visiting birds and animals because they may destroy nodes. For maximizing connectivity with little cost and

Table 1.2 Comparative analysis of routing protocols in IoMT (Part a).

Paper	Architecture	Entities	Communication mode	Link establishment criteria	Suitability criteria	Optimal path finding	Noiseless path selection	Environment adaption	Goal	Methodology
[29]	Cluster	Sensor nodes, Sink nodes	Device-to-device	Transmission energy consumption, Distance	Transmission Cost, energy cost	Integer Linear Programming model solved using IBM ILOG CPLEX Optimizer	×	×	Designed a join based method which can perform preverification of fault tolerance, fault tolerant load distribution and fault tolerant cost minimization.	A cluster based fault tolerant routing which minimize total energy consumption during packet transmission. The proposed approach uses virtual clustering and flow graph modeling to handle faulty CH.
[30]	General	IoT multimedia storage devices, vehicle (bus and taxi type), DC	Vehicle-to-vehicle	Message establishment, time spent on road segment, density, speed, transmission range	Prioritize over probability of delivery at nearest data centers	Designed a threshold based optimal route policy algorithm	×	×	Goal is reduction of packet drop of low priority data and task delivery within deadline.	Proposed routing policy based on delivery delay and delivery prioritization of data.
[31]	General	IoT device, Base station (BS)	Device-to-device	Distance, density	Rank based Trusted Connectivity Probability (T-CP)	Dijkstra Algorithm	✓	×	Carried out an analysis for cases at known and unknown channel state information.	Author proposed an approach of Social relationship based on T-CP for multihop routing path selection using tight lower bound and exact closed bound with decode and forward mechanism.

(continued on next page)

Table 1.2 *(continued)*

Paper	Architecture	Entities	Communication mode	Link establishment criteria	Suitability criteria	Optimal path finding	Noiseless path selection	Environment adaption	Goal	Methodology
[26]	Tree topology	Multimedia device		Cumulative path carbon footprints, cumulative path delay, cumulative path link energy, battery status, idle time for parent node	Delay constraints, battery consumption, type of energy resource	–	✓	✓	An adaptive and dynamic optimization based RPL which can support multimedia communication.	Proposed a Green-RPL routing to minimize carbon footprint emissions and energy consumption using tree like network topology.
[27]	Tree topology	IoT devices	Point-to-point, multipoint to point, point to multipoint	Remaining power status, queue utilization for parents to root chain	Rank based Context aware routing metric (CARF)	–	x	x	Loop free and less packet loss as well less power depletion for RPL under heavy and dynamic load, eliminates thundering hard and equality illusion problem	Context aware load balance RPL with DODAG structure find route using two steps as: 1) CARF, 2) parent selection mechanism.
[32]	General	Sensors, router	Device-to device	Distance, battery voltage, packet length, data rate,	Energy consumption during listening, transmitting, receiving and sleeping	Dijkstra Algorithm	x	✓	Aimed to maximize life-time in an energy efficient manner.	Proposed a Kalman Filtering (KF) adaptive energy harvesting routing by enhancing energy back-off parameter. If harvested energy is high then use current energy and store the excess energy

Table 1.3 Comparative analysis of routing protocols in IoMT (Part b).

Paper	Merits	Demerits	Evaluation metrics	Simulation details	Assumption	Complexity	Designed approach	Comparison	Tool
[29]	Average backup energy consumption for new CH without aggregation is 0.1856 mJ while proposed approach is around 0.0098 mJ	Average failure affected as energy consumption is reduced around 10% at 0.2 failure rate while 4% increasing at failure rate up to 0.5	Average backup energy consumption, Average failure-affected energy consumption, Network lifespan after fault tolerance	Sensor nodes (1000 with 10% powerful sensor and 90% sink node), Sensed data (500 bytes), failure threshold (0.2–0.5), Simulation run (50)	Same data size sensed by each member	–	Virtual CH, Flow bipartite graph techniques	New CH without aggregation, New CH with aggregation, distributed join CH, distributed join CH with multiple factor	NS2
[30]	Average delay during data re-collection is reduced around 17.3% for general and 41.8% for suburban and average data delivery ratio improved up to 16.9% with respect to OVDF.	Delay decreased percentage graph turns towards deteriorate around 11.11% at 140 with respect to 120 in its own scheme.	Average delay, Delivery ratio, Data packet impartiality	Multidataset of Beijing city of "T-drive trajectory", Data packet size (512 bytes), Data storage size (10M), Communication range (150 meters), vehicles (30–150)	–	–	DDSV (delay and delivery ratio for sensing vehicles)	OVDF	–
[31]	Optimal path route by proposed algorithm is same as ES.	Optimal path selection is depends on BS position	Connectivity Probability	Nodes (20), noise variance (1), Path loss exponent (4)	Node equipped with half duplex mode single antenna.	O((log n-2)!)	Trusted Connectivity Routing Algorithm	Exhaustive search (ES)	Monte Carlo simulation

(continued on next page)

Table 1.3 *(continued)*

Paper	Merits	Demerits	Evaluation metrics	Simulation details	Assumption	Complexity	Designed approach	Comparison	Tool
[26]	Average packet delay is 20.4 msec, while ETX and OF0 requires 47 msec and 26.2 msec.	Energy consumption depends on path link quality. At poor link successful delivery using single hop probably reduced and causes retransmission.	Number of packet delivered, energy consumed.	ICMPv6, packet size (128 byte), datarate (25 pkt/sec)	–	–	Green-RPL	ETX, OF0	Cooja Simulator for Contiki-OS
[27]	RPL experience 32% packet loss while proposed approach packet loss is 15% at 100 nodes.	Instability and inconsistency increase at new node joining and frequent messaging.	Queue loss ratio, packet loss ratio, lifetime, overhead, Runtime	Number of nodes (50), Traffic rate (15–100 pkt/min), Packet format (IPv6), Transmission power (15 dBm), FIFO queue size (20 pkts), Range (100),	–	–	CLRPL	RPL	Cooja Simulator for Contiki-OS
[32]	Less energy consumption wrt R-MPRT-Mod (12.54%) and AODV-EHA (52.3%).	Works for very small packet size (1000 bits)	Average consumed energy, lifetime, Alive nodes, Packet loss ratio, end-to-end delay	Data-rate (250 Kbps), Simulation period (12 hr), End-to-end delay (0.5%), Throughput (5–80 pkt/s)	–	–	EHARA	R-MPRT-Mod, AODV-EHA	MATLAB®

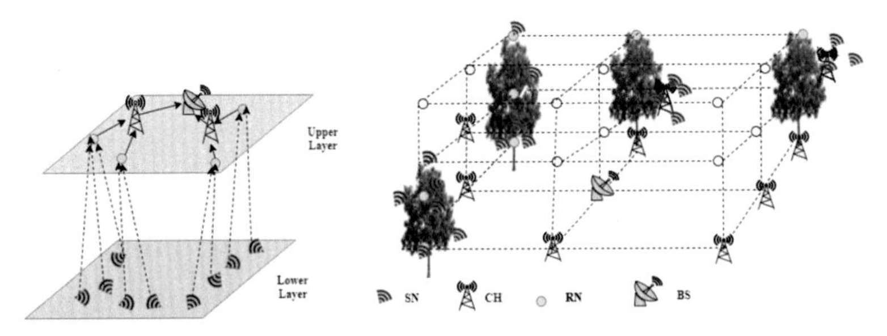

FIGURE 1.7

A two layered hierarchical and 3-D grid based hierarchical cluster based arrangement of sensor nodes.

long-life sensing devices, installing and placing 3-D relay nodes is the most optimal solution to handle the challenges of the wireless network in a harsh environment [33].

For improving performance, a two layered hierarchical setup can be assumed in large-scaled environment. Lower layer nodes (SN) gather data and send it cluster heads (CH) in the upper layer. The upper layer contains relay nodes (RN) for better transmission. Cluster nodes summarize gather data from the lower layer, and then with the help of relay node, it sends to the base station (BS) in the upper layer. In a 3-D cubic grid-based model, the target environment is assumed partitioned in a grid-based cluster in hierarchical manners. Cluster heads (CH) are placed in the most appropriate vertices surrounded by the largest number of sensor nodes, and the base station is placed on the fixed position as per requirement. For efficient connectivity between cluster heads and base station (BS), it is mandatory to seek proper placement of relay nodes in grid subject to minimize cost and maximize connectivity. Fig. 1.7 which is taken from [33], shows a two-layered hierarchical and 3-D grid-based hierarchical cluster-based arrangement of sensor nodes, respectively.

The main agenda is to search the proper placement of RNs so that cost of deployment could be minimized by reducing the total counts of CHs/RNs in model with the constraint of strong connectivity, low cost, and longer network lifetime. The overall cost can be minimizing by reducing the costlier equipment for the network scenario. Since the cost of equipment is dependent on the functionality (RNs in this case because of additional storage and forwarding capacity) and thus, cost can be minimized by adjusting the number of RNs. The strong connectivity can be achieved by optimal design and reducing obstacles between communicating devices, and therefore, the grid model is more suitable for this purpose.

Algorithm description

Turjman et al. [33] have introduced an Optimized 3-D deployment with Lifetime Constraint (O3DwLC) algorithm to achieve the objective of minimal cost, long network lifetime, and high connectivity. This algorithm works in two phases. In the first phase, they designed a connected backbone structure using First Phase Relay Nodes (FPRN). These FPRNs are placed in an optimized manner so that the requirements of FPRNs become minimum for connecting cluster heads to the base station. Thus, a minimum cost spanning tree (MST) is used between vertices in the grid model. MST can be constructed using the 3-D grid vertices pointing location of nodes. Further, the algorithm seeks the two closest nodes in the set of nodes, CHs and BS. In case the closet two nodes are not adjacent on the 3-D grid, then it introduces again a minimum number of grid vertices on which the relays have to be positioned to establish a path between these two nodes. After connecting the closest two nodes repeatedly, the algorithm looks for the next closest nodes and so on. The extra relay nodes (SPRN) can be introduced in optimal manner to maximize network backbone connectivity. The connectivity can be computed by analyzing the consecutiveness of graph nodes in the grid. The algorithm locates optimized SPRNs position in the second phase of processing to maximize lifetime and minimize the backbone generated in the first phase.

For routing multimedia data, the algorithm's performance is measured based on backbone connectivity level, the number of CHs, RNs, BS nodes used, and the number of turns. The number of turns measures the total turns for the deployed network in which it can stay operational.

Discussion. There are many cluster-based [29], general (non-cluster-based) [30,31, 27] and tree-topology-based [25,26] routing protocols are discussed in Table 1.2 and 1.3. In [29], the proposed work avoids backup workload shifting to new CH by re-joining new cluster according to energy suitability. Although in this research work sensed data size were same for all sensor and no deadline, which is an irrelevant assumption. To prevent the challenge of [29], a deadline prioritize delivery routing protocol is designed by Li et al. [30] and a dynamic load balanced routing protocol on rank is proposed by Taghizadeh et al. [27]. Moreover, both approaches retransmit packet to reduce packet drop rate which has no threshold during frequent messaging which may cause overhead retransmit broadcast. Here, majority of research work minimize delay, packet drop while some approach gives attention energy consumption by designing green routing [26] and Energy-harvesting-aware routing [27]. The proposed work of [26,27] based on "energy-store and energy-use" protocol by utilizing the current available energy consumption of node and "excess-energy" for future use. Although, the suitability metric in [26] avoids battery level which results to small lifetime and paper [27] ignores network heterogeneity. To overcome the network heterogeneity a social-behavioral multihop optimal routing proposed for device-to-device communication for dense IoMT environment in [31]. The proposed work finds path based on rank according to distance and density for fixed base station, while ignoring routing at outage period. The routing algorithms discussed here attempt to improve the efficiency of the IoMT network, energy-saving, fault-tolerant capability, optimal connection or link probability, etc. The energy-harvesting-aware (EHA) routing protocol discussed the quality issues of the IoMT services and the routing procedure. The other routing protocols specially designed for QoS are illustrated in the next subsequent section.

1.3 Quality of Service (QoS) routing in IoMT

In IoMT, there are two types of applications in general – video-based applications, which can tolerate some amount of packet-loss but requires timely delivery of a packet, and other applications that are sensitive to packet loss due to their constrained nature. The other category of applications relies upon retransmission to guarantee message delivery. Both application categories require QoS to provide a better experience. The QoS protocols are discussed below in detail.

1.3.1 Traffic-aware QoS routing protocol

The traffic-aware QoS routing protocol [34] is designed to solve the packet delay and packet loss issues. The protocol fulfills the QoS requirement of applications by introducing a greedy-based approach based on Yen's K-shortest paths algorithm for computation of the optimal path. The routing protocol calculates the QoS path for all the incoming flows in an SDN controller by taking all the conditions into account of the SDN rule capacity constraint. The protocol considers the particular type of traffic and the associated QoS requirement like packet loss and packet delay or both. The routing protocol takes a graph and set of flows with their associated QoS requirement as input. The user must prioritize delay-sensitive (ds) and loss-sensitive (ls) to each flow. The protocol provides an output set of paths on which ds-flows and ls-flows can be routed. The protocol performs the following operation:

- The protocol takes the cost of each link in terms of delay or packet loss as inputs and returns a feasible routing for all the ds and ls flows.
- The ds and ls paths are considered with respect to assign priority with each application.
- The ds and ls path are routed according to the path calculated, with the help of the K-Shortest path algorithm.
- The last GET-QoS-Path function checks all the delay, loss, bandwidth, and check rules if all are satisfied, then returns TRUE, else FALSE.

1.3.2 QoS-aware and Heterogeneously Clustered Routing (QHCR) protocol

The QoS-aware and Heterogeneously Clustered Routing (QHCR) protocol [35] is an energy-efficient routing protocol that provides the dedicated routes for the bandwidth-hungry, delay-sensitive, real-time, and QoS-aware applications and saves energy in the network. The QHCR protocol works in the following phases: Information gathering phase, cluster head (CH) election phase, node association phase, and intracluster communication. In the information-gathering phase, every node sends and receives the broadcast messages to and from their neighbors to collect the energy levels and other information. After exchanging the information from their neighbors, each node maintains a neighbor table and updates the base station. The neighbor table consists of information such as the number of neighbor nodes, initial energy levels of neighbor nodes, the distance of neighbor nodes from the base station, and distance between one neighbor node to another. This protocol uses CSMA/CD method to minimize the collision in the network when two nodes broadcast at the same time. QHCR is a centralized protocol in which the base station collects all information from the nodes in the network and uses that information to elect the cluster head. In the CH election phase, the cluster head is elected based on the cost value at each energy level. This cost value is computed using the information stored in the node's neighbor table. A node having the lowest cost value is elected as the cluster head at every energy level. The initial energy of nodes can be divided into four energy levels. The nodes can be clustered based on these energy levels such that the nodes having the same energy levels belong to the same cluster. In the node association phase, the elected cluster head starts sending the broadcast messages to other nonCHs nodes and member nodes of the clusters in the network. When a nonCHs node does not get any broadcast message from any cluster head, they elect themselves as cluster heads. During intracluster communication, when the distance between nonCH nodes and CH or BS is very long, the nonCH nodes use intermediate nodes for forwarding the messages to them or other nodes. Thus, the main goal of the QHCR protocol is to select the best route for transmitting the messages with minimum delay by using the intermediate nodes lie on multiple paths between the sending node and the cluster head or base station.

1.3.3 QoS in wireless multimedia sensor network based IoMT

The IoMT applications use sensor nodes to gather data for subsequent processing. The sensor nodes formulate an interconnected network to provide better coordination for input data sharing. This network uses wireless links to exchange multimedia data, popularly known as Wireless Multimedia Sensor Network (WMSN). The WMSN is an integral part of IoMT networks. The objective of WMSN is to provide the real-time delivery and standard QoS of multimedia data within constrained energy requirements. The key QoS parameters are minimum packet drop, reduced latency, in-order, and prioritized multimedia data delivery. Hamid and Hussain [36] presented a detailed survey on layered and cross-layered approaches used to achieve QoS in WSMN based IoMT. Here, the QoS requirements to deliver multimedia content are explicitly examined on application, transport, routing, MAC, and physical layer. They illustrate the use of less complex multimedia data coding schemes at the application layer; Unequal Error Protection (UEP) method to provide reliability, priority-based rate control methods to manage congestions, and optimized packet scheduling to tolerate the processing delay at the transport layer; geographic routing protocol [37] to handle delay constrained multimedia data and multipath routing [38] to solve the load balancing issue on aggregator sensor node at the routing layer; dynamic adjustment of contention window (CW) to handle heterogeneous multimedia services, Multiple Queuing methods to satisfy the scheduling requirements of different of multimedia data flows and variable contention period to better channel access at the MAC layer; dynamic control of beacon interval and use of Ultra-Wideband (UWB) technology to achieve reliable listening and a high rate of data transmission at the physical layer. A special investigation on cross-layer approaches, i.e., optimized cooperation of different layers, to satisfy the QoS requirements of multimedia data is presented. The cross-layer protocols in WMSN successfully achieve the real-time delay constraint, rate adaptation, reliability, and optimized energy requirements.

1.3.4 SAMS framework for WMSN-based IoMT

With the increase of the Internet of Things (IoT) and real-time adjustment development, people's quality of life is improving. Multimedia wireless sensor nodes and devices that make the Internet of Multimedia Things (IoMT) are essential for IoT applications, which are diverse in nature. These constrained nodes and devices depend on standardized communication stacks and effective routing protocols that form the Wireless Multimedia Sensor Network (WMSN). Quick response services, traffic control, smart cities, smart hospitals, smart agriculture, smart buildings, criminal inspection, security systems, Internet of bodies (IoB), and Industrial IoT (IIoT) are examples of real-world multimedia technologies. IoMT devices need high protection, higher bandwidth, memory, and faster computing resources to process data. This study suggests a model for a cluster-based hierarchical WMSN called Seamless and Authorized Streaming (SAMS) [39]. There are two phases for building secure clusters of a SAMS: set-up and steady-state. During the set-up phase, two-level authentication (i.e., between BS and elected CHs; and formation of clusters) is performed, and during the steady-state phase, seamless data transmission is achieved. Only legal nodes may transmit captured data to their Cluster Heads (CHs) after these clusters have been established. During mutual authentication, each multimedia node in the cluster regularly senses the environment and stores any data received in a buffer with a predefined limit.

Each multimedia node waits for its turn to transmit the capture data using its allocated timeslot/channel that are prescribed by its respective cluster head. This waiting can cause excessive packet loss and end-to-end delay in wireless channels. A novel channel allocation technique is used to keep the WMSNs' Quality of Service (QoS) at an acceptable scale to resolve this problem.

A member node in one cluster moves to a nearby CH if the latter has an available channel for allocation in the event of a buffer overflow. Fig. 1.8 which is taken from [39], depicts the SAMS framework's network architecture. The timely and accurate distribution of data is a key characteristic of IoMT. As a result, the SAMS outperforms existing schemes in terms of different QoS metrics like average packet loss and average end-to-end delay while also providing robust protection against various attacks (such as tampering, sinkhole, insider, Sybil, and password guessing).

Discussion. The research work reviewed in this section are QoS oriented protocols, which improve the seamless real-time servicing for IoMT. The comparative analysis of QoS protocols for IoMT is summarized in Table 1.4 and 1.5. The quality of service routing protocol in [34] focuses on throughput, jitter and reliability for two types of applications such as delay and loss sensitive using greedy heuristic based k-shortest

Table 1.4 Comparative analysis of QoS protocols for IoMT (Part a).

Paper	Architecture	Entities	Communication mode	Link establishment criteria	Suitability criteria	Optimal path finding	Noiseless path selection	Environment adaption	Goal	Methodology
[34]	General	IoT sensors, traffic lights	Device-to-device	Delay duration, packet-loss probability, current bandwidth	Delay sensitive cost function and loss sensitive cost function	K-shortest path	x	X	Author motivation to design a network routing protocol which focuses QoS (throughput, jitter, reliability etc.) for delay and loss sensitive application.	Proposed approach utilize a programmatic SDN and routes delay and loss sensitive data using greedy heuristic based k-shortest path algorithm.
[35]	Cluster	WSN, sensor node	Device to device	Energy consumption at sensing, processing and radio communication, number of bits transmitted, distance	Initial and current energy cost function.	–	x	X	The aim is to design a QoS-aware routing protocol to deliver data for real-time and nonrealtime multimedia service with balanced load and fault tolerable.	Proposed a cluster based energy efficient routing according to delay sensitivity, bandwidth, and energy fluctuation. CH is selected according to path metric (initial energy).
[37]	Distributed	Image Sensor nodes	Device-to-device	Distance, queue size, angular bearing towards destination	Relative location based cost function	–	✓	X	Aims to propose a best effort QoS support, low delay with avoidable in-node routing table and information related to outdated states of sensor.	Designed a routing algorithm for event driven real-time geographical distributed environment based on greedy approach.

(continued on next page)

Table 1.4 *(continued)*

Paper	Architecture	Entities	Communication mode	Link establishment criteria	Suitability criteria	Optimal path finding	Noiseless path selection	Environment adaption	Goal	Methodology
[38]	General	Sensors	Device-to-device	Size of real time and nonrealtime traffic.	Probability of successful data delivery	K-path finding	x	x	Aims to maximize network lifetime through energy balancing across node according to acceptable delay.	Propose a QoS and energy aware routing technique which balance fault tolerance using XOR based forward error correction method.
[39]	Cluster	BS, sensors	X to Y where X,Y ε {BS, sensors}	Payload, distance	buffer overflow	–	x	x	Goal is to propose secure intercluster communication framework to reduce packet loss due excessive waiting for its turn to send captured data.	Proposed a lightweight secure AES based streaming for seamless delivery. After successful authentication Cluster formation initiated according to time slot. Here, CM moves according to buffer overflow.

Table 1.5 Comparative analysis of QoS protocols for IoMT (Part b).

Paper	Merits	Demerits	Evaluation metrics	Simulation details	Assumption	Designed approach	Comparison	Tool
[34]	The proposed approach improves QoS Violated flows around 15%, 14%, and 13% (using AttMpls) and 39%, 38%, and 37% (using Goodnet) wrt MRC, SPD, and LARP respectively.	Forwarding without fine grained QoS	End-to-end delay, runtime, QoS violated flow, Active Node	Topology (AttMpls, Goodnet), bandwidth (0.20–0.40 kbps), packet size (94–699 bytes)	Assume constant jitter	SWAY	Minimum Rule Capacity (MRC), Shortest path delay (SPD), LAgrangian Relaxation Aggregated cost(LARAC)	–
[35]	Lifetime ends for proposed algorithm at 2750 rounds while ECHERP is 1300, PASCCC is 2000, and CCWM is 2100 rounds.	Due to no energy harvesting feature the proposed work conserve no energy from renewable source.	Lifetime, stability period, throughput, average energy consumption, End-to-end delay	Sensors (100), buffer size (256 k-bytes), Transmit power (20 mW), receiving power (15 mW)	Fixed sensing nodes, same packet size,	QHCR	CCWM, PASCCC, ECHERP	MATLAB
[37]	The proposed scheme increases the life-time from 29% to 43% compared to angle based approach	In-accurate location information leads to retransmission and packet loss.	Number of alive node, delay	Radio range (75 m), Nodes(196), payload size (1), energy (200–400 units)	Payload size is fixed	Hybrid scheme of angle and distance	Distance based scheme, angle based scheme	–
[38]	Due to use of forward error correction mechanism the proposed routing outperforms than MCMP to retrieve original message.	MCMP protocol outperform better than proposed model with nonreal-data traffic because of low overhead by queuing.	Energy consumption, delivery ratio, average delay	Sensor node (300), transmission range (25 m), buffer size (256 K-bytes), transmit and receive power (13, 12 mW)	Fully connected environment, sensors must be failure free		MCMP	NS2
[39]	Secure against replay, DoS, Insider, Sybil, Eavesdrop	Every time at cluster formation CH authenticates with BS	Overhead, energy cost, resilience, cluster size, end-to-end delay	Buffer size (60), AES-128	–	SAMS	Das approach, Amin-biswas approach	–

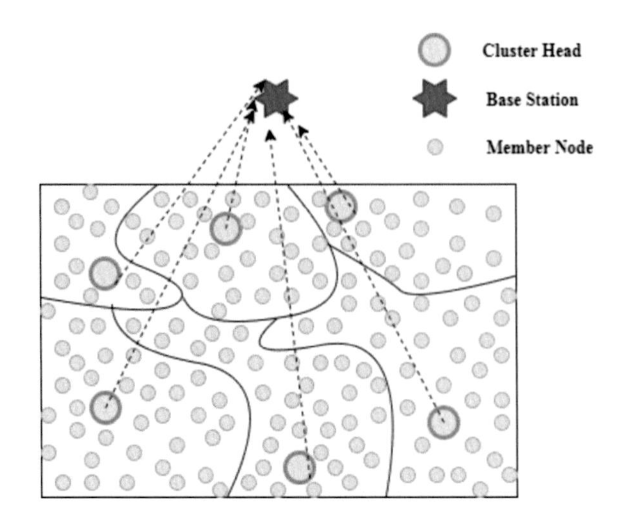

FIGURE 1.8

Wireless network model.

path algorithm to overcome the bandwidth consumption while ignoring storage of outdated states of sensors which is resolved by Savidge et al. in [37]. Here, the authors presented a purely location-based QoS routing which can become as a faulty network due to sharing of in-accurate location information by attacker node. To overcome the above issue, an AES-128 authentication based secure and cluster-based approach is designed in [39]. Although, here is an issue of frequent authentication at the time of fault-tolerance between CH to BS can leads to overhead at fault-tolerance. Some other cluster based approach has been designed with fault tolerance by selecting CH and CM according to available bandwidth, delay as well as energy fluctuation for real-time and nonrealtime IoT application with fixed payload size [35]. Moreover, the approach is limited due to no energy consumption from renewable resource. In [38], Othman and Yahya gave an energy-aware protocol based on balanced fault tolerance and forward error correction method which shows significant overhead in queuing process.

1.4 Conclusion and future directions

The Internet of Multimedia Thing (IoMT) has shown its applications in every corner of today's day-to-day life. The multimedia data sharing among the users through the IoMT network improves the user experiences at maximum height. However, the challenges in IoMT like real-time data delivery, throughput, device energy constraint, cross-layer issues, etc., still need to be handled in efficient manner. This chapter discusses the routing and QoS protocols to tackle the challenge of the IoMT network. The design of routing and QoS protocols primarily concentrated on the low energy consumption of the devices, minimum communication cost over the links, harvesting energy to reuse, and providing real-time experience to the user. The in-depth discussion on each of the protocols is presented lucidly. Further, in future, we need to collectively focus on Quality of Experience (QoE) parameters like inter disruption time between subsequent packet deliveries, packet buffering interval on intermediate nodes, and user satisfaction feedback with discussed QoS parameters such as packet loss, end-to-end delay, error rate to evaluate the performance of IoMT applications. An improved cross-layer model can be designed in view of the QoE as mentioned above and QoS parameters. The smart routing protocol based on machine learning and optimization techniques to discover the energy-efficient route for bandwidth-hungry IoMT application need to be further explored. The storage requirements for Big Multimedia Data (BMD) open the door to reinvestigate the energy harvesting-based routing and link layer protocol and strengthen the provision of green routing protocol for IoMT networks. The Software Defined Network (SDN) architecture can be incorporated to improve the routing capabilities by decoupling network control from forwarding decisions in IoMT applications. Moreover, some use cases can be taken for different types of IoMT applications to exemplify identified challenges.

References

[1] L. Atzori, A. Iera, G. Morabito, The internet of things: a survey, Computer Networks 54 (15) (2010) 2787–2805.

[2] S. Greengard, The Internet of Things, MIT Press, 2021.

[3] https://www.statista.com/statistics/1101442/iot-number-of-connected-devices-worldwide/. (Accessed 3 October 2021).

[4] F. Samie, L. Bauer, J. Henkel, Hierarchical classification for constrained IoT devices: a case study on human activity recognition, IEEE Internet of Things Journal 7 (9) (2020) 8287–8295.

[5] Y.B. Zikria, M.K. Afzal, F. Ishmanov, S.W. Kim, H. Yu, A survey on routing protocols supported by the Contiki Internet of things operating system, Future Generation Computer Systems 82 (2018) 200–219.

[6] F. Javed, M.K. Afzal, M. Sharif, B.S. Kim, Internet of Things (IoT) operating systems support, networking technologies, applications, and challenges: a comparative review, IEEE Communications Surveys and Tutorials 20 (3) (2018) 2062–2100.

[7] A.K. Maurya, D. Singh, IoT enabling platforms, in: Internet of Things, An Indian Adaptation: Concepts and Applications, Wiley, 2021, pp. 93–122.

[8] S. Li, L. Da Xu, S. Zhao, 5G Internet of Things: a survey, Journal of Industrial Information Integration 10 (2018) 1–9.

[9] G.A. Akpakwu, B.J. Silva, G.P. Hancke, A.M. Abu-Mahfouz, A survey on 5G networks for the Internet of Things: communication technologies and challenges, IEEE Access 6 (2017) 3619–3647.

[10] T. Vaiyapuri, V.S. Parvathy, V. Manikandan, N. Krishnaraj, D. Gupta, K. Shankar, A novel hybrid optimization for cluster-based routing protocol in information-centric wireless sensor networks for IoT based mobile edge computing, Wireless Personal Communications (2021) 1–24.

[11] A.K. Maurya, D. Singh, Median predictor based data compression algorithm for wireless sensor network, International Journal of Computer Applications 975 (2011) 8887.
[12] W. Sun, Z. Wang, G. Zhang, A QoS-guaranteed intelligent routing mechanism in software-defined networks, Computer Networks 185 (2021) 107709.1.
[13] V.K. Quy, V.H. Nam, D.M. Linh, N.T. Ban, N.D. Han, A survey of QoS-aware routing protocols for the MANET-WSN convergence scenarios in IoT networks, Wireless Personal Communications (2021) 1–14.
[14] A.K. Maurya, D. Singh, A. Kumar, R. Maurya, Random waypoint mobility model based performance estimation of On-Demand routing protocols in MANET for CBR applications, in: 2014 International Conference on Computing for Sustainable Global Development (INDIACom), IEEE, 2014, pp. 835–839.
[15] D. Singh, A.K. Maurya, A.K. Sarje, Comparative performance analysis of LANMAR, LAR1, DYMO and ZRP routing protocols in MANET using Random Waypoint Mobility Model, in: 2011 3rd International Conference on Electronics Computer Technology, IEEE, 2011, pp. 62–66.
[16] A.K. Maurya, D. Singh, A. Kumar, Performance comparison of DSR, OLSR and FSR routing protocols in MANET using random waypoint mobility model, International Journal of Information and Electronics Engineering 3 (5) (2013) 440–443.
[17] A. Katiyar, D. Singh, R.S. Yadav, State-of-the-art approach to clustering protocols in vanet: a survey, Wireless Networks 26 (7) (2020) 5307–5336.
[18] G. Sun, Y. Zhang, H. Yu, X. Du, M. Guizani, Intersection fog-based distributed routing for V2V communication in urban vehicular ad hoc networks, IEEE Transactions on Intelligent Transportation Systems 21 (6) (2019) 2409–2426.
[19] G. Premsankar, M. Di Francesco, T. Taleb, Edge computing for the Internet of Things: a case study, IEEE Internet of Things Journal 5 (2) (2018) 1275–1284.
[20] D. Kumar, A.K. Maurya, G. Baranwal, IoT services in healthcare industry with fog/edge and cloud computing, in: IoT-Based Data Analytics for the Healthcare Industry, Elsevier, Amsterdam, The Netherlands, 2021, pp. 81–103.
[21] A.K. Maurya, A.K. Tripathi, Deadline-constrained algorithms for scheduling of bag-of-tasks and workflows in cloud computing environments, in: Proceedings of the 2nd International Conference on High Performance Compilation, Computing and Communications, 2018, pp. 6–10.
[22] Y.B. Zikria, M.K. Afzal, S.W. Kim, Internet of Multimedia Things (IoMT): opportunities, challenges and solutions, Sensors (Basel) 20 (8) (2020) 2334.
[23] A. Nauman, Y.A. Qadri, M. Amjad, Y.B. Zikria, M.K. Afzal, S.W. Kim, Multimedia Internet of Things: a comprehensive survey, IEEE Access 8 (2020) 8202–8250.
[24] Quality of Service Technical White Paper, Microsoft, Redmond, WA, USA, 1999.
[25] T. Winter, P. Thubert, A. Brandt, J.W. Hui, R. Kelsey, P. Levis, K. Pister, R. Struik, J. Vasseur, R.K. Alexander, RPL: IPv6 Routing Protocol for Low-Power and Lossy Networks, rfc, 6550, 1–157, 2012.
[26] S.A. Alvi, G.A. Shah, W. Mahmood, Energy efficient green routing protocol for internet of multimedia things, in: 2015 IEEE Tenth International Conference on Intelligent Sensors, Sensor Networks and Information Processing (ISSNIP), IEEE, 2015, pp. 1–6.
[27] S. Taghizadeh, H. Bobarshad, H. Elbiaze, CLRPL: context-aware and load balancing RPL for IoT networks under heavy and highly dynamic load, IEEE Access 6 (2018) 23277–23291.
[28] H. Bouzebiba, M. Lehsaini, FreeBW-RPL: a new RPL protocol objective function for internet of multimedia things, Wireless Personal Communications 112 (2020) 1003–1023.

[29] J.W. Lin, P.R. Chelliah, M.C. Hsu, J.X. Hou, Efficient fault-tolerant routing in IoT wireless sensor networks based on bipartite-flow graph modeling, IEEE Access 7 (2019) 14022–14034.
[30] T. Li, S. Tian, A. Liu, H. Liu, T. Pei, DDSV: optimizing delay and delivery ratio for multimedia big data collection in mobile sensing vehicles, IEEE Internet of Things Journal 5 (5) (2018) 3474–3486.
[31] G. Chen, J. Tang, J.P. Coon, Optimal routing for multihop social-based D2D communications in the Internet of Things, IEEE Internet of Things Journal 5 (3) (2018) 1880–1889.
[32] T.D. Nguyen, J.Y. Khan, D.T. Ngo, A distributed energy-harvesting-aware routing algorithm for heterogeneous IoT networks, IEEE Transactions on Green Communications and Networking 2 (4) (2018) 1115–1127.
[33] F.M. Al-Turjman, H.S. Hassanein, M.A. Ibnkahla, Efficient deployment of wireless sensor networks targeting environment monitoring applications, Computer Communications 36 (2) (2013) 135–148.
[34] N. Saha, S. Bera, S. Misra, Sway: traffic-aware QoS routing in software-defined IoT, IEEE Transactions on Emerging Topics in Computing (2018).
[35] M. Amjad, M.K. Afzal, T. Umer, B.S. Kim, QoS-aware and heterogeneously clustered routing protocol for wireless sensor networks, IEEE Access 5 (2017) 10250–10262.
[36] Z. Hamid, F.B. Hussain, QoS in wireless multimedia sensor networks: a layered and cross-layered approach, Wireless Personal Communications 75 (1) (2014) 729–757.
[37] L. Savidge, H. Lee, H. Aghajan, A. Goldsmith, QoS-based geographic routing for event-driven image sensor networks, in: 2nd International Conference on Broadband Networks, IEEE, 2005, pp. 991–1000.
[38] J. Ben-Othman, B. Yahya, Energy efficient and QoS based routing protocol for wireless sensor networks, Journal of Parallel and Distributed Computing 70 (8) (2010) 849–857.
[39] M.A. Jan, M. Usman, X. He, A.U. Rehman, SAMS: a seamless and authorized multimedia streaming framework for WMSN-based IoMT, IEEE Internet of Things Journal 6 (2) (2018) 1576–1583.

CHAPTER

2 Energy efficient data communication in Internet of Multimedia Things (IoMT)

Shailendra Shukla[a], Asheesh Kumar Mani Tripathi[b], and Amit Kumar Singh[c]

[a]*Computer Science and Engineering Department, MNNIT Allahabad, Allahabad, India*

[b]*Information and Cyber Security Services, HCL, Noida, India*

[c]*Computer Science and Engineering Department, NIT Patna, Patna, India*

2.1 Introduction

In the last couple of years, the advancement of low power and low-cost small-scaled devices such as Micro Electro Mechanical Systems (MEMS) [1] has given the explosive growth in the number of IoT devices [2] and increased scalability. IoT is useful for the application in home automation, security, automated applications to provide useful contextual information frequently to help with their works, and decision making. Internet-enabled sensors and actuators' operations can be rapid, efficient, and more economical. Multimedia data in IoT [3] is bulky and specially meant for the real-time application requires real-time communication. Few examples are real-time multimedia-based surveillance systems [4] for smart home, office, at critical infrastructure, telemedicine service in the smart hospital, remote patient monitoring, transportation management system, remote multimedia based monitoring, etc. Multimedia services require higher processing and memory resources, which raises the need for different modus operandi for multimedia-based IoT.

The IoT devices are mainly equipped with a mic and camera. The major task of equipped devices is to sense the multimedia content and send it to sink nodes. For accurate information to be communicated, knowledge of the exact location of the sensor nodes in WSN [5] is mandatory. However, this location knowledge is possible with the integration of GPS devices to the sensor devices, but an increase in cost and reduction in battery life of sensor nodes are some of the demerits of this approach. So an alternate approach for this is to detect the most important nodes [6] in the networks.

Critical node detection aka *boundary node* (B-N) [6] actually means to find the nodes, which reside in those positions of the network, which if removed then causes the partition in the networks. Hence, with the detection of boundary nodes, we can achieve connectivity and coverage of the network at a minimal cost. Till now, there

Internet of Multimedia Things (IoMT). https://doi.org/10.1016/B978-0-32-385845-8.00007-1

are various methods have been proposed which focus on the detection of boundary nodes in the sensor networks. But to the best of our knowledge, no method provides the virtual coordinates [7] simultaneously to the detected boundary nodes. So in this chapter, we propose an algorithm, which not only detects the boundary nodes but also provides the virtual coordinates to them, which later contributes to providing the virtual coordinates to all the remaining nonboundary nodes in the wireless sensor networks. In IoMT the nodes are required to be capable of highly resourceful which is not possible as IoT devices are low power and fewer storage devices. Hence the detection of boundary nodes and virtual coordinate optimize the resource requirements.

The main contribution of this chapter is to develop a distributed algorithm to detect the boundary nodes and then virtual coordinate in IoT networks using the central node (ς). The proposed algorithm is free from any hardware dependability and flooding-based communications, it simply uses the RSSI value and cosine rules of trigonometry and. Finally, the proposed algorithm is deployed over the RPL protocol. Remaining chapter has been organized as follows: Section 2.2 covers the related work. Section 2.3 deals with system model. Our proposed method is described in section 2.4. Section 2.5 deals with the results and comparison analysis with other proposed approach. And, finally we conclude our chapter in section 2.6.

2.2 Related work

This section of the chapter is divided into subparts in the first part presents the reported work on IoMT content dissemination protocol and the second part presents the various boundary detection and virtual coordinate algorithms.

2.2.1 Routing protocol for IoMT

Reported work on WSN in multimedia IoT (IoMT) shows various protocols which support the low power consumption and heterogeneity in IoT [8–11]. The WSN deployed for IoMT uses CoRE protocol rather than HTTP for the Internet and a lightweight OMA-DM to support low power and low resource IoMT devices. RPL (Routing Protocol for Low-Power and Lossy Networks) [12] is a routing protocol for resource-constrained devices that uses Destination Oriented Directed Acyclic Graph (DODAG) to maintain network topology. This DAG comprises multihop routes from leaf nodes to the root node. RPL optimizes an objective function and chooses the best path by selecting the desired predecessor nodes starting from leaf nodes. However, they are not developed for large multimedia data transmission.

The Green-RPL [13] is an energy-efficient routing protocol for IoMT. It is an enhanced version of the RPL protocol [12]. The previous RPL protocol implementations are not feasible for IoMT and do not consider multimedia data. In contrast, the Green-RPL protocol considers the data generated from multimedia devices. The Green-RPL protocol reduces energy consumption and carbon footprint emissions together with QoS requirements of applications. To guarantee QoS for a particular multimedia application, it determines the delay bound for the application. For example, in VoIP applications, the delay limitation usually is 120 msec. To ensure energy efficiency, the protocol considers the features of all the intermediary links between the leaf nodes and the root node and estimates the energy consumption by the chosen immediate predecessor node to support traffic needs for one more immediate successor node. An optimization model is given for this protocol based on various requirements and constraints. In Green-RPL routing protocol, the desired immediate predecessor node is selected according to the objective function minimizing the emissions of cumulative path carbon footprints on all the links from the immediate predecessor node to the root node, and satisfying the constraints such as cumulative path link energy, cumulative path delay, idle time, and battery status of the immediate predecessor node. When more than one immediate predecessor node satisfies the given constraints, then the node that provides the most greener path is chosen as the desired immediate predecessor node. In cumulative path link energy constraint, energy consumption by a node is calculated based on the quality of links in the selected route. The cumulative path delay constraint specifies that the predefined delay threshold should not be increased. According to the idle time constraint, an immediate predecessor node should be chosen as the desired node only if its idle time is sufficient to support another child node. In the last constraint battery status, a child node should choose an immediate predecessor node as the desired one if the battery level of its energy resources is greater than the predefined threshold. A modified version of RPL, known as Context-aware and load balancing RPL under heavy and highly dynamic load for IoT and IoMT applications is discussed in [14]. It addresses the power depletion and packet loss problems associated with RPL. Paper [15] proposes an efficient protocol CoUDP for real-time IoT multimedia communications. Paper [16] proposes a four-layer approach physical devices layer, network layer, combination layer, and context layer for IoMT data communications. Caching policy for popularity-aware video caching in the topology-aware content-centric network is proposed in the [17]. The paper [17] uses scalable video coding for fast video delivery and cached each video layer. Paper [18] proposed an approach to achieve reliability and reduce delay in IoMT networks. To enhance the QoS metrics, and efficient data gathering scheme is designed in [19]. Cross layer protocols for IoMT are proposed in [20] and [21]. The major limitation of reported work is most of them do not consider the energy efficiency and fault-tolerant approach for critical IoMT applications.

2.2.2 Boundary detection and virtual coordinate algorithms

According to [22], Boundary detection algorithms can be classified into 3 categories, geometrical approach, statistical approach, and topological approach. Paper [23] is considered as a geometrical approach as it is based on the exact geometric location identification using global positioning devices such as GPS. Using the location information, sensor nodes identify the voids to create an alternative path for data traffic. The author has used the term stuck node for the nodes which are closer to the destination than its one-hop neighbor. Paper [24] is based on some graph techniques and computational geometry, to find out the closed polygonal boundaries of the voids. It is also based on the assumption that every node must be able to obtain its location. The main drawback with the geometrical approach is the requirement of an external hardware device to find out the geographical position, which is not only costly but also doesn't work indoor and in bad weather conditions.

The statistical approach is based on some complex mathematical and statistical functions and categorizes the nodes into the interior and boundary nodes. Paper [25] uses the restricted stress centrality measure which is a subclass of centrality indices and makes a comparison of it with a defined threshold value that is whether it is above or below the threshold value, based on which it categorizes the nodes into the above-mentioned classes. The main drawback of this approach is its impractical node distribution and high-density requirement. Paper [26] utilized the fact that the degree of nodes that are closer to the boundary of the network will be less in comparison to the node interior nodes. No requirement of global positioning devices is one of the major advantages of the statistical approaches, but these methods work only with the denser regions. Moreover, these methods require a node degree of more than 100 or higher.

The third category is the topological approaches which use only the topological information available to sensor nodes in the network. Paper [27] describes a two-step algorithm boundary recognition and topology extraction. A combinatorial structure called flower and augmented cycles are searched in this approach which is the base of this algorithm, but it is not always the case where one can find a flower-like structure in a randomly deployed network which is the major drawback of this approach. Paper [28] builds a shortest path tree and finds out cut nodes by flooding the network. Using the cut nodes, it finds out the network boundary. Paper [29] and paper [30] use the iso-contours to recognize the network boundary. Iso-contours are build using hop-counts from a root node and the end nodes of these iso-contours are treated as the boundary nodes. The drawback with this approach is its high-density requirement and flooding of more than 3 different nodes. Paper [31] describes a 3 step approach for the detection of boundary nodes. In the first step, which is the information collection step, it builds an x-hop neighbor list. 2nd step is the path construction, in which the link between the x-hop neighbors is identified by building the first set and it uses a recursive procedure to construct the path and finally, the last set is made which contains the node which cannot be extended further. 3rd step is the path checking phase which categorizes the node as an interior or boundary node according to the intersection of neighbors of the first set and last set. According to my understanding, a lot of computation is done over a single node recursively for the path construction, which is the only drawback of this approach.

2.3 System model

An IoT network is assumed where sensor nodes are homogeneously spread into a region L in such a way that it forms a connected graph G(V, E). In a connected graph

G(V, E), V represents the set of nodes and E represents the link between the nodes. The Nbr(v) represents the neighbor set of a node v ($v \in V$). Each node is equipped with a small range camera and has a communication range 'R', also each node can determine the distance to its direct one hope neighbors using ranging techniques like RSSI, TOA, or TDOA [33] also they are not aware of their location. In consonance with Internet-Connected multimedia, we envisage our network broadly consists of the following components: **Central Node** (ς)**:** A node $\varsigma \in V$ is said to be a central node if its graph centrality is maximum [34]. **Boundary Node** (B-N)**:** A node $B - N \in V$ is said to be a Boundary node if it lies on the critical position of the network graph G. **Network Boundary:** Network boundary is the imaginary circle formed by the Boundary nodes after the first phase of the algorithm. **Interior Node** ($I - N$)**:** A node $I - N \in V$ is said to be an interior node if it lies inside the network boundary and not a Boundary nodes (B-N).

2.3.1 Problem description

The problem can be formulated as follows: To find an IoT node from a set Min($B - N$) ∈ V such that it lies at the most critical places of the networks, detection of a central node (ς) in the given graph G(V, E) which is at a minimum distance (min-dis) from all the nodes of networks. Our second problem is to find the nodes which are at the maximum distance from the central node and provide them some virtual coordinates so that a virtual overlay coordinate is a form for the data communications. The main contribution of this chapter is to develop a distributed algorithm to detect the virtual coordinate which is based on graph centrality (ς). Developed an algorithm to find the network boundary in IoT networks using the central node and simple cosine rules of trigonometry and finally, deploying the algorithms over the RPL protocol.

2.4 Proposed approach

2.4.1 Working of algorithm

The idea behind the proposed approach is to detect the boundary nodes and then to assign virtual coordinates to these boundary nodes (B-N) using the graph centrality. To assign virtual coordinates it uses RSSI (Received signal strength indicator), $P_r(RI(N_v[i]))$, to detect the distance between target nodes (mobile nodes present in the network). Since the boundary nodes are detected hence the anchor nodes [35] are detected from boundary nodes. Finally, trilateration is used to detect the virtual coordinate and critical nodes.

We have divided our proposed approach into 2 phases. In phase-I, the boundary nodes detection algorithm also acts as anchor nodes. In phases II-a and II-b, the distance estimation is used to detect the center node, and then a virtual coordinate algorithm is proposed.

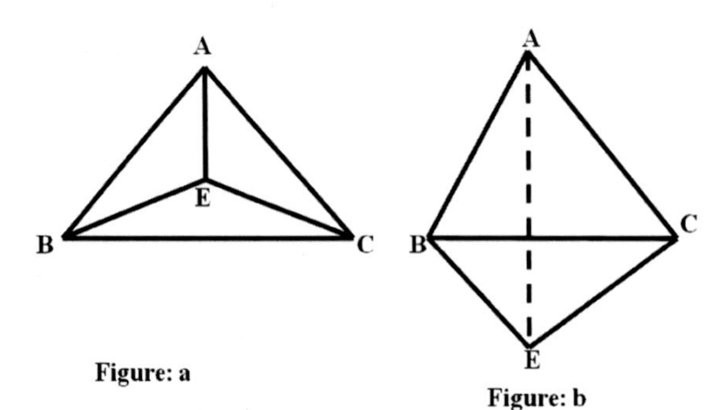

FIGURE 2.1

Node E as interior node (figure a) and boundary node (figure b).

2.4.2 Phase I: boundary detection algorithms

Assume a smaller sub graph of four nodes A, B, C, E. Suppose node E wants to test whether it is a boundary or interior node. Node E tests the *coverness* information i.e. it is covered by at least three nodes. Fig. 2.1 a and b shows the placement of node E as interior node and boundary node. For any set of four noncollinear nodes with complete connectivity, as illustrated in Fig. 2.1. If node E lies inside the triangle then it is interior node otherwise it is a boundary node. The distance between the nodes (E, C), (E, B), and (E, A) is represented as x_3, x_2, and x_1. We have used Lemma 1 to derive the relation to find the interior and boundary nodes and used HCR's inverse Cosine Formula along with the property of inscribing a center point in a triangle to prove it.

Lemma 1. *A given node is interior node if it satisfies three relationships in terms of angle.*

1. $\angle EAC < \angle CAB$

$$\cos^{-1}\left(\frac{b_1^2 + c^2 - x_3^2}{2x_1 c}\right) < \cos^{-1}\left(\frac{c^2 + b^2 - a^2}{2cb}\right) \tag{2.1}$$

2. $\angle EBC < \angle ABC$

$$\cos^{-1}\left(\frac{a^2 + x_2^2 - x3^2}{2ax_2}\right) < \cos^{-1}\left(\frac{a^2 + c^2 - b^2}{2ac}\right) \tag{2.2}$$

3. $\angle ECA < \angle ACB$

$$\cos^{-1}\left(\frac{x_3^2 + b^2 - x_1^2}{2bx_3}\right) < \cos^{-1}\left(\frac{a^2 + b^2 - c^2}{2ab}\right) \tag{2.3}$$

Proof. Angle divided into two parts will always be greater than the divided individual parts. For the boundary node this does not hold as illustrated in Fig. 2.1. Using Angle

Addition Postulate, Angle Addition Postulate states that if a point S lies in the interior of $\angle BAC$, then $\angle EAC + \angle EAB = \angle BAC$ from Fig. 2.1. □

A node N_v wants to test its boundary condition and initialize Algorithm 1. Node N_v reads the RSSI value of its neighbor and stores the value in $RI(N_v[i])$ to compute the distance $d(v_i)$ of its neighbor. Node N_v tests the boundary condition. If the number of a neighbor is greater than 3 and dissatisfies Lemma 1 then mark the B-N value as 1. Else the node is an interior node and mark B-N as 0, finally send the message (contains the value of B-N) to the sink node.

2.4.3 Phase II: assignment of virtual coordinates to boundary nodes

The initial task of the phase-II algorithm is to assign virtual coordinates to Boundary nodes B-N. To do this, all the nodes of the network are divided into two categories, Boundary nodes (B-N) and interior nodes as stated in Algorithm 1. Boundary nodes have to be mapped to an imaginary circle to form a network boundary. This task is done by the central node (ς). So the first task of this approach is to elect the central node which is done based on graph centrality $C_G(v)$ (Fig. 2.2).

Algorithm 1 Pseudocode to detect boundary node (B-N).

1: Initialize the node N_v
2: Read the $RSSI$ value from neighbor nodes.
3: $RI(N_v[i]) \leftarrow RI(N_v[i])$ + gain
4: $d(v_i) \leftarrow P_r(RI(N_v[i]))$
5: **if** $Nb(v) \geq 3$ && Lemma $1 = TRUE$ **then**
6: $\quad v$ is Interior node, mark $B - N_v = 0$.
7: **else**
8: $\quad v$ is Boundary node and mark $B - N_v = 1$.
9: **end if**
10: Send MSG[$B - N_v$] to Sink node

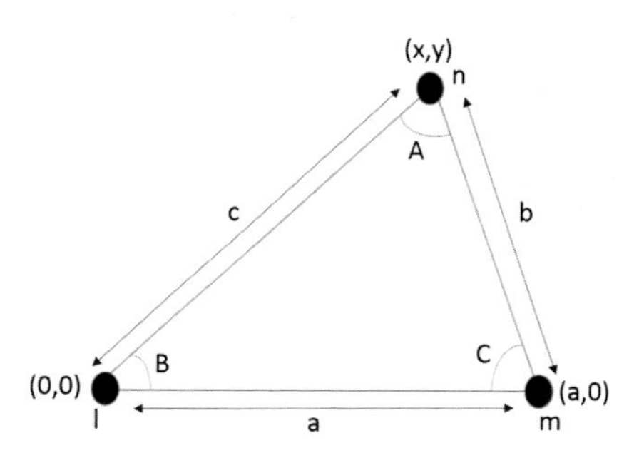

FIGURE 2.2

A triangle showing 3 nodes with their known distances.

1. Phase: II-a election of central node

In phase-II, each node v in the network broadcasts a hello message which is piggybacked with a hop counter. The piggybacked (hop count) calculates the hop distance between the nodes in networks. With the help of received piggybacked, Nodes know the hop count distance to each node in the network. Once piggybacked is received, then each node calculates its graph centrality. Each node compares its maximum hop count distance using $C_G(v)$:

$$C_G(v) = 1/\max_{t \in V} d_G(v, t) \quad \text{(graph centrality)} \tag{2.4}$$

The node with the maximum graph centrality will be elected as the central node and assigned with coordinates $(0, 0)$.

Algorithm 2 Phase: II-a election of central node (ς).

1: **Input:** Nodes know their one hop neighbor
2: **Output:** Central node (ς)
3: For each node v Broadcast a message < *node_id, hop_counter* > in the network, where node_id is id of each node from where the message is initialized and hop_counter is the counter of each hop.
4: At each node, build a distance vector D containing distances to all nodes
5: For each node
Select max_hop_count from their distance vector D
6: For each node
Calculate Graph centrality C_G and broadcast it into the network.
7: Return Maximum C_G as the central node (ς) and assign it coordinates $(0, 0)$.

2. Phase: II-b virtual coordinates to perimeter nodes

In the first step, all the nodes know the hop count distance to each node. So a distance vector D is build at each node. Central node will select its maximum hop count (max_d) distance node and assign it virtual coordinates $(max\_d, 0)$. Again select next maximum hop count node from the center and assign the virtual coordinates to it using cosine rule as described below.

According to cosine rule, we have

$$a^2 = b^2 + c^2 - 2bc\cos(A) \tag{2.5}$$

$$b^2 = a^2 + c^2 - 2ac\cos(B) \tag{2.6}$$

$$c^2 = a^2 + b^2 - 2ab\cos(C) \tag{2.7}$$

These equations can be converted to find the angles as follows:

$$A = \cos^{-1}(b^2 + c^2 - a^2/2bc) \tag{2.8}$$

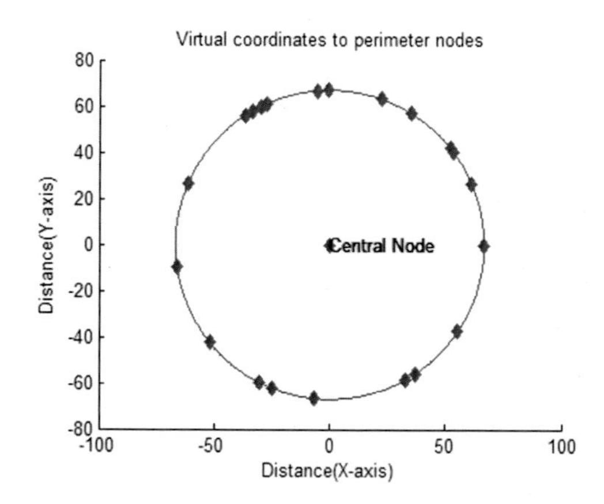

FIGURE 2.3

Circle representing the boundary nodes.

$$B = \cos^{-1}(a^2 + c^2 - b^2/2ac) \tag{2.9}$$

$$C = \cos^{-1}(a^2 + b^2 - c^2/2ab) \tag{2.10}$$

here it is needed to calculate the angle made at the central node, which is $\angle B$ calculated by Eqs. (2.5)–(2.10). Now as we know the angle made at the central node, we can assign coordinates to node $n(x, y)$ as $(a\cos(B), a\sin(B))$. In the similar way, we assign coordinates to remaining $k - 2$ perimeter nodes by assuming central node (ς) and first boundary node as the reference nodes shown in Fig. 2.3.

This is how, we assign coordinates to k perimeter nodes by using the central node. Now the next task is to assign coordinates to remaining $n - k$ nonperimeter nodes, which is explained in the next section.

Algorithm 3 Phase: II-b virtual coordinate assignment.

1: **Input:** Select max_hop_count from the Distance Vector D of central node (ς).
2: **Output:** Assign it coordinates (max_hop_count, 0)
3: repeat 4 to 5 $k - 1$ times
4: At each node, build a distance vector D containing distances to all nodes
5: For each node
Select max_hop_count from their distance vector D
6: Select next maximum hop_count from D of ς and find its coordinates (x, y) as follows:
7: Calculate internal angle made at ς using cosine rule i.e. Eqs. (2.5)–(2.10).
8: assign $x = max\_hop\_count * \cos(B)$
9: assign $y = max\_hop\_count * \sin(B)$
10: Return (x,y) as coordinates of this boundary node (B-N).

2.4.3.1 Complexity

The complexity of our proposed VC-RPL protocol depends on the number of neighbors and its 2D list intersection computation ($\mathcal{O}(Nb(N_v))$). Since it is a 2D list so the time complexity of intersection will take almost $\mathcal{O}(N^2)$, and computation of boundary nodes and virtual coordinate assignment requires $\mathcal{O}(\frac{(N)!}{3!})$. So the overall complexity will be around $\mathcal{O}(N^2) + \mathcal{O}(\frac{(N)!}{3!}) = \mathcal{O}(N^3)$. The communication cost is $\mathcal{O}(1)$ as it is a distributed algorithm and depends only on 1-hop information of Algorithm 1.

2.5 Implementation and results

The proposed algorithm is implemented over the RPL protocol.

2.5.1 Preliminaries

RPL [12] is considered as one of the most prominent routing protocols for data delivery in IoT networks. The limitations of IoT networks are they have limited resources and lousy connections. RPL routing protocol provides both P2MP (Point-to-Multipoint) and MP2P (Multipoint-to-Point) packets forwarding. RPL protocols form a Destination-Oriented Directed Acyclic Graph (DODAG) to remove the cycle in the networks and more than one path to the sink or root nodes. RPL protocol uses three types of ICMPv6 [32] messages for DODAG formation: DIO (DODAG Information Object) is used to form the networks by advertising whereas the DIS (DODAG Information Solicitation) is used for the new nodes to join the networks by soliciting for DIO messages. Once the DODAG network is formed then the node sends the DAO (Destination Advertisement Object) message to the information in the up direction.

The initial step of the RPL protocol is to build the RANK for each of the nodes in the networks. The purpose of the RANK is to help the DIO/DAO message to create and maintain the Directed Acyclic Graph (DAG) root. RANK is assigned in such a way that the node's rank increases as we move away from the root and decreases when moved away it increases. To generate and construct the RANK an Objective Function (OF) is used. Objective Function (OF) is the network administrator-defined routing metrics like $RSSI$, Expected Transmission Count (ETX), residual energy, hop count, link quality, etc. Once the RANK is assigned to all the nodes in the networks then by the help of the DIO/DAO message assign itself as the candidate as the root of neighboring nodes. An overlay RSSI-based network of Boundary Node (B-N) is created over RPL-DAG.

2.5.2 Simulation setup

To evaluate the effect of our proposed virtual coordinate detection algorithm, we have considered the Cooja simulator of Contiki 3. It is a lightweight and open-source operating system designed specifically for IoT devices. We have picked Tmote Sky (MSP430-based board) motes that uses IEEE specified 802.15.4 compliant CC2420 radio chip standards over the MAC layer and mobility based RP L [12] protocol for routing at the network layer.

Each node is randomly (uniform random distribution) deployed in the two-dimensional simulation area. The simulation scenario consists of 20 to 50 nodes in an area of 200×200 units^2. One node is acting as a sink node and as discussed in [12] DIO-Minimum-Interval, DIO-IntervalDoublings, and Radio-Duty-Cycling intervals are set to 12, 16, and 16, respectively. Performance is measured by taking the median of 20 experimental results with different seeds to statistically validate the results. The simulation is performed for 60 minutes with a packet size of 1 KB and a data rate of 250 kbps. The data is generated at the rate of 25 pkts/sec. Since the work is only limited to the performance of the routing protocol so we have to ignore the MAC layer. For comparison, the network topology is kept the same for all three cases, i.e. VC-RPL(proposed), Green-RPL, and ETX-based RPL approach.

2.5.2.1 Boundary and Internal nodes

The results in this subsection present the performance of our proposed algorithm at different communication ranges. The number of deployed sensor nodes is 150 and 300 in an area of 1000×1000 m^2 square. Results in Fig. 2.4 and Fig. 2.5 show the boundary nodes detected using Algorithm 1. We can observe that there is a decrement in the percentage boundary nodes from Fig. 2.4 to Fig. 2.5. This is because of the increase in density of nodes in the networks which helps Algorithm 1 to implement the cosine rule.

2.5.2.2 Percentage of correctly detected Boundary nodes

The result in Fig. 2.6 shows the correctly detected Boundary nodes (N-N) with respect to the increase in the number of nodes. We have compared the boundary nodes detected after Algorithm 1 with theoretical boundary detection. In this simulation, the number of nodes in the networks is increased from 50 to 300 then the correctly detected boundary nodes (B-N) are estimated. Since the boundary nodes are essential to be detected to know as they are the most important nodes for multimedia cache storage. The result shows that our proposed algorithm has a slight false positivity, it shows that Algorithm 1 produces 4% to 40% extra boundary nodes with the increase in the number of nodes in networks.

2.5.2.3 Number of packets delivered to the sink node

In this simulation, we compared three approaches (ETX-RPL, Green-RPL, and VC-RPL) to analyze the behavior in multimedia-based networks. The result in Fig. 2.7 illustrates the successful delivery of our packets to the root node in 20 nodes networks. The simulation is run for almost one hour and the result clearly shows that in our proposed case VC-RPL algorithm provides a significant improvement in the number of successful packet transmissions. The result shows in Fig. 2.7 that our proposed algorithm outperforms the ETX-RPL approach as it delivers the 3 times more

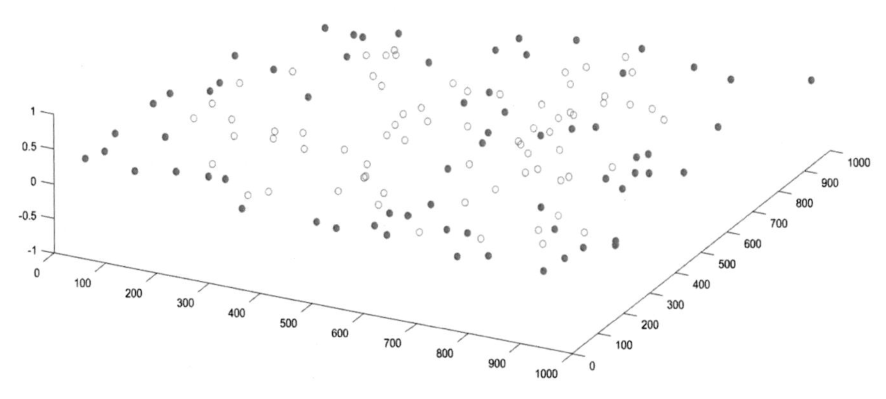

FIGURE 2.4

B-N at 150 node (using Algorithm 1).

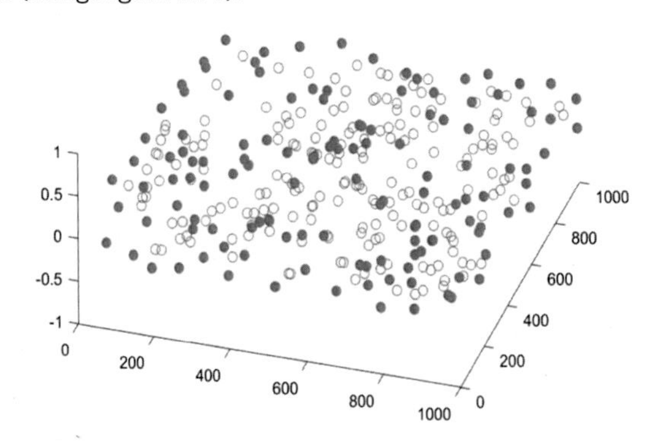

FIGURE 2.5

B-N at 300 node (using Algorithm 1).

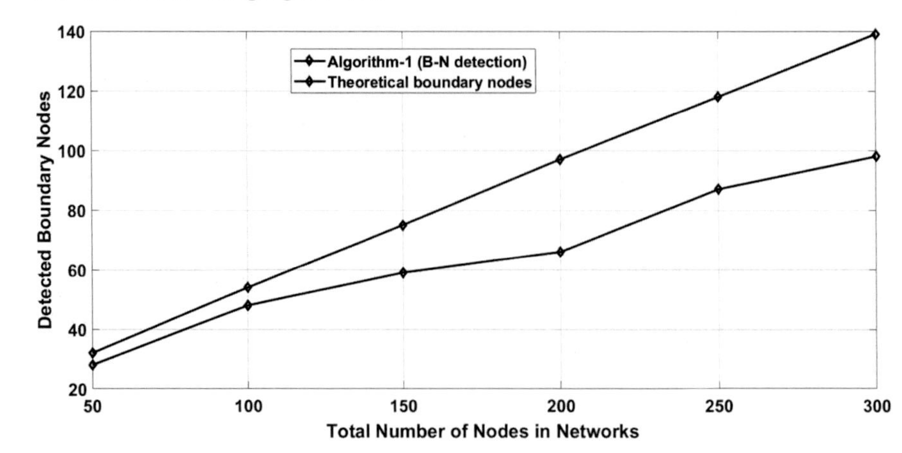

FIGURE 2.6

Percentage of correctly detected boundary nodes vs. total number of nodes.

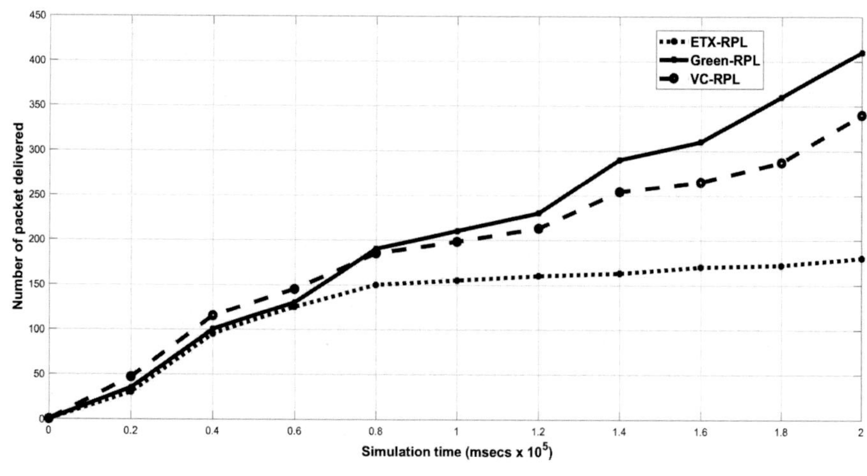

FIGURE 2.7

Number of packets delivered to the sink node.

data in IoT networks however the performance of VC-RPL is slightly less by 1.5 times when compared with the Green-RPL.

2.5.2.4 Energy consumptions with respect to time

Cooja simulator provides the time information of motes activity/state like transit, listen, etc. In our simulation, we have considered the TMote specification for the computation of energy consumption. According to the specification voltage of 3.6 V is required for the TMote and current values transmit (21 mA), listen (23 mA), low power mode (0.21 mA), ideal (1.2 mA), CPU state (2.4 mA). The comparative analysis of the Green-RPL, ETX-RPL, and VC-RPL is illustrated in Fig. 2.8. The cumulative energy consumed by all the networks node is shown in the resulting Fig. 2.8. The result shows that our proposed algorithm initially requires high energy as it has to compute the virtual coordinates first and then the normal RPL gets on the execution. The result shows that our proposed algorithm required almost 50% less energy when compared to the green RPL. The higher energy consumption in Green-RPL is due to the higher number of packet generation as can be seen in Fig. 2.7. ETX-RPL is the basic RPL protocol with objective function as ETX, Since its delivery ratio is low hence the energy consumption is also low.

2.6 Conclusion

This chapter presents the problem of energy-efficient multimedia content delivery in IoT networks. To solve this problem, the chapter considers the problem of Boundary node (critical node) selection in the IoMT scenario. To select boundary nodes, a spa-

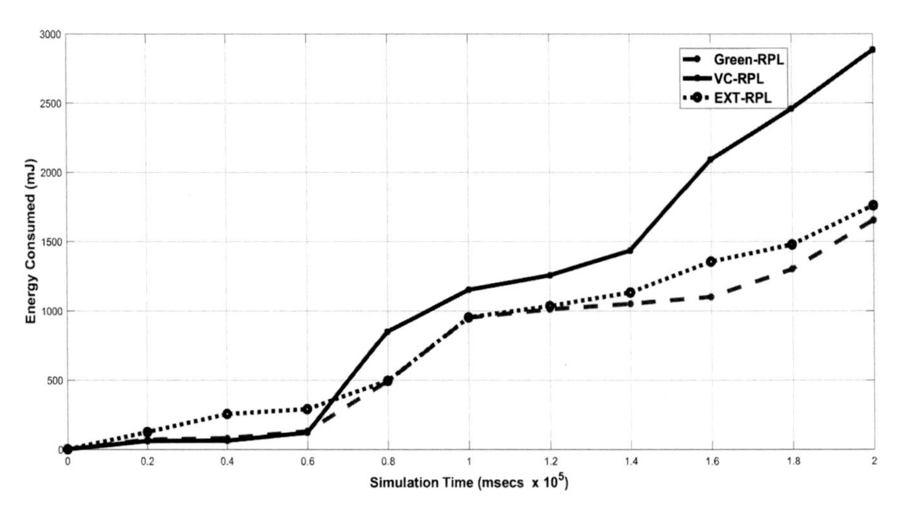

FIGURE 2.8

Energy consumed with respect to time.

tial correlation measurement-based virtual coordinate algorithm that depends only upon the local connectivity is proposed. The proposed correlation degree is a spatial correlation measurement that measures the correlation between an IoT device's position and its neighboring position. We have compared our proposed algorithm with the existing IoMT algorithm in terms of delivery ratio and energy consumption. The result shows that our proposed algorithm required almost 50% less energy when compared to the green RPL and delivers 3 times more data than ETX-RPL. For future work, we are planning to implement our proposed approach in a dynamic IoT network and independent of RPL protocols. Also, the boundary node can also act as the cache-store, which is not much explored in the chapter.

References

[1] Chih-Ming Ho, Yu-Chong Tai, Micro-electro-mechanical-systems (MEMS) and fluid flows, Annual Review of Fluid Mechanics 30 (1) (1998) 579–612.

[2] M.W. Woo, J. Lee, K. Park, A reliable IoT system for personal healthcare devices, Future Generation Computer Systems 78 (2018) 626–640.

[3] B. Jiang, J. Yang, H. Xu, H. Song, G. Zheng, Multimedia data throughput maximization in Internet-of-Things system based on optimization of cache-enabled UAV, IEEE Internet of Things Journal 6 (2) (2018) 3525–3532.

[4] G. Valenzise, L. Gerosa, M. Tagliasacchi, F. Antonacci, A. Sarti, Scream and gunshot detection and localization for audio-surveillance systems, in: 2007 IEEE Conference on Advanced Video and Signal Based Surveillance, IEEE, 2007, September, pp. 21–26.

[5] P. Maheshwari, A.K. Sharma, K. Verma, Energy efficient cluster based routing protocol for WSN using butterfly optimization algorithm and ant colony optimization, Ad Hoc

Networks 110 (2021) 102317.
[6] K.K. Pattanaik, A. Trivedi, A dynamic distributed boundary node detection algorithm for management zone delineation in precision agriculture, Journal of Network and Computer Applications 167 (2020) 102712.
[7] Y. Zhang, W. Shen, A novel particle swarm optimization algorithm for k-coverage problems in wireless sensor networks, in: 2021 IEEE 24th International Conference on Computer Supported Cooperative Work in Design (CSCWD), IEEE, 2021, May, pp. 831–836.
[8] Q. Du, L. Sun, H. Song, P. Ren, Security enhancement for wireless multimedia communications by fountain code, in: IEEE COMSOC MMTC Commun. Front., vol. 11, no. 2, Mar. 2016, pp. 47–51.
[9] S.H. Ahmed, S.H. Bouk, H. Song, Multimedia streaming in named data networks and 5G networks, IEEE COMSOC MMTC Commun. Front. 11 (2) (Mar. 2016) 57–61.
[10] J. Yang, et al., Multimedia recommendation and transmission system based on cloud platform, Future Generation Computer Systems 70 (May 2017) 94–103.
[11] F. Siddiqui, J. Beley, S. Zeadally, G. Braught, Secure and lightweight communication in heterogeneous IoT environments, Internet of Things 14 (2021) 100093.
[12] T. Winter, P. Thubert, A. Brandt, J.W. Hui, R. Kelsey, P. Levis, et al., RPL: IPv6 Routing Protocol for Low-Power and Lossy Networks, rfc, 6550, 2012, 1–157.
[13] S.A. Alvi, G.A. Shah, W. Mahmood, Energy efficient green routing protocol for internet of multimedia things, in: 2015 IEEE Tenth International Conference on Intelligent Sensors, Sensor Networks and Information Processing (ISSNIP), IEEE, 2015, April, pp. 1–6.
[14] S. Taghizadeh, H. Bobarshad, H. Elbiaze, CLRPL: context-aware and load balancing RPL for IoT networks under heavy and highly dynamic load, IEEE Access 6 (2018) 23277–23291.
[15] W. Jiang, L. Meng, Design of real time multimedia platform and protocol to the Internet of Things, in: Proc. IEEE 11th Int. Conf. Trust Security Privacy Comput. Commun. (TrustCom), Liverpool, UK, 2012, pp. 1805–1810.
[16] A. Floris, L. Atzori, Quality of experience in the multimedia Internet of Things: definition and practical use-cases, in: Proc. IEEE Int. Conf. Commun. Workshop (ICCW), London, UK, 2015, pp. 1747–1752.
[17] Z. Liu, et al., Fast-start video delivery in future Internet architectures with intra-domain caching, Mobile Networks and Applications 22 (1) (Feb. 2017) 98–112.
[18] M. Dong, K. Ota, A. Liu, M. Guo, Joint optimization of lifetime and transport delay under reliability constraint wireless sensor networks, IEEE Transactions on Parallel and Distributed Systems 27 (1) (Jan. 2016) 225–236.
[19] J. Long, M. Dong, K. Ota, A. Liu, S. Hai, Reliability guaranteed efficient data gathering in wireless sensor networks, IEEE Access 3 (2015) 430–444.
[20] C. Han, J.M. Jornet, E. Fadel, I.F. Akyildiz, A cross-layer communication module for the Internet of Things, Computer Networks 57 (3) (2013) 622–633.
[21] I. Al-Anbagi, M. Erol-Kantarci, H.T. Mouftah, A survey on crosslayer quality-of-service approaches in WSNs for delay and reliabilityaware applications, IEEE Communications Surveys & Tutorials 18 (1) (1st Quart. 2016) 525–552.
[22] I. Khan, H. Mokhtar, M. Merabti, A survey of boundary detection algorithms for sensor networks, in: Proceedings of the 9th Annual Postgraduate Symposium on the Convergence of Telecommunications, Networking and Broadcasting, 2008.
[23] Q. Fang, J. Gao, L.J. Guibas, Locating and bypassing holes in sensor networks, Mobile Networks and Applications 11 (2) (2006) 187–200.
[24] A. Shirsat, B. Bhargava, Local geometric algorithm for hole boundary detection in sensor networks, Security and Communication Networks 4 (9) (2011) 1003–1012.

[25] S.P. Fekete, M. Kaufmann, A. Kroller, K. Lehmann, A new approach for boundary recognition in geometric sensor networks, in: Proc. 17th Canadian Conference on Computational Geometry (CCCG'05), Windsor, 2005, pp. 82–85.
[26] S.P. Fekete, A. Kroller, D. Pfisterer, S. Fischer, C. Buschmann, Neighborhood-based topology recognition in sensor networks, in: Proc. 1st International Workshop on Algorithmic Aspects of Wireless Sensor Networks (ALGOSENSORS'04), Turku, Finland, July 2004.
[27] A. Kroller, S.P. Fekete, D. Pfisterer, S. Fischer, Deterministic boundary recognition and topology extraction for large sensor networks, in: Proc. 17th Annual ACM-SIAM Symposium on Discrete Algorithms (SODA'06), 2006, pp. 1000–1009.
[28] Y. Wang, J. Gao, Joseph S.B. Mitchell, Boundary recognition in sensor networks by topological methods, in: Proc. 12th Annual International Conference on Mobile Computing and Networking (Mobi-Com'06), Los Angeles, California, September 2006.
[29] S. Funke, Topological hole detection in wireless sensor networks and its applications, in: Proc. of 3rd ACM/SIGMOBILE International Workshop on Foundations of Mobile Computing (DIALM-POMC), Cologne, 2005, pp. 44–53.
[30] S. Funke, C. Klein, Hole detection or: "How much geometry hides in connectivity?", in: Proc. 22nd ACM Symposium on Computational Geometry (SoCG), Sedona, 2006, pp. 377–385.
[31] I.M. Khan, N. Jabeur, S. Zeadally, Hop-based approach for holes and boundary detection in wireless sensor networks, IET Wireless Sensor Systems 2 (4) (2012).
[32] A. Conta, S. Deering, M. Gupta, Internet control message protocol (icmpv6) for the Internet protocol version 6 (ipv6) specification, RFC 2463, December 1998.
[33] M. Laaraiedh, L. Yu, S. Avrillon, B. Uguen, Comparison of hybrid localization schemes using RSSI, TOA, and TDOA, in: 17th European Wireless 2011 – Sustainable Wireless Technologies, VDE, 2011, April, pp. 1–5.
[34] S.P. Borgatti, M.G. Everett, A graph-theoretic perspective on centrality, Social Networks 28 (4) (2006) 466–484.
[35] Taha Bouchoucha, Zhi Ding, Anchor selection for topology inference and routing in wireless sensor networks, Journal of Communications and Information Networks 5 (3) (2020) 318–323.

CHAPTER

3 Visual information processing and transmission in Wireless Multimedia Sensor Networks: a deep learning based practical approach

Yasar Abbas Ur Rehman[a] **and Muhammad Tariq**[b]

[a]*TCL Corporate Research Hong Kong, Hong Kong*

[b]*National University of Computer and Emerging Sciences (NUCES), Peshawar, Pakistan*

3.1 Introduction

The integration of low-cost Complementary Metal-Oxide Semiconductor (CMOS) cameras and microphones to the traditional wireless sensing devices has opened new gateways to harvest multimedia information from the physical environment. This integration enables, the now called, Wireless Multimedia Sensor Networks (WMSN) to simultaneously capture audio, video, still images, and scalar sensors data. As a result, it also attracts both the research and industrial communities from multiple domains in signal processing, computer vision, and communication networks.

Nowadays, WMSN has found a wide array of applications that include the health sector, border security, seismic activity monitoring, weather prediction, surveillance, industrial automation, and smart cities [70], etc. The Wireless Multimedia Sensor (WMSens) nodes work collaboratively with other wireless sensor nodes in WMSN, which could be scalar or visual, homogenous or heterogenous, to process and transmit the information from the acquisition end to the receiving end [1].

Visual data carries complex and a multitude of information compared to data from other media. As a result, the visual data in WMSens (equipped with visual sensors) and WMSN as a whole enables potential new applications and improves the existing applications in industrial automation, environmental monitoring, telemedicine, and wearable sensors. For example, in habitat monitoring, vision systems (visual sensors and visual processing units) are being deployed underwater to study the marine fauna [74]. In road monitoring, vision systems provide speed checks, the number of cars passing through highways, and automatic accident detection and reporting. In telehealth [7], vision systems are utilized in monitoring and measuring the flow of

Internet of Multimedia Things (IoMT). https://doi.org/10.1016/B978-0-32-385845-8.00008-3

Intravenous (IV) infusions [21]. Also, remote photoplethysmography (rPPG) signals from the face image of the person are utilized to measure the heart rate and blood pressure [80]. In automatic surveillance, vision systems provide continuous monitoring and reporting of any unusual incident or movement. In industrial automation, the vision systems help in identifying the defects in the chips during manufacturing, automatic packaging, and placement. There are also potential benefits to the WMSN by utilizing vision systems. Incorporating vision systems in WMSN can provide an enhanced view in multiple resolutions thus improving analyses and getting time-critical information that other sensors may not provide [13].

The integration of the vision system to traditional wireless-based scalar sensor nodes has added additional capabilities to wireless sensor nodes (now called wireless visual nodes or WMSens nodes). However, integrating the vision system to the conventional wireless-based sensor platform requires the modification in traditional image processing and computer vision techniques, which would be applicable, compatible, and efficient once deployed in these wireless-based sensor nodes [87]. This is because visual sensors generate a huge amount of data – unlike scalar data. In contrast, wireless sensor platforms offer limited bandwidth, battery power, computational and storage capabilities, respectively.

In addition to the modification of the traditional computer vision and image processing techniques required for the vision systems in WMSens, an additional challenge is provided by the fast-paced development in computer vision algorithms utilizing deep learning and artificial neural networks. With the successful incarnation of Deep Neural Networks (DNN) in 2012 [38], computer vision algorithms have been divided into 2 categories. Hand-crafted features-based algorithms and deep feature-based algorithms. The former category of algorithms has been developed for various Internet of Things (IoT) applications. However, the accuracy of these algorithms degrades within the complex environment as these algorithms were not trained on a sufficient amount of data. On the other hand, the deep feature-based algorithms have seen tremendous success in vision-based applications, such as autonomous driving assistance, scene understanding, semantic segmentation, object detection. However, deploying the same in the WMSens node would require intelligent modification to maintain the required level of accuracy and energy cost, respectively.

The objectives for these modifications can be summarized as follows:

1. To enable the WMSN nodes to capture the key visual information, along with other scalar data, in their vicinity.
2. To enable the deployment of state-of-the-art computer vision algorithms based on hand-crafted features and deep learning in WMSens for enhanced performance and to add additional capabilities to the WMSN.
3. To increase the lifetime of a wireless sensor node by minimizing the in-node and subsequently the in-network energy cost.
4. To build autonomous systems, that are self-sufficient, and can respond timely to user-based queries.

Different from other works and to the best of our knowledge, this chapter is the first to discuss a variety of computer vision algorithms related to object detection, semantic segmentation, low-light image restoration, semantic colorization, superresolution, and their implementation in WMSens and WMSN. We believe that it would provide people in both the research and industrial community with new insights into the design and realization of novel potential applications in WMSens and WMSN.

This chapter first provides an overview of the vision systems in WMSens nodes, the protocols developed to transmit the visual information from the WMSens nodes to the base station, hand-crafted feature-based algorithms, and deep feature-based algorithms deployed in WMSN, respectively in section 3.2. We also give a brief introduction of the deep neural network architectures to make this chapter self-sufficient for the readers in section 3.3. A variety of hand-crafted feature-based algorithms and deep feature-based algorithms are utilized and those which can be potential candidate algorithms in WMSN for various applications are discussed in section 3.4. Finally, the chapter is concluded with a conclusion and future work in section 3.5.

3.2 Literature review

Vision system in WMSens motes, incorporating smart camera or combination of cameras, has been around for more than a decade, such as Mesheye [29], CITRIC [12], and SensEye [39]. The vision module in these WMSN motes is capable of processing the object geometry, position, and distance from the WMSN mote, etc. They are also capable of performing some low-level object detection, object classification, and semantic labeling [15]. Often the vision module in these WMSens nodes incorporates quite low-resolution cameras, (usually monochrome), to conserve the WMSens node energy during the acquisition and processing of visual information. The high-resolution camera is incorporated and activated in case of acquiring more details of the object [37].

The visual information, whether in raw format or processed format, has to be transmitted from the source WMSens to the base station or the receiving node for further processing [101]. Visual information processing in the WMSens node has a relationship with the in-node energy cost and in-network energy cost. For example, the WMSens in-node energy is consumed in activating the sensor to capture the information, processing the information, and transmitting the processed information to the nearby WMSens nodes or Wireless Sensor (WSens) nodes. On the other hand, the in-network energy is consumed during network topology maintenance, further processing of the received information, and transmitting it to the base station [3].

The conservation of in-node energy in WMSens largely depends upon the visual processing algorithm and the corresponding protocols developed for processing the incoming information and transmitting it to the nearby homogenous or heterogeneous sensor nodes [87]. The question now is: How much information should be processed at the sensor node to prolong the WMSens node lifetime and consequently the WMSN?

A WMSens node consumes energy during the processing of information using built-in systems that execute a variety of algorithms and during the transmission of information to the nearby WMSens node or base station [4]. The energy consumed by the WMSens node due to the transmission of visual information has a direct relationship with the visual processing algorithm within the WMSens node and the amount of information (visual/semantic) that needs to be transmitted. Additionally, the energy consumption during information processing is comparatively lower than energy consumption during the transmission of information [53]. As a result, the processed information needs to be compressed inside the WMSens before transmitting it to the base station [41]. Several techniques have been proposed in the literature to conserve in-node energy while processing visual information [5,73,85,88].

This leads us to design the visual processing system in WMsens with two key ingredients: vision algorithms and compression algorithms. For example, a visual processing system in WMSens may consist of object detection, using background subtraction or other schemes, to detect and localize the stationary or moving object in its vicinity [66]. This information can then be further processed to understand the object geometry, perform object classification, and semantic labeling. As a result, the information processed by the visual processing system can be classified into low-level, medium-level, and high-level information, respectively [98]. The information at different levels can then be compressed using various encoding schemes (lossless encoding and lossy encoding) to reduce the amount of information being transmitted, thus decreasing the transmission energy cost.

In some cases, the visual algorithm alone can provide some form of compression, e.g., generating semantic labels from input images and transmitting them to the base station [25]. However, this would put much burden on the WMSens node that processes this information, particularly when the battery power to WMSens is limited. In the worse case, if the visual algorithm is not robust to environmental changes, false detection could happen and can cause a swift drain out of energy, both in the WMSens node and WMSN as a whole, due to perpetual transmission.

This leads us to the second question of how much we can rely on the visual processing algorithm deployed in the WMSens node to ensure the prolonged battery life of WMSens. Since a WMSens comes with stringent computing capabilities, therefore the low-level computer vision algorithms based on hand-crafted features have been the choice of processing visual information. However, these algorithms were often designed by experts in the field and cannot cover every possible situation. Thus these algorithms lack robustness in the case of a dynamic environment. Nonetheless, they are much compatible with WMSens nodes compared with the high-level computer vision algorithms [2,15,36,87]. Table 3.1 shows the comparison of the methods proposed for visual information processing in WMSens and WMSN. The methods are compared based on the application type, consideration of in-node processing, in-node energy minimization, in-network energy minimization, compression method, data transmission mode, and the type of data transmitted. Additionally, Table 3.2 shows the comparison of various schemes based on the type of features, hand-crafted or deep features utilized in WMSN.

Table 3.1 Comparison of various methods and the target application for WM-Sens.

Method	Target application	In-node processing	In-node energy minimization	In-network energy minimization	Compression method	Data transmission mode	Data transmission type
[37]	Object detection and classification	Yes	Yes	No	No	Wireless	Text
[101]	Object detection and classification	No	No	No	No	Wired	Image
[87]	Object detection and image transmission	Yes	Yes	Yes	2D-DWT	Wireless	Image
[4]	Object detection	Yes	Yes	No	No	Wireless	Image
[41]	Image transmission scheme	Yes	Yes	Yes	2D-DWT	Wireless	Image
[5]	Image transmission scheme	Yes	Yes	Yes	SPIHT	Wireless	Image
[73]	Image transmission scheme	Yes	Yes	Yes	EALAIC	Wireless	Image
[15]	Object detection and classification	Yes	Yes	No	No	Wireless	Text
[2]	Object Identification	Yes	Yes	No	No	Wireless	Text
[36]	Object detection and image transmission	Yes	Yes	No	Discret-Tchebichef Transform	Wireless	Image
[25]	Object detection and Image Transmission	Yes	No	No	Semantic encoding	Wireless	Video stream

3.3 Deep learning based practical approach for WMSN

Recently, the remarkable success of Deep Neural Network (DNN) in various categories of computer vision algorithms such as object recognition [38] and object detection [100] has attracted the attention of both the research community and industry. A variety of computer vision algorithms have found improvement and robustness utilizing DNN, such as semantic segmentation [92], instance segmentation [90], semantic matting [76], Super-Resolution [91], Image Quality Assessment [46], etc. A prominent characteristic of these DNN is that they can directly learn robust features from the large dataset using end-to-end learning powered by a back-propagation algorithm.

DNNs are characterized by hierarchical learning, i.e., each layer of the neural network learns a certain level of information. For example given an input image, the

Table 3.2 Various algorithms proposed for WMSN and the feature extraction type.

Method	Target application	Features type	Features extraction method	Classification/regression
[37]	Object detection and classification	Hand-crafted	Shape-based features, width/height ratio, compactness, total blob area/frame area ratio, SURF Features	SVM, KNN
[101]	Object detection and classification	Deep features	VGG16	Classification, Regression
[87]	Object detection and image transmission	Hand-crafted	Probabilist histogram-based features	–
[4]	Object detection	Hand-crafted	Background subtraction based thresholding	–
[15]	Object detection and classification	Hand-crafted	Shape-based features, width/height ratio, compactness, total blob area/frame area ratio, SURF Features	SVM, KNN
[2]	Object Identification	Hand-crafted	Background subtraction, centroid distance shape signature	–
[36]	Object detection and image transmission	Hand-crafted	Sum of absolute difference	–
[25]	Object detection and image transmission	Deep features	STFGAN	–

initial layers of the deep neural network learn about the object's low-level features such as edges. The middle layers learn mid-level features such as textures and the high-level layers learn abstract information such as object's shapes, etc. The features can be learned directly from the data using a back-propagation algorithm using end-to-end learning [20,43]. The algorithm performance can be determined using intra-database and cross-database testing [67]. In the following subsections, we provide a brief overview of deep learning techniques and learning methods that can be utilized to build state-of-the-art deep models for WMSN.

3.3.1 Convolutional Neural Networks

The Convolutional Neural Network (CNN) has been extensively utilized for nearly all computer vision tasks. The CNN can be directly applied to images without performing any transformation. The CNNs are characterized by local connectivity, weight-sharing, and pooling. Given an input image containing the object of interest (OOI), the CNN network can directly output the probability vector for the whole image or each pixel [42].

Let suppose the input image x_i containing an object of class j from a given dataset S is fed to the CNN parametrized by weights θ. Then the output probability y_i for the

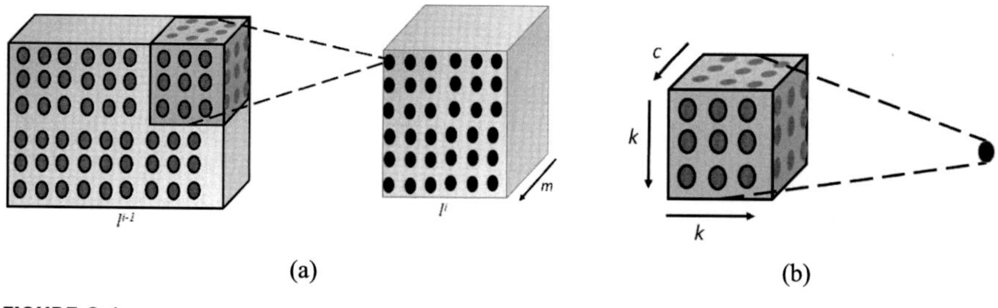

FIGURE 3.1

Illustration of how feature maps are formed through a convolution operation.

input image x_i can be represented as a softmax function:

$$p(y_i = j \mid x_i, \theta) = \frac{e^{\theta x_i + b}}{\sum_N e^{\theta x_i + b}}. \tag{3.1}$$

Imagine that a CNN contains n layers. To generate m feature maps at ith layer l^i ($l \in \{1, ..., n\}$), m unique windows of $k \times k \times c$ (c denotes the channel at layer l^{i-1}) kernels are slide on the input at layer l^{i-1}, recursively. Each neuron in the feature map at layer l is then considered to be locally connected to $k \times k \times c$ volume in the layer l^{i-1} (Fig. 3.1).

To generate a slice at the layer l^i, we need to slide a single $k \times k \times c$ kernel through the feature map at the layer l^{i-1}. Thus the total number of parameters at the layer l^i are $m \times k \times k \times c$ (excluding the m biases). Variants of this approach exist in the literature, such as separable convolution [14], and dilated convolution [11], etc.

3.3.2 Activation units

After performing convolution, the feature maps are passed through activation functions. One of the most commonly used activation functions is Rectified Linear Unit (ReLU) [57] or its variant Leaky ReLU [63]. The rectified linear unit has more similarity with the human biological response as it gives maximum response when the value of the neuron is greater than some threshold. Other types of activation functions include sigmoid, tanh, parametric rectified linear unit (PReLU), Exponential Linear Unit (ELU), and soft plus [26].

3.3.3 Max pooling

Normally, the deeper layers in the CNN have more feature maps than the initial layers. As a result, the number of parameters can quickly become large if we keep the resolution of the feature map unchanged throughout the layers. To reduce the number of parameters, it is necessary to reduce the resolution of the feature maps. However, this may incur the loss of information thus reducing the generalization of the CNN network. To overcome this, max-pooling has been utilized to select only that neuron in a certain $k \times k$ window, which can give a maximum response [56]. This not only reduces the number of parameters but inherently introduces a translation invariance. Max-pooling can be replaced by average-pooling [81] or gated-pooling [44]. Most state-of-the-art object recognition CNN benchmarks have utilized average-pooling before the final classification layer.

3.3.4 Batch normalization

Batch Normalization (BN) was introduced to reduce the internal covariate shift and to improve the training of the CNN. The BN is represented using the following equations [33]:

$$y_i \leftarrow \gamma \hat{x}_i + \beta, \tag{3.2}$$

$$\hat{x}_i = \frac{x_i - \mu_B}{\sqrt{\sigma_B^2 + \epsilon}}. \tag{3.3}$$

In BN, each scalar feature in the CNN layer is normalized to zero mean and unit variance, using the statistics of a minibatch. This also provides some sort of regularization and minimizes the reliance on careful weight initialization, allowing higher learning rates during training. There are other forms of normalization proposed in the literature, such as instance normalization [86], group normalization [93], and layer normalization [94].

3.3.5 Regularization

The DNNs have a large number of learnable parameters which increases their capacity. Asa result, they tend to overfit (a.k.a. remembering the training data) on the training data thus reducing the generalization to unseen samples. To overcome this limitation, dropout was introduced to randomly remove or zero out some of the neurons in the hidden layer [82]. Another way to prevent the DNN from overfitting is to generate perturbed copies of the training samples, also called data augmentation. Data augmentation can be random [79] or objective-based [97]. This includes flipping, cropping, scaling, translation, rotation, brightness, adding noise, transforming the color space [52].

3.3.6 Transfer learning

Often the domain-specific data is quite low to train the CNN. To overcome this difficulty, a transfer learning approach has been widely adopted. The deeper layers in the CNN learn high-level and abstract features specific to the data, whereas the lower layers learn low-level general features such as edges and blobs etc. [96]. In transfer learning, the last few layers of a CNN, pretrained on a large-scale dataset such as ImageNet [69], are retrained on the target data [84].

3.3.7 Recurrent Neural Networks

Recurrent Neural Networks (RNN) are widely utilized for sequential data that include time-series, temporal data, text, and speech. In such data, there is a dependency of the data at any point on previous data [72]. The input to the RNN model is the data at the current time t and the hidden state at time $t-1$, to output a target value and a new hidden state. However, they suffer from exploding and vanishing gradients problems. To resolve this issue, Long Short-Term Memory (LSTM) was introduced [30]. LSTM architecture consists of three gates, the input gate, output gate, and forget gate.

3.3.8 Auto-Encoders

Auto-Encoders (AE) models learn the mapping of data from an input domain to the target domain utilizing two networks connected in cascade. The encoder stage project the data to a latent-space representation that is decoded by using the decoder stage. They have been applied in a variety of tasks such as semantic segmentation, image colorization, superresolution, 3D-construction, presentation attack detection [48,58, 67,76], etc.

3.3.9 Generative Adversarial Networks

A Generative Adversarial Network (GAN) consists of a generator and discriminator. The generator usually consists of either an encoder-decoder network that maps the input data to such a representation that could fool the discriminator. The GANs are trained using the minimax game. The generator aims to maximize the discriminator error, whereas the discriminator tries to minimize the classification error [23].

3.3.10 Mobile neural networks

The requirement of large memory footprints and high computational resources in DNN may not be suitable for mobile and resource-constrained applications, which are constrained by limited memory footprints, power, and computational resources, respectively. Furthermore, latency is also an issue in these resource-constrained applications. For example, an algorithm running on a portable device should not take more than 3 to 4 seconds to process input information. In most cases, it is possible to reduce the parameter space of the DNN without largely sacrificing the performance. For example, the MobileNets proposed in [31] utilized depth-wise separable convolution to reduce the parameter space. This architecture was later improved with the introduction of inverted residuals [71] and quantizations [77]. Commercially, Neural Architectural Search (NAS) has been adopted to find device-aware models that can find the trade-off between memory footprints, power, and computational resources [19,55].

Deep feature-based algorithms utilizing DNN architectures provide unprecedented improvements mainly in computer vision applications. However, certain limitations hinder the potential scope of deep feature-based algorithms. Table 3.3 depicts some of the merits and demerits of deep feature-based algorithms.

Table 3.3 Merits and demerits of deep feature-based algorithms.

Deep feature-based algorithms	Merits	Demerits
CNN	• Weight sharing, local connectivity, pooling • Works better for image data • Requires no transformation of the image • Translation invariant with max-pooling	• Overfitting on the small dataset • A large number of parameters • Rotation variant • Susceptible to noise and adversarial examples
Transfer Learning	• Utilized the pretrained networks on the target task • Works better for training small scale dataset on deep neural nets	• Works poorly when the domain gap is large • May tend to overfit on small scale data • Requires a multiple hyper-parameters searches and layer changes
RNN	• Works better for sequential and time-series data	• Tend to overfit • Sometimes unstable due to exploding and vanishing gradients
AE	• Useful for learning latent features • A hierarchical unsupervised feature extractor • Can be utilized to initialize a neural network before supervise training	• Tends to provide smooth outputs which may not be desirable in some applications • Tends to overfit on the data
GAN	• Provide better and sharp results than Auto-Encoders. • Learns better representations for reconstruction tasks.	• Hard to train • No set rules when to stop training except analyzing the quality of generated images

3.4 Computer vision algorithms in WMSN

This section briefly discusses the computer vision algorithms utilizing hand-crafted features and deep-features being implemented in WMSN. In particular, we discuss generic object detection and semantic segmentation as they are vastly utilized in WMSens and WMSN. We also discuss image restoration, image colorization, and superresolution as they are also the candidate research areas in WMSN.

3.4.1 Object detection

Object detection refers to either finding the spatial location of the object in an image or finding the salient object in an image. The former detection task can be achieved by using bounding box regression. While the later detection task can be achieved by utilizing segmentation at the pixel level. The detection task provides the object spatial location which can be utilized to isolate the object from the rest of the image and subsequently classified into one of the representative classes.

The detection and classification of objects into different classes are important from the perspective of WMSN. The vision system in WMSens nodes incorporates the algorithms to classify the objects based on their geometry etc. The computer vision algorithms perform this task by either extracting features using generic feature extraction algorithms or using DNN, often followed by utilizing Support Vector Machines (SVM) [10] or MLPs [24,47]. The generic features extraction algorithms are independent of the dataset in considerations. Common algorithms include Histogram of Oriented Gradients (HOG) [17], Gabor filters [64], Scale Invariant Feature Transform (SIFT) [51], Speeded Up Robust Features (SURF) [6]. On the other hand, features extracted from DNNs are often data specific. Common algorithms utilized for object recognition are AlexNet [38], VGGNet [78], ResNet [28], GoogleNet [83], DenseNets [32], HR-Nets [89] and their variants. To execute the task of object detection the common algorithms include, Region-based Convolutional Neural Network (RCNN) [22], Faster-RCNN [68], R-FCN [16], Mask R-CNN [27], Single Scale MultiBox Detector (SSD) [49], YOLO [65] and their variants.

The object detection and classification algorithm can be implemented inside the WMSens nodes or at the base station. However, the implementation is typically based on the application scenario. This also depends upon the computing resources and battery power provided to a particular WMSens node. For example, a visual module in the WMSens node for object detection and classification deployed in a subway station for monitoring purposes may have abundant compute resources and energy supply compared to the same WMSens node deployed at the remote location.

To execute the task of object detection and classification at the sensor node, traditional computer vision algorithms are a natural choice owing to their lower computational complexity compared to DNN based computer vision algorithms. However as discussed before, they might work better in a controlled scenario and generate many false positives and false negatives in dynamic scenarios due to factors like camera movement, temperature, fog, rain, light variations, object orientation, and clutter, occlusion, etc. Moreover, vision systems designed using traditional computer vision algorithms are trained on a very small-scale dataset using hand-crafted feature extractors. As a result, they may not cover every possible situation (unseen examples outside the dataset), which limits the scope of their implementation in resource-constrained WMSens nodes deployed in remote regions for activity monitoring. The DNN algorithms on the other hand are trained on large-scale datasets and learn features directly from the data can provide good performance in dynamic scenarios, compared to traditional computer vision algorithms. The requirement of high computational and energy resources in DNN can be mitigated by either designing specific hardware, design of neural network layers that have low parameters, or utilizing device-aware NAS [34,45,54].

3.4.2 Semantic segmentation

Semantic segmentation refers to the pixel-based localization of objects and scenes in a given image. Unlike the object classification task, the segmentation task classifies each pixel into one of the defined categories. Each pixel in an image may belong to one of its representative classes such as trees, buildings, cars, people, sky, and so on. Segmentation algorithms give a class label to each pixel in the image [52]. Segmentation can be either binary-class, such as classifying each pixel in an image as belonging to foreground or background, or multiclass which classify each pixel in an image to one of the defined classes.

A simple segmentation algorithm is the semantic detection and isolation of objects of interest from the background using the thresholding operations [59]. Most in-node algorithms in WMSN for moving object detection utilize a thresholding-based strategy to separate foreground objects from the background. Often these algorithms are based on the assumption of the static background, which is updated overtime. This is followed by object feature extraction and classification. After classification, either the image (whole image or part of the image) or its contextual representation is transmitted to the base station depending upon the application [92]. Other algorithms include region growing [60], K-means clustering [18], active contours [35], graph cuts [8], Conditional Markov random fields [62].

Fully Convolutional Neural Network (FCN) was one of the first DNN proposed for semantic segmentation [75]. This was later improved by deep architectures such as ParseNet [50], SegNet, U-Net, Feature Pyramid Network (FPN), Pyramid Scene Parsing Network (PSP) [40,52].

Instead of transmitting the whole image, it is possible to transmit only the segmentation results to the base station. As discussed previously in this case, the visual algorithm provides some sort of compression as instead of transmitting the whole image, a portion of the image containing Region of Interest (RoI) is transmitted to the base station. An extension of our method, proposed in [87], is the transmission of only the RoI in the image segments rather than the image segments [36]. However, designing a low-cost and efficient segmentation algorithm is still an open issue in WMSN.

3.4.3 Image restoration, superresolution, and semantic colorization

Often the WMSens nodes are equipped with a low-resolution camera, usually monochrome, to capture the image or video of the vicinity in which they are deployed. Once captured, the image/video data is first compressed and then sent to the base station via a wired or multihop wireless network [25]. The received image often gets degraded depending upon the camera quality, encoding scheme [95], and channel distortion. To extract semantics and meaningful information from the low-resolution, monochrome, and degraded image, it is necessary to restore the details. Additionally, further improvement in the semantics can be achieved by upscaling the restored image using superresolution [89] and semantic image colorization techniques [99], respectively.

To the best of our knowledge, this area of computer vision has not been sufficiently researched in the context of WMSN and there are still open problems that require attention. For example, one problem could be the restoration of the image of the scene captured under low-light [9], such as nighttime. The image captured under low-light contains very negligible details to be effectively utilized for extracting any semantic information. For example, it is hard if not impossible to get a semantic mask or localize the object from the image captured under low-light using monochrome or RGB camera, respectively. As such, it is fairly rationale to incorporate low-light image restoration either in the WMSens node or at the base station, whichever is feasible, before applying further generic or domain-specific computer vision algorithms for the extraction of semantic information from the image.

Since the WMSens node has stringent resources it is not possible to get a high resolution such as HD, 4K, or even higher resolution image. However, once the image is received at the base station it is possible to generate a high-resolution image of the scene using superresolution techniques. Including superresolution techniques in the WMSN systems design could greatly reduce the requirement of high-end and sophisticated cameras and consequently reduces the in-node computational budget of the WMSens node.

Applying the image colorization algorithms on the monochrome images provide distinct color to the various objects and add more semantic meaning to the image respectively [61]. Colorization information is indispensable for intra-class object recognition. The image colorization in WMSens can encourage the utilization of monochrome images to 1) reduce the cost of utilizing high-end cameras, 2) reduces the burden on in-node processes, and provide some sort of compression, as only a single channel image information is transmitted by the WMSens node. However, assigning a proper color to each object in an image is still an open research issue in computer vision.

3.5 Conclusion and future considerations

In this chapter, we first gave a brief overview of the visual processing system in WMSens. The visual processing algorithms, based on hand-crafted features and deep-features, in the WMSens, controls the amount of information that needs to be transmitted to the base station and subsequently regulates the energy consumption within the WMSens node and WMSN. Afterward, we gave a brief introduction of DNN, in particular the deep CNN. We also discussed various areas of computer vision such as object detection and classification, semantic segmentation, image restoration, superresolution, and image colorization and their applicability, compatibility, and issues in WMSens and WMSN. Finally, we believe that this chapter would help the researcher to open a broad new spectrum of research solutions and areas in WMSN.

With the rapid advancement in miniaturization technology, the Internet of Multimedia Things (IoMT), and deep learning, it is inevitable to design techniques and systems by utilizing cross-disciplinary research to fill the technological gaps that will enable better applications and smart products. As a result, the modification in traditional computer vision algorithms is indispensable to cope with the stringent requirements offered by the nature of WMSN devices. For example, Neural Architectural Search (NAS) techniques can be utilized to design DNN architectures for WMSN devices that will keep the balance between the performance and computational resources while enabling a variety of vision algorithms deployment in WMSN devices. The combination of image enhancement and restoration techniques like low-light enhancement, blur removal, image colorization, and superresolution can enable potential applications in the study of habitat monitoring and the study of ocean ecology using WMSN. Underwater images are corrupted with many factors simultaneously such as low light, dispersion due to water, polarized imaging, etc. Designing WMSN that can perform distributed processing on these images and sending important and useful data to the base station are still open issues.

References

[1] I.F. Akyildiz, T. Melodia, K.R. Chowdhury, A survey on wireless multimedia sensor networks, Computer Networks 51 (4) (2007) 921–960, https://doi.org/10.1016/j.comnet.2006.10.002.

[2] M.S. Alhilal, A. Soudani, A. Al-Dhelaan, A shape-based object identification scheme in wireless multimedia sensor networks, Advances in Intelligent Systems and Computing 314 (2015) 251–259, https://doi.org/10.1007/978-3-319-10383-9_23.

[3] I.T. Almalkawi, M.G. Zapata, J.N. al-Karaki, J. Morillo-Pozo, Wireless multimedia sensor networks: current trends and future directions, Sensors 10 (7) (2010) 6662–6717, https://doi.org/10.3390/s100706662.

[4] S.M. Aziz, D.M. Pham, Energy efficient image transmission in wireless multimedia sensor networks, IEEE Communications Letters 17 (6) (2013) 1084–1087, https://doi.org/10.1109/LCOMM.2013.050313.121933.

[5] N. Bano, M. Alam, S. Ahmad, Energy-efficient, low memory listless SPIHT coder for wireless multimedia sensor networks, Advances in Wireless and Mobile Communications 10 (5) (2017) 871–883.

[6] H. Bay, T. Tuytelaars, L. Van Gool, SURF: speeded up robust features, in: A. Leonardis, H. Bischof, A. Pinz (Eds.), Computer Vision – ECCV 2006, in: Lecture Notes in Computer Science, vol. 3951, 2006, pp. 404–417.

[7] H. Bolhasani, M. Mohseni, A.M. Rahmani, Deep learning applications for IoT in health care: a systematic review, Informatics in Medicine Unlocked 23 (December 2021) 100550, https://doi.org/10.1016/j.imu.2021.100550.

[8] Y. Boykov, O. Veksler, R. Zabih, Fast approximate energy minimization via graph cuts, IEEE Transactions on Pattern Analysis and Machine Intelligence 23 (11) (2001) 1222–1239, https://doi.org/10.1109/34.969114.

[9] L. Bu, Z. Xu, G. Zhang, Z. Zhang, Night-light image restoration method based on night scattering model for Luojia 1-01 satellite, Sensors (Switzerland) 19 (17) (2019), https://doi.org/10.3390/s19173761.

[10] C.-C. Chang, C.-J. Lin, Libsvm, ACM Transactions on Intelligent Systems and Technology 2 (3) (2011) 1–27, https://doi.org/10.1145/1961189.1961199.

[11] L.C. Chen, G. Papandreou, I. Kokkinos, K. Murphy, A.L. Yuille, DeepLab: semantic image segmentation with deep convolutional nets, atrous convolution, and fully connected CRFs, IEEE Transactions on Pattern Analysis and Machine Intelligence 40 (4) (2018) 834–848, https://doi.org/10.1109/TPAMI.2017.2699184.

[12] P. Chen, P. Ahammad, C. Boyer, S. Huang, L. Lin, E. Lobaton, et al., CITRIC: a low-bandwidth wireless camera network platform, in: ACM/IEEE Conference on Distributed Smart Cameras (ICDSC), September, 2008, pp. 1–10, Electrical Engineering and Computer Sciences, University of California, Berkeley, CA 94720, USA; Electrical Engineering and Computer Sciences, University of California, Merced, CA 95344, USA.

[13] Q. Chen, Y. Xie, S. Guo, J. Bai, Q. Shu, Sensing system of environmental perception technologies for driverless vehicle: a review of state of the art and challenges, Sensors and Actuators A, Physical 319 (2021) 112566, https://doi.org/10.1016/j.sna.2021.112566.

[14] F. Chollet, Xception: deep learning with depthwise separable convolutions, in: IEEE Conference on Computer Vision and Pattern Recognition (CVPR), 2017, pp. 1251–1258.

[15] M. Civelek, A. Yazici, Automated moving object classification in wireless multimedia sensor networks, IEEE Sensors Journal 17 (4) (2017) 1116–1131, https://doi.org/10.1109/JSEN.2016.2638853.

[16] J. Dai, Y. Li, K. He, J. Sun, R-FCN: object detection via region-based fully convolutional networks, in: Advances in Neural Information Processing Systems (Vol. 29), 2016, retrieved from https://github.com/daijifeng001/r-fcn.

[17] N. Dalal, B. Triggs, Histograms of oriented gradients for human detection, in: Proceedings – 2005 IEEE Computer Society Conference on Computer Vision and Pattern Recognition, CVPR 2005, I, 2005, pp. 886–893.

[18] N. Dhanachandra, K. Manglem, Y.J. Chanu, Image segmentation using K-means clustering algorithm and subtractive clustering algorithm, Procedia Computer Science 54 (2015) 764–771, https://doi.org/10.1016/j.procs.2015.06.090.

[19] J.D. Dong, A.C. Cheng, D.C. Juan, W. Wei, M. Sun, DPP-Net: device-aware progressive search for pareto-optimal neural architectures, in: V. Ferrari, M. Hebert, C. Sminchisescu, Y. Weiss (Eds.), Computer Vision – ECCV 2018, in: Lecture Notes in Computer Science, vol. 11215, 2018, pp. 540–555.

[20] K. Fukushima, S. Miyake, T. Ito, Neocognitron: a neural network model for a mechanism of visual pattern recognition, IEEE Transactions on Systems, Man and Cybernetics SMC-13 (5) (1983) 826–834, https://doi.org/10.1109/TSMC.1983.6313076.

[21] N. Giaquinto, M. Scarpetta, M. Spadavecchia, G. Andria, Deep learning-based computer vision for real time intravenous drip infusion monitoring, IEEE Sensors Journal 21 (13) (2021) 14148–14154, https://doi.org/10.1109/JSEN.2020.3039009.Deep.

[22] R. Girshick, J. Donahue, T. Darrell, J. Malik, Region-based convolutional networks for accurate object detection and segmentation, IEEE Transactions on Pattern Analysis and Machine Intelligence 38 (1) (2016) 142–158, https://doi.org/10.1109/TPAMI.2015.2437384.

[23] I.J. Goodfellow, J. Pouget-Abadie, M. Mirza, B. Xu, D. Warde-Farley, S. Ozair, et al., Generative adversarial nets, in: Advances in Neural Information Processing Systems (Vol. 27), 2014, retrieved from http://www.github.com/goodfeli/adversarial.
[24] I.J. Goodfellow, D. Warde-Farley, M. Mirza, A. Courville, Y. Bengio, Maxout networks, retrieved from http://arxiv.org/abs/1302.4389, 2013.
[25] J. Guo, X. Gong, W. Wang, X. Que, J. Liu, SASRT: semantic-aware super-resolution transmission for adaptive video streaming over wireless multimedia sensor networks, Sensors 19 (14) (2019) 3121, https://doi.org/10.3390/s19143121.
[26] S. Hayou, A. Doucet, J. Rousseau, On the impact of the activation function on deep neural networks training, in: International Conference on Machine Learning, 2019, pp. 2672–2680, retrieved from http://proceedings.mlr.press/v97/hayou19a.html.
[27] K. He, G. Gkioxari, P. Dollár, R. Girshick, Mask R-CNN, IEEE Transactions on Pattern Analysis and Machine Intelligence 42 (2) (2020) 386–397, https://doi.org/10.1109/TPAMI.2018.2844175.
[28] K. He, X. Zhang, S. Ren, J. Sun, Deep residual learning for image recognition, in: 2016 IEEE Conference on Computer Vision and Pattern Recognition (CVPR), 2016, pp. 770–778.
[29] S. Hengstler, D. Prashanth, S. Fong, H. Aghajan, MeshEye: a hybrid-resolution smart camera mote for applications in distributed intelligent surveillance, in: IPSN 2007: Proceedings of the Sixth International Symposium on Information Processing in Sensor Networks, 2007, pp. 360–369.
[30] S. Hochreiter, J. Schmidhuber, Long short-term memory, Neural Computation 9 (8) (1997) 1735–1780, https://doi.org/10.1162/neco.1997.9.8.1735.
[31] A. Howard, M. Zhu, B. Chen, et al., Mobilenets: efficient convolutional neural networks for mobile vision applications, arXiv, retrieved from https://arxiv.org/abs/1704.04861.
[32] G. Huang, Z. Liu, L. Van Der Maaten, K.Q. Weinberger, Densely connected convolutional networks, in: Proceedings – 30th IEEE Conference on Computer Vision and Pattern Recognition, CVPR 2017, 2017-Jan, pp. 2261–2269.
[33] S. Ioffe, C. Szegedy, Batch normalization: accelerating deep network training by reducing internal covariate shift, in: Proceedings of the 32nd International Conference on Machine Learning, 2015, pp. 448–456, retrieved from http://proceedings.mlr.press/v37/ioffe15.html.
[34] G. Jose, A. Kumar, S. Kruthiventi, S. Saha, H. Muralidhara, Real-time object detection on low power embedded platforms, in: Proceedings of the IEEE/CVF International Conference on Computer Vision (ICCV), 2019, pp. 2485–2492.
[35] M. Kass, A. Witkin, D. Terzopoulos, Snakes: active contour models, International Journal of Computer Vision 1 (4) (1988) 321–331, https://doi.org/10.1007/BF00133570.
[36] N. Kouadria, K. Mechouek, S. Harize, N. Doghmane, Region-of-interest based image compression using the discrete Tchebichef transform in wireless visual sensor networks, Computers and Electrical Engineering 73 (2019) 194–208, https://doi.org/10.1016/j.compeleceng.2018.11.010.
[37] M. Koyuncu, A. Yazici, M. Civelek, A. Cosar, M. Sert, Visual and auditory data fusion for energy-efficient and improved object recognition in wireless multimedia sensor networks, IEEE Sensors Journal 19 (5) (2019) 1839–1849, https://doi.org/10.1109/JSEN.2018.2885281.
[38] A. Krizhevsky, I. Sutskever, G.E. Hinton, Imagenet classification with deep convolutional neural networks, Advances in Neural Information Processing Systems (2012) 1097–1105.

[39] P. Kulkarni, D. Ganesan, P. Shenoy, The case for multi-tier camera sensor networks, in: Proceedings of the International Workshop on Network and Operating System Support for Digital Audio and Video, 2005, pp. 141–146.
[40] F. Lateef, Y. Ruichek, Survey on semantic segmentation using deep learning techniques, Neurocomputing 338 (2019) 321–348, https://doi.org/10.1016/j.neucom.2019.02.003.
[41] V. Lecuire, C. Duran-Faundez, N. Krommenacker, Energy-efficient transmission of wavelet-based images in wireless sensor networks, EURASIP Journal on Image and Video Processing 2007 (2007), https://doi.org/10.1155/2007/47345.
[42] Y. Lecun, Y. Bengio, G. Hinton, Deep learning, Nature 521 (7553) (2015) 436–444, https://doi.org/10.1038/nature14539.
[43] Y. LeCun, B. Boser, J.S. Denker, D. Henderson, R.E. Howard, W. Hubbard, L.D. Jackel, Backpropagation applied to handwritten zip code recognition, Neural Computation 1 (4) (1989) 541–551, https://doi.org/10.1162/neco.1989.1.4.541.
[44] C.Y. Lee, P. Gallagher, Z. Tu, Generalizing pooling functions in CNNs: mixed, gated, and tree, IEEE Transactions on Pattern Analysis and Machine Intelligence 40 (4) (2018) 863–875, https://doi.org/10.1109/TPAMI.2017.2703082.
[45] F. Li, Z. Mo, P. Wang, Z. Liu, J. Zhang, G. Li, et al., A system-level solution for low-power object detection, in: Proceedings – 2019 International Conference on Computer Vision Workshop, ICCVW 2019, 2019, pp. 2461–2468, retrieved from http://arxiv.org/abs/1909.10964.
[46] Y. Li, L.-M. Po, L. Feng, F. Yuan, No-reference image quality assessment with deep convolutional neural networks, in: 2016 IEEE International Conference on Digital Signal Processing (DSP), 0 2016, pp. 685–689.
[47] M. Lin, Q. Chen, S. Yan, Network in network, https://doi.org/10.1109/ASRU.2015.7404828, 1–10.
[48] J. Liu, Y. Yao, W. Hou, M. Cui, X. Xie, C. Zhang, X. Hua, Boosting semantic human matting with coarse annotations, pp. 8563–8572, retrieved from http://arxiv.org/abs/2004.04955, 2020.
[49] W. Liu, D. Anguelov, D. Erhan, C. Szegedy, S. Reed, C.Y. Fu, A.C. Berg, SSD: single shot multibox detector, in: B. Leibe, J. Matas, N. Sebe, M. Welling (Eds.), Computer Vision – ECCV 2016, in: Lecture Notes in Computer Science, vol. 9905, 2016, pp. 21–37.
[50] W. Liu, A. Rabinovich, A.C. Berg, ParseNet: looking wider to see better, arXiv:1506.04579, retrieved from https://github.com/, 2015.
[51] D.G. Lowe, Object recognition from local scale-invariant features, in: Proceedings of the IEEE International Conference on Computer Vision, 2, 1999, pp. 1150–1157.
[52] S. Minaee, Y. Boykov, F. Porikli, A. Plaza, N. Kehtarnavaz, D. Terzopoulos, Image segmentation using deep learning: a survey, preprint arXiv, retrieved from http://arxiv.org/abs/2001.05566, 2020.
[53] R.A.F. Mini, A.A.F. Loureiro, Energy in wireless sensor networks, Middleware for Network Eccentric and Mobile Applications 3 (24) (2009), https://doi.org/10.1007/978-3-540-89707-1_1.
[54] M. Możejko, T. Latkowski, Ł. Treszczotko, M. Szafraniuk, K. Trojanowski (n.d.), Superkernel neural architecture search for image denoising.
[55] M. Mozejko, T. Latkowski, L. Treszczotko, M. Szafraniuk, K. Trojanowski, Superkernel neural architecture search for image denoising, in: IEEE Computer Society Conference on Computer Vision and Pattern Recognition Workshops, 2020-June, pp. 2002–2011.
[56] N. Murray, F. Perronnin, Generalized max pooling, in: Proceedings of the IEEE Computer Society Conference on Computer Vision and Pattern Recognition, 2014, pp. 2473–2480.

[57] V. Nair, G.H. On, Rectified linear units improve restricted Boltzmann machines, in: ICML'10: Proceedings of the 27th International Conference on International Conference on Machine Learning, pp. 807–814, retrieved from https://dl.acm.org/doi/abs/10.5555/3104322.3104425.
[58] O. Nikisins, A. George, S. Marcel, Domain adaptation in multi-channel autoencoder based features for robust face anti-spoofing, in: IEEE International Conference on Biometrics, 2019.
[59] Otsu Nobuyuki, A threshold selection method from gray-level histograms, IEEE Transactions on Systems, Man and Cybernetics 9 (1) (1979) 62–66.
[60] R. Nock, F. Nielsen, Statistical region merging, IEEE Transactions on Pattern Analysis and Machine Intelligence 26 (11) (2004) 1452–1458, https://doi.org/10.1109/TPAMI.2004.110.
[61] S. Pahal, P. Sehrawat, Image colorization with deep convolutional neural networks, in: G. Hura, A. Singh, L. Siong Hoe (Eds.), Advances in Communication and Computational Technology, in: Lecture Notes in Electrical Engineering, vol. 668, 2021, pp. 45–56.
[62] N. Plath, M. Toussaint, S. Nakajima, Multi-class image segmentation using conditional random fields and global classification, ACM International Conference Proceeding Series 382 (2009) 1–8, https://doi.org/10.1145/1553374.1553479.
[63] D. Ranjan Nayak, D. Das, R. Dash, S. Majhi, B. Majhi, Deep extreme learning machine with leaky rectified linear unit for multiclass classification of pathological brain images, Multimedia Tools and Applications 1 (16) (2019), https://doi.org/10.1007/s11042-019-7233-0.
[64] N.K. Ratha, A.K. Jain, S. Lakshmanan, Object detection in the presence of clutter using Gabor filters, in: A.G. Tescher (Ed.), spiedigitallibrary.org, 1994, pp. 612–623.
[65] J. Redmon, S. Divvala, R. Girshick, A. Farhadi, You only look once: unified, real-time object detection, in: Proceedings of the IEEE Computer Society Conference on Computer Vision and Pattern Recognition, 2016-Dec, pp. 779–788.
[66] Y.A.U. Rehman, M. Tariq, O.U. Khan, Improved object localization using accurate distance estimation in wireless multimedia sensor networks, PLoS ONE 10 (11) (2015), https://doi.org/10.1371/journal.pone.0141558.
[67] Yasar Abbas Ur Rehman, L. Po, M. Liu, Z. Zou, W. Ou, Y. Zhao, Face liveness detection using convolutional-features fusion of real and deep network generated face images, Journal of Visual Communication and Image Representation 59 (2019) 574–582, https://doi.org/10.1016/j.jvcir.2019.02.014.
[68] S. Ren, K. He, R. Girshick, J. Sun, Faster R-CNN: towards real-time object detection with region proposal networks, Advances in Neural Information Processing Systems 28 (2015) 91–99, retrieved from https://github.com/.
[69] O. Russakovsky, J. Deng, H. Su, J. Krause, S. Satheesh, S. Ma, et al., ImageNet large scale visual recognition challenge, International Journal of Computer Vision 115 (3) (2015) 211–252, https://doi.org/10.1007/s11263-015-0816-y.
[70] O. Said, A. Tolba, Accurate performance prediction of IoT communication systems for smart cities: an efficient deep learning based solution, Sustainable Cities and Society 69 (2021) 102830, https://doi.org/10.1016/j.scs.2021.102830.
[71] M. Sandler, A. Howard, M. Zhu, A. Zhmoginov, L.-C. Chen, MobileNetV2: inverted residuals and linear bottlenecks, in: Proceedings of the IEEE Conference on Computer Vision and Pattern Recognition (CVPR), 2018, pp. 4510–4520.
[72] M. Schuster, K.K. Paliwal, Bidirectional recurrent neural networks, IEEE Transactions on Signal Processing 45 (11) (1997) 2673–2681, https://doi.org/10.1109/78.650093.

[73] A. Senturk, R. Kara, I. Ozcelik, Fuzzy logic and image compression based energy efficient application layer algorithm for wireless multimedia sensor networks, Computer Science and Information Systems 17 (2) (2020) 509–536, https://doi.org/10.2298/CSIS191124008S.

[74] E.V. Sheehan, D. Bridger, S.J. Nancollas, S.J. Pittman, PelagiCam: a novel underwater imaging system with computer vision for semi-automated monitoring of mobile marine fauna at offshore structures, Environmental Monitoring and Assessment 192 (1) (2020), https://doi.org/10.1007/s10661-019-7980-4.

[75] E. Shelhamer, J. Long, T. Darrell, Fully convolutional networks for semantic segmentation, IEEE Transactions on Pattern Analysis and Machine Intelligence 39 (4) (2017) 640–651, https://doi.org/10.1109/TPAMI.2016.2572683.

[76] X. Shen, X. Tao, H. Gao, C. Zhou, J. Jia, Deep automatic portrait matting, in: B. Leibe, J. Matas, N. Sebe, M. Welling (Eds.), Eccv, 2016, pp. 92–107.

[77] T. Sheng, C. Feng, S. Zhuo, X. Zhang, L. Shen, M. Aleksic, A quantization-friendly separable convolution for MobileNets, in: Proceedings – 1st Workshop on Energy Efficient Machine Learning and Cognitive Computing for Embedded Applications, EMC2 2018, 2018, pp. 14–18.

[78] K. Simonyan, A. Zisserman, Very deep convolutional networks for large-scale image recognition, in: International Conference on Learning Representations (ICLR), 2015, 2015, p. 14, retrieved from https://arxiv.org/abs/1409.1556.pdf.

[79] E.A. Smirnov, D.M. Timoshenko, S.N. Andrianov, Comparison of regularization methods for ImageNet classification with deep convolutional neural networks, AASRI Procedia 6 (2014) 89–94, https://doi.org/10.1016/j.aasri.2014.05.013.

[80] R. Song, H. Chen, J. Cheng, C. Li, Y. Liu, X. Chen, PulseGAN: learning to generate realistic pulse waveforms in remote photoplethysmography, IEEE Journal of Biomedical and Health Informatics 2194(c) (2021), https://doi.org/10.1109/JBHI.2021.3051176.

[81] J.T. Springenberg, A. Dosovitskiy, T. Brox, M. Riedmiller, Striving for simplicity: the all convolutional net, in: 3rd International Conference on Learning Representations, ICLR 2015 – Workshop Track Proceedings, 2015.

[82] N. Srivastava, G. Hinton, A. Krizhevsky, R. Salakhutdinov, Dropout: a simple way to prevent neural networks from overfitting, Journal of Machine Learning Research 15 (2014) 1929–1958, retrieved from https://www.jmlr.org/papers/volume15/srivastava14a/srivastava14a.pdf?utm_campaign=buffer&utm_content=buffer79b43&utm_medium=social&utm_source=twitter.com.

[83] C. Szegedy, W. Liu, Y. Jia, P. Sermanet, S. Reed, D. Anguelov, et al., Going deeper with convolutions, in: Proceedings of the IEEE Conference on Computer Vision and Pattern Recognition, 2015, pp. 1–9.

[84] C. Tan, F. Sun, T. Kong, W. Zhang, C. Yang, C. Liu, A survey on deep transfer learning, in: V. Kůrková, Y. Manolopoulos, B. Hammer, L. Iliadis, I. Maglogiannis (Eds.), Artificial Neural Networks and Machine Learning – ICANN 2018, in: Lecture Notes in Computer Science, vol. 11141 (1), 2018, pp. 270–279.

[85] M. Tausif, A. Jain, E. Khan, M. Hasan, Memory-efficient architecture for FrWF-based DWT of high-resolution images for IoMT applications, Multimedia Tools and Applications (2021), https://doi.org/10.1007/s11042-020-10258-0.

[86] D. Ulyanov, A. Vedaldi, V. Lempitsky, Instance normalization: the missing ingredient for fast stylization, retrieved from http://arxiv.org/abs/1607.08022, 2016.

[87] Y.A. Ur Rehman, M. Tariq, T. Sato, A novel energy efficient object detection and image transmission approach for wireless multimedia sensor networks, IEEE Sensors Journal 16 (15) (2016) 5942–5949, https://doi.org/10.1109/JSEN.2016.2574989.

[88] H. Wang, J. Xie, User preference based energy-aware mobile AR system with edge computing, in: Proceedings – IEEE INFOCOM, 2020-July, 2020, pp. 1379–1388.

[89] J. Wang, K. Sun, T. Cheng, B. Jiang, C. Deng, Y. Zhao, et al., Deep high-resolution representation learning for visual recognition, IEEE Transactions on Pattern Analysis and Machine Intelligence (2020), https://doi.org/10.1109/TPAMI.2020.2983686, 1–1.

[90] X. Wang, T. Kong, C. Shen, Y. Jiang, L. Li, SOLO: segmenting objects by locations, retrieved from http://arxiv.org/abs/1912.04488, 2019.

[91] Z. Wang, J. Chen, S.C.H. Hoi, Deep learning for image super-resolution: a survey, IEEE Transactions on Pattern Analysis and Machine Intelligence (2020), https://doi.org/10.1109/TPAMI.2020.2982166, 1–1.

[92] C. Wei, X. Jiang, Z. Tang, W. Qian, N. Fan, Context-based global multi-class semantic image segmentation by wireless multimedia sensor networks, Artificial Intelligence Review 43 (4) (2015) 579–591, https://doi.org/10.1007/s10462-013-9394-y.

[93] Y. Wu, K. He, Group normalization, in: Proceedings of the European Conference on Computer Vision (ECCV), 2018, pp. 3–19.

[94] R. Xiong, Y. Yang, D. He, K. Zheng, S. Zheng, C. Xing, et al., On layer normalization in the transformer architecture, in: Proceedings of the 37th International Conference on Machine Learning, PMLR, 2020, pp. 10524–10533, retrieved from http://proceedings.mlr.press/v119/xiong20b.html.

[95] H. Zaineldin, M.A. Elhosseini, H.A. Ali, Image compression algorithms in wireless multimedia sensor networks: a survey, Ain Shams Engineering Journal 6 (2) (2015) 481–490, https://doi.org/10.1016/j.asej.2014.11.001.

[96] M.D. Zeiler, R. Fergus, Visualizing and understanding convolutional networks, in: Lecture Notes in Computer Science (including subseries Lecture Notes in Artificial Intelligence and Lecture Notes in Bioinformatics), vol. 8689, 2014, pp. 818–833.

[97] Y. Zhang, L. Man Po, M. Liu, Y.A. Ur Rehman, W. Ou, Y. Zhao, Data-level information enhancement: motion-patch-based Siamese Convolutional Neural Networks for human activity recognition in videos, Expert Systems with Applications 147 (2020) 113203, https://doi.org/10.1016/j.eswa.2020.113203.

[98] Z. Zhang, C. Lai, H. Chao, A green data transmission mechanism for wireless multimedia sensor networks, (August 2014) 14–19.

[99] Y. Zhao, L.M. Po, K.W. Cheung, W.Y. Yu, Y.A.U. Rehman, SCGAN: saliency map-guided colorization with generative adversarial network, IEEE Transactions on Circuits and Systems for Video Technology (2020), https://doi.org/10.1109/TCSVT.2020.3037688.

[100] B. Zhou, A. Khosla, A. Lapedriza, A. Oliva, A. Torralba, Learning deep features for discriminative localization, in: 2016 IEEE Conference on Computer Vision and Pattern Recognition (CVPR), 2016, pp. 2921–2929.

[101] X. Zhu, B. Ding, W. Li, L. Gu, Y. Yang, On development of security monitoring system via wireless sensing network, EURASIP Journal on Wireless Communications and Networking 2018 (1) (2018) 221, https://doi.org/10.1186/s13638-018-1235-x.

CHAPTER

Cognitive radio-medium access control protocol for Internet of Multimedia Things (IoMT)

4

Shweta Pandit[a], Prabhat Thakur[b], Alok Kumar[a], and Ghanshyam Singh[c]

[a] *Department of Electronics and Communication Engineering, Jaypee University of Information Technology, Solan, India*

[b] *Department of Electrical and Electronics Engineering Science, University of Johannesburg, Johannesburg, South Africa*

[c] *Centre for Smart Information and Communication Systems, Department of Electrical and Electronics Engineering Science, University of Johannesburg, Johannesburg, South Africa*

4.1 Introduction

Connected devices to the internet are increasing day-by-day. The multimedia content which includes text, image, graphic, animation, audio, and video, is the traffic traveling over the internet in terms of emails, chat in social networking sites, blogs, forums, games, music, video, online shopping, online study etc. The human's penetration into the internet world is exponentially increasing and as per the prediction of CISCO, the multimedia content over the internet covers 49 Exabyte per month data by the year of 2022, out of which video content would be 38.14 Exabyte [1]. The text will include only around 2% of internet multimedia data out of these multimedia contents. In addition, the impact of video-based internet multimedia data can be seen from the Statistics-2019, which claims to have around 300 hours of video upload on YouTube every minute by around two billion YouTube users and each day around 5 billion videos are watched on YouTube [1]. These studies have shown the higher impact of Internet of Multimedia Things (IoMT) in the technology world and is defined as the technology which allows smart multimedia things or devices to interact and communicate with one another over the internet to assist the multimedia-based services and applications [2]. The IoMT is basically integration of IoT network with multimedia contents however the stringent requirement of large spectrum, least delay, more memory space and faster processing differentiate it from the conventional internet of things (IoT) network. Kevin Ashtonhas first coined the term IoT in 1999 [3] and is referred as "*things having identities and virtual personalities operating in smart spaces using intelligent interfaces to connect and communicate within social, environmental, and user contexts via the Internet*" [4]. Therefore, we may say that IoMT is an

Internet of Multimedia Things (IoMT). https://doi.org/10.1016/B978-0-32-385845-8.00009-5

IoT network with strict quality of service (QoS) and quality of experience (QoE) requirements. The prominent applications of the IoMT cover a vast range of areas like smart home, smart wearables, smart city, smart grids, connected vehicles or smart cars, environment monitoring, intelligent agriculture, smart supply chain, smart retail, connected health etc. Currently, all these applications are integral part of human life.

Further, the worldwide pandemic situation created by COVID-19 has shifted the dependency of human beings to work-from-home through real-time multimedia applications. The online-learning platform has also seen drastic improvement and modernization in this pandemic and with the help of real-time multimedia over the internet it has been made possible. The children education dependency and their learning have been totally dependent on the real-time multimedia transfer over the internet. Therefore, we can say that it has become the essential commodity of this generation. Now in near future, after the pandemic emergency, the working model through the internet will be new normal and will certainly have significant research contributions in the field of IoMT. The respective multimedia devices in IoMT which require being less expensive and energy-efficient, are basically accessible over the internet through a unique IP address as that of the computer system had a unique IP address to be accessed over the internet. Therefore, these multimedia devices of the IoMT are no different than the individual entity of the IP network. However, the multimedia devices of the conventional multimedia network only collect data from its surroundings without being assigned an IP address and need a device that transfers the content over the internet. Therefore, the multimedia devices of the conventional multimedia network are not the entity directly communicating over the IP network. Further, this data or content may be streamed to the user or stored at the cloud for processing or for later, on-demand retrieval by the end-user [2]. Further, we may say that it provides heterogeneity support where its multimedia devices interact over the internet with any kind of device (not necessarily multimedia device) or system in addition to taking/giving commands from/to other homogeneous or heterogeneous devices. This feature is lacking in the conventional multimedia network where multimedia devices can interact with only homogeneous multimedia devices and are not smart enough as that in IoMT to prompt any action based on its interaction or perception [2]. In addition, the multimedia data is unstructured data and need some feature extraction [5]. Further, the IoMT takes QoE into consideration along with QoS and therefore IoMT network parameters are equally important as that of multimedia device's hardware capabilities for IoMT network performance [5]. Moreover, the QoE is the performance measure from the user-centric way however, the QoS is a performance measure from the network-centric way.

4.2 Difference in Internet of Things and Internet-of-Multimedia-Things

IoT architecture or network faces difficulty in supporting multimedia traffic due to its delay-sensitive nature and more bandwidth requirement, therefore this architecture needs to be modified for IoMT. The data produced by IoT device is scalar such as measured temperature, pressure, humidity, distance etc by the respective sensors and requires limited processing speed, bandwidth, and memory e.g. in bytes to kilobytes. In contrast, the multimedia data in IoMT device especially of real-time communication is bulky, requires higher processing speed and bandwidth e.g. memory requirement is in Megabyte to Gigabyte. The data produced by the multimedia application are from immersive 360 degrees, distributed gaming, free-viewpoint video, augmented reality (AR), virtual reality (VR), and extended reality (XR) etc. [6]. Further, the processor's specification for IoT network is in kHz to MHz while for IoMT network, it is in MHz to GHz [5]. Moreover, the IoT data is periodic while IoMT data is continuous. In addition, the main differentiating factors among IoT and IoMT are described below.

4.2.1 Architecture

IoT architecture is said to have a three-layered structure as shown in Fig. 4.1. It consists of the sensor layer, network layer, and application layer. The sensor layer interacts with the environment as that of the physical layer of OSI (open system interconnection) model. It acquires the data from the environment, process data locally at the IoT device or sends it for processing to the centralized system or to the sink node. Further, the network layer function is to gather the data from the sensor layer and process it. In addition, the network layer provides the interconnection between different IoT devices using network protocols and responsible for determining the best route which is mostly dependent on the least energy cost [5]. Further, the third layer of IoT architecture is the application layer which directly deals with the end IoT device. In addition, the information management, data mining, data analytics, and decision-making services are also provided by the application layer of the IoT

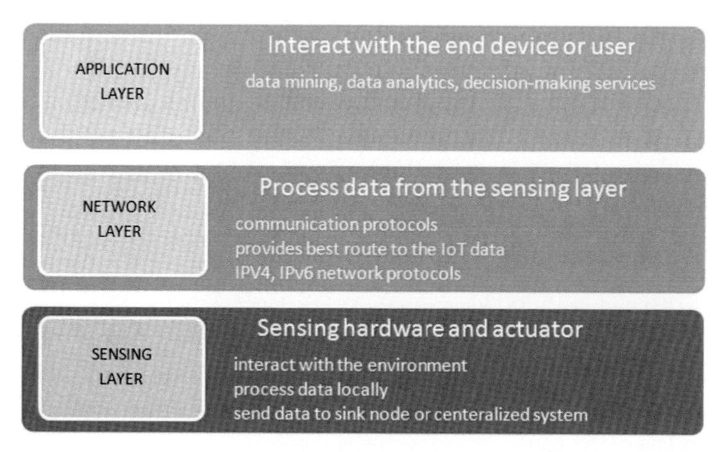

FIGURE 4.1

The IoT architecture [5].

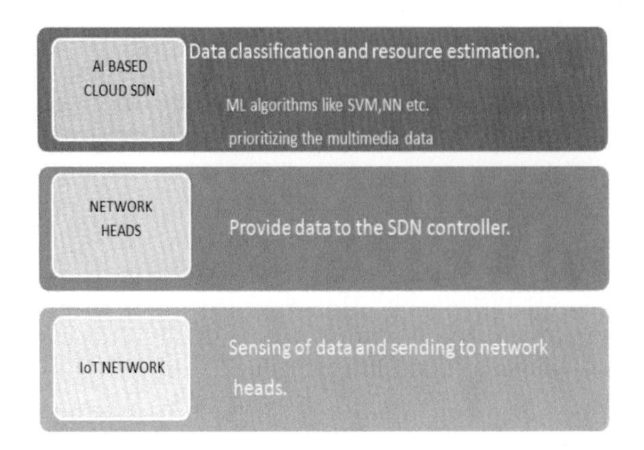

FIGURE 4.2

AI-based software-defined IoMT architecture [7].

architecture [5]. However, the IoMT architecture needs to provide services to heterogeneous multimedia traffic which poses different service requirements at different time. In this direction, various researchers have proposed their architecture out of which the artificial intelligence (AI)-based Software-Defined IoMT Architecture [7] and Big Data Layered architecture [8] are presented in this chapter. However, these articles have presented the name of IoT, still, the traffic versatility is that of IoMT network, due to which these works have been presented here, the discussion of which is presented further.

The authors in [7] have proposed architecture for video surveillance system based on AI which provides QoS in terms of least delay, loss rate, and jitter and is shown in Fig. 4.2. The software-defined networking (SDN) is integrated with the AI, which after interaction with the sensed information, send a command to the AI module to classify or scale the data in the most sensitive to a least sensitive category within the scale of 5 to 1, respectively. For the classification of the multimedia data of IoMT network, various machine learning (ML) algorithms like Support Vector Machine (SVM), Neural Network (NN), and statistic method (Kernel) etc. are exploited. Once after the classification of the data, the resource estimation for the multimedia traffic is performed.

Further, the authors in [8] have proposed big data layered architecture for multimedia IoT and its layered architecture is illustrated in Fig. 4.3. The big data layered architecture comprises six different layers namely, the identification layer, physical layer, communication layer, middleware layer, multimodal computational layer, and application layer. The hardware components required for the computation and real-time operating system (RTOS) are provided by the multimodal computation layer as shown in Fig. 4.3.

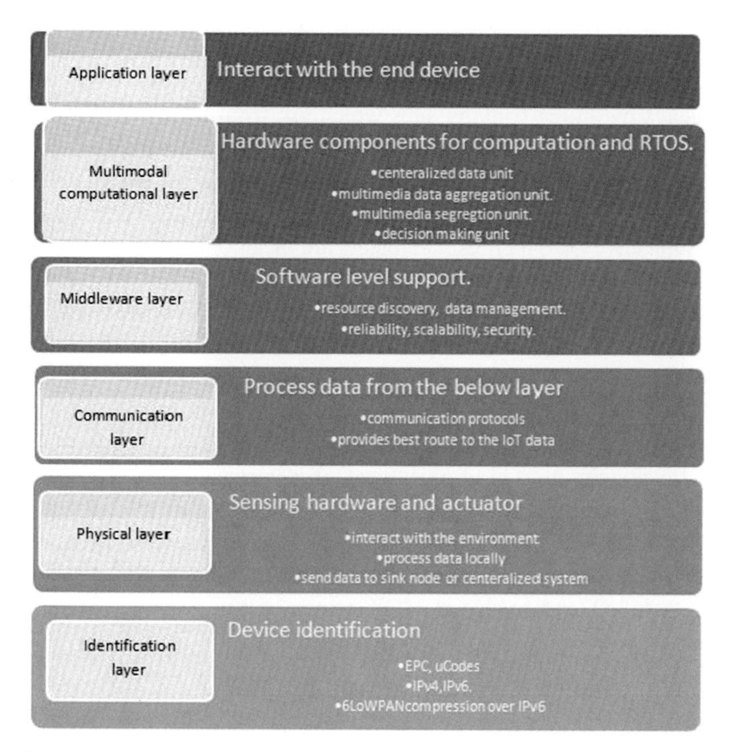

FIGURE 4.3

Big data layered IoMT architecture [8].

4.2.2 Performance parameters

As it is also described above in this chapter that the IoMT network performance is measured with QoE parameters which are user-centric along with QoS parameters which are network-centric. Therefore, the user satisfaction parameters which may be either subjective or objective, are equally important to be considered along with network-centric parameters like throughput, delay, jitter, peak signal-to-noise ratio, outage probability, data loss rate, bit error rate etc in the IoMT network in comparison to the IoT network. As per Microsoft, "Network QoS refers to the ability of the network to handle this traffic such that it meets the service needs of certain applications" [9]. However, the International Telecommunication Union (ITU) defines the QoE as follows: "The overall acceptability of an application or service, as perceived subjectively by the end-user" [10]. With the aspect of QoS, it is reported by Conviva that viewers of video content over the internet expect high-quality multimedia service and 75% of them leave the poor-quality video in just 4 minutes [1]. Therefore, the importance of high-quality multimedia content is clearly reflected by the drop-out rate of the users of the service/application.

One of the relatable parameters of QoE is the rating from the user asked by the particular application or service after its use e.g. as on the scale of one to five stars done nowadays by online banking sites which ask from the customer about online banking experience. However, in the IoT network, the QoS parameters are only considered. The user satisfaction in QoE is generally, based on the user perception, user experience, and expectation from the application or service [1]. The most commonly employed subjective QoE metrics parameters are mean opinion score (MOS), degradation MOS (DMOS), and the user satisfaction where MOS is obtained as the average of absolute ratings collected for each delivered multimedia content, and the DMOS is obtained as the average of the arithmetic difference between the ratings of the delivered multimedia content and the original multimedia content [11]. Further, the user engagement or satisfaction depends upon the playtime of the video content or the number of views of the multimedia content [12] and visual signal-to-noise ratio (VSNR) is the popular metrics of objective QoE parameters [13]. Khan et al. in [14] have mapped the quantifying parameter peak signal-to-noise ratio (PSNR) to MOS and said that the PSNR greater than 37 dB value corresponds to MOS value of 5, however, 31–36.9 dB, 25–30.9 dB, 20–24.9 dB, and less than 19.9 dB PSNR values correspond to 4, 3, 2, and 1 value of MOS, respectively.

4.2.3 Scalar data of IoT and big data of IoMT

The IoT only deals with the scalar data where the sensors only send the sensed information, which is generally, a single numeric value. The information is then sent to the sink node or some central entity or gateway for processing and control operation. IoT data is machine-generated like data from the different sensors, for example, the sensors employed at the appliances of smart home, the sensor at our health monitoring wrist bands etc. Therefore, the volume of data is small in comparison to the big volume data produced in IoMT. In addition, the multimedia content is human-generated, for example, emails, content in video streaming, images, graphics, gif, animations, contents shared in social networking sites etc. Therefore, we can say that IoMT is generating big data. Further, the data content produced by IoT through sensors has some specific format and the data produced is therefore called structured data. However, the multimedia data in IoMT does not necessarily have a format and is generating both structured and unstructured traffic and data. Further, the multimedia data in IoMT may be delay-sensitive however others might be delay tolerant. For example, online games, interactive video sessions are delay-sensitive application while streaming of stored audio and video content are somewhat tolerant to a certain amount of delay. In addition, few multimedia applications may tolerate the loss of few bits from the information while others might not. For example, if a certain frame from the video sequence that is GOP (group of pictures) is lost due to severe network congestion, the user's eye may not be able to recognize the loss of that frame. However, the text information e.g. in email, the data loss is not tolerable and affect the perceived service by the user.

4.3 Internet-of-Multimedia-Things and cognitive radio

4.3.1 Feasibility of cognitive radio to multimedia traffic

In order to support the demand of multimedia traffic in the IoMT network, the multimedia device needs to be bandwidth/spectral efficient and should have the capability to support green communication. Further, an excellent higher processing capability requirement and ability to support different multimedia content through IoMT is a challenging task. As it is evident that the multimedia transmission is bandwidth-hungry due to the large volume or bulky nature of the data, therefore, one of the efficient solutions to support IoMT is with the help of a cognitive radio network [15]. The cognitive radio (CR) concept came into light by Mitola in 2000 in his thesis [16], where it is defined as a device that utilizes the spectrum white spaces or underutilized spectrum of the licensed services. The advantage of cognitive radio is its reconfigurability and adaptability as per the environment scenario. It can change its modulation technique, transmit power, carrier frequency and can operate at different licensed bands where they serve as a secondary user of that band. This results in efficient spectrum utilization of the primary/licensed spectrum which earlier after allocation has not been utilized fully and due to poor allocation strategy and planning, has created spectrum scarcity. Therefore, we can say that to support a large volume of multimedia traffic, as per the conventional fixed allocation policy, spectrum or bandwidth is limited, however, in actual the allocated spectrum is unable to be utilized efficiently and enough bandwidth for multimedia applications is available [17]. Instead of a fixed spectrum allocation policy [18] for spectrum access, which is employed in traditional communication services, the cognitive radio employs dynamic spectrum access (DSA) [19] policy for spectrum access which provide flexibility to the user to access the spectrum. The cognitive device also has the freedom to aggregate the carriers through channel bonding of licensed and unlicensed bands to handle the variable traffic demand. Further, based on the QoS demand by the application, the cognitive device can be assigned with the best available channel of the spectrum. Mitola itself in [20] has claimed the use of cognitive radio for supporting multimedia traffic. Hence, based on the above-mentioned services provided by the cognitive radio, multimedia traffic support is very much possible in the cognitive radio-based network.

4.3.2 Why cognitive radio is a potential candidate for Internet-of-Multimedia-Things?

The answer to this question is provided below which maps the IoMT traffic requirement with the cognitive radio capabilities. Basically, the multimedia content requires large bandwidth for its operation, therefore, the amount of bandwidth available within an assigned licensed band of a particular service is difficult to grab. Since CR provides provision of its operation in different licensed band and support reconfigurability, therefore a large chunk of bandwidth availability is possible for CR user operating over different licensed and unlicensed bands. Following points describes the traffic requirement or demands put up by IoMT and the cognitive radio capability to handle those demands:

- The carrier aggregation is possible in the cognitive radio network, whereas per the user demand the different carriers available in the non-contiguous spectrum may be combined to fulfill the user demand. Therefore, the multimedia traffic like text and audio may operate on a single available channel however for video streaming in IoMT, multiple carriers need to be aggregated to support video traffic and this aid can be provided by cognitive radio.
- The delay-sensitive multimedia applications require quick attention from the network in the allocation of resources, and cognitive radio can provide the provision for handling this traffic over the spectrum white spaces or can simultaneously operate with the licensed user using spectrum underlay mode [19] of CR operation.
- The variable bit rate multimedia traffic can be served efficiently with high user experience and better QoS by allocating the best detected idle channels. The high data rate requirement by the IoMT traffic will also be handled efficiently by the detected idle licensed channels with better quality.

Further, several researchers have explored cognitive radio with respect to multimedia traffic support. However, before that, we can highlight below the practical use of CR in various applications in order to claim this technology to be a full proof technology to be implemented for IoMT.

- The cognitive radio-based IEEE 802.22 WRANs (Wireless Regional Area Networks) standard is discussed in [21] which allows the coexistence of television users and cognitive radio users for wireless internet access. The cognitive radio users use the unutilized television band of rural area for Internet applications which is advantageous to provide broadband internet access over these television white spaces (TVWS), otherwise separate broadband network deployment could be difficult and costly in the rural areas.
- The implementation of CR in the 5G network is already well acknowledged and is there in the Release-16 of 3GPP with the name 5G New Radio-Unlicensed (NR-U) spectrum [1].
- IEEE 802.11af is again a CR based standard that uses Wi-Fi i.e. IEEE 802.11 for TV band operations called "WiFi extension to TV white space (TVWS)" [22].
- CR based IEEE 802.16h standard also providing broadband internet access over Wi-Max band called Wi-Max extension to TVWS [23].
- IEEE 802.19.1 is working on "Co-existence of several white space systems" [24] which is again a cognitive radio-based standard.

All these above-mentioned technologies prove that CR is just not a theoretical concept, but its practicality is well known. Apart from the above standards, there are many other standards or working groups which are using CR technology for different applications.

4.3.3 Cognitive radio in IoMT network

Further, one of the ways to provide multimedia traffic support in cognitive radio network is provided by Popescu et al. in the [25] chapter, where they have employed a support node for this purpose. The architecture proposed for multimedia traffic support in CRN has a middleware, SDR and overlays, where the middleware is a software entity communicating with underlying components. With this architecture support, the varying QoS demand of multimedia traffic is met by cross-layer communication and finding suitable routing paths [25].

In addition, the reported literature utilizing cognitive radio for multimedia traffic has mostly provided the methods and schemes to carry the video multimedia content over the cognitive radio network which put the maximum resource demand and computation with respect to the other multimedia content. Therefore, if the cognitive radio network can provide service for the most challenging traffic, then it ensures the meeting of QoS and QoE requirements of other multimedia content over the IoMT network. Bocus et al. in [26] have considered SVC (scalable video coding) in a cognitive radio environment where the interference limit of primary users and two different networks need to be considered. They have performed subcarrier, bit, and power allocation in orthogonal frequency division multiple access (OFDMA) based cognitive radio network for transmission of video content through SVC.

Another cognitive radio network with multimedia video traffic support is considered by Aripin et al. in [27]. The cognitive radio network is working over the ultra-wide band (UWB) ranging from 3.1–10.6 GHz called cognitive UWB network. Further, a medium access control (MAC) centric cross-layer approach is described where the MAC layer takes the QoS requirements for the video content to be delivered from the topmost application layer of the OSI model and the channel condition and sensing information from the underneath physical layer. Further, the MAC layer in this cross-layer interaction-based model, finalizes various actions to be communicated to the respective layers for implementation. For example, adaptive modulation and coding format are communicated to the physical layer in order to achieve the target data rate and tolerable error rate etc. Therefore, the MAC layer is the main controller of the proposed scheme. In addition, researchers have mentioned cooperative sensing schemes to be implemented in the cognitive UWB network for packet reception rate-based channel allocation scheme. However, these authors in their work have assumed homogeneous multimedia traffic with the same QoS requirements like error rate and packet reception rate.

In addition, the digital fountain codes are implemented in the cognitive radio network to distribute the video content over the subcarriers/subchannels of the spectrum bands [28], where the distribution scheme has to take care of the primary user traffic pattern. Further, the fountain codes use also combat the channel impairments and primary user interference as they act as channel coding method. The method is also proposed to set access priority of channels based on the primary user's arrival rate, fading and noise in the channel. It is proposed to access the highest priority channel as long as all the subcarriers of that channel are not exhausted in order to get the maximum spectrum efficiency.

In addition, Libiao et al. in [29] have implemented heterogeneous i.e. both delay-sensitive and delay-tolerant multimedia traffic over the cognitive ad-hoc network and resource scheduling scheme for the multimedia traffic has been performed. The authors have tried to get the best user experience of the multimedia traffic over the cognitive ad-hoc network, wherein the earlier literature mostly sensing accuracy and spectrum utilization has been the main focus. The scheme for OFDM subcarrier assignment, power allocation and adaptive modulation and coding has been devised to maximize or optimize the utility function of the user which has expected delay and throughput of the multimedia traffic at the cognitive radio receiver in the optimization function. However, the utility function maximization has been done by maintaining interference at the primary user below a tolerable limit and constraining the transmit power of cognitive users over the OFDM subcarriers. The authors have proposed a method to select the optimal power and modulation strategy for different subcarriers based on their quality and correspondingly allocation of these subcarriers will be done to the multimedia traffic.

Further, Jiang et al. in [30] have provided the scheme for channel allocation for transmission of images and H.264 video in cognitive radio network which has subjected to different QoE requirements. The multimedia traffic needs to be served by the cognitive network over primary channels are either delay-sensitive application or delay-tolerant application based on which the primary channels with low switching or dropping probability will be assigned to the delay-sensitive application. The mean opinion score (MOS) which is one of the QoE measuring metrics, has been calculated by taking into account the sender bitrate (SBR), dropping probability (DP), and frame rate (FR).

In all the previously mentioned literature, the cognitive radio user can only transfer the multimedia traffic over the white spaces of licensed users only if the contiguous single white space is available for video transmission. Therefore, as per the above-mentioned proposed schemes by the researchers, even if non-contiguous white spaces are available and sensed by the cognitive users whose total bandwidth is greater than that of the required bandwidth for multimedia video traffic transmission, it could not be implemented for video transmission until the single sufficient white space is available. Therefore, Bhattacharya et al. in [31] have proposed a method to implement non-contiguous bands for multimedia traffic transmission over the cognitive network. Two-channel allocation schemes have been proposed, one based on frequency division multiplexing-frequency division multiple access (FDM-FDMA) using non-overlapping channels and the other is the orthogonal frequency division multiplexing (OFDM)-FDMA using overlapping subcarriers of OFDM. In the proposed protocol, the performance parameter is given as the average number of attempts made by the proposed scheme for acquiring the required number of channels from the FDM or OFDM scheme in order to communicate a given multimedia signal. The proposed protocol has been compared with the first-fit and best-fit channel allocation schemes which have been outperformed by their proposed schemes.

Further, Fig. 4.4 represents the cognitive radio based IoMT network, where the cognitive users are utilizing the available licensed channel of the primary user net-

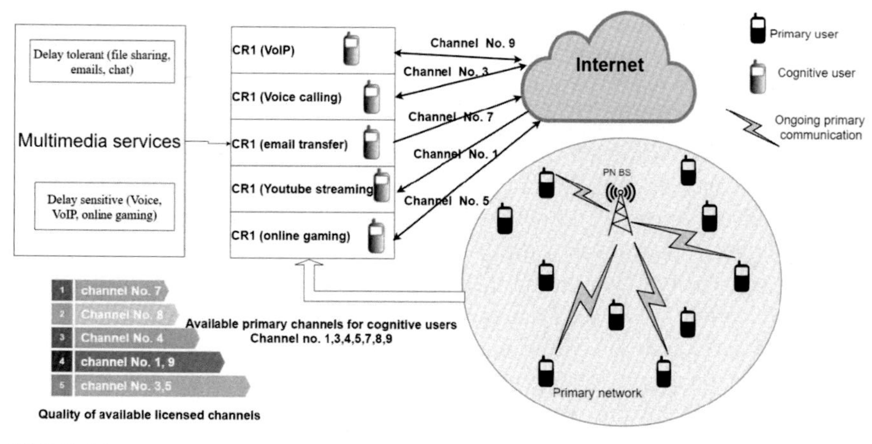

FIGURE 4.4

Cognitive radio based IoMT network.

work. The accession of the available licensed channels by the cognitive users for their multimedia traffic is shown to be dependent on the quality of the channels.

The quality of the channel is defined in terms of the delay sensitivity of the applications, that is highly delay-sensitive service needs a high-quality channel in comparison to the delay-tolerant application. In the figure, we have shown the highest quality channel with number 5 and the lowest one with number 1.

4.4 CR-MAC protocols for Internet-of-Multimedia-Things

The MAC protocol in the cognitive radio network plays an important role and there are some additional tasks and features which are enabled by the cognitive MAC protocol in comparison to the conventional MAC protocol [32]. The main features of the cognitive-MAC protocol employed in the IoMT network are highlighted below:

- MAC protocol of the cognitive radio has to distinguish among the primary and cognitive-users transmission in order to withdraw from its current channel access when the primary user resumes its transmission and continue transmission if another same priority cognitive user tries to access. However, the conventional MAC protocols are designed for only a single type of user all having the same access rights.
- The cognitive-MAC protocol has to work on a multichannel multiuser framework where the wideband channel access support is provided by the cognitive user.
- Allocation of the spectrum/channel to the cognitive users over a wide spectrum band and coordination among the users for accessing the spectrum is provided by the cognitive-MAC.

- The spectrum mobility support is provided by the cognitive MAC in comparison to the conventional MAC. It allows the cognitive nodes to switch to different available channels when the primary user resumes its transmission. Therefore, in-band and out-of-band spectrum sensing task need to be performed by the cognitive user. In-band sensing allows the cognitive node to know the return of the primary user on that channel and out-of-band sensing allows the cognitive nodes to find another available licensed channel that may be used in case current secondary user transmission is interrupted by the primary user.
- The heterogeneity of the traffic in IoMT and the volume of data is huge. The CR-MAC protocol for IoMT can operate on a wide range of spectrum and is not confined to the assigned spectrum as that of traditional MAC protocol, therefore the facility to support heterogeneity and amount of traffic carried over the CR-MAC is provided by the protocol.

Different cognitive radio-based MAC protocols for multimedia data have been reviewed by the researchers and are presented further. These MAC protocols are designed for multimedia traffic communication over the cognitive radio network and the comparison of these protocols is shown in Table 4.1. The centralized and distributed cognitive MAC protocol for multimedia transmission has been surveyed in [12]. In the centralized cognitive MAC protocol, a central entity is there to coordinate and schedule sensing and data transmission operation of different cognitive users over the licensed channels while meeting the interference constraint at the primary user. While in the distributed or ad-hoc cognitive MAC protocol, there is no central physical entity. However, the coordination among the cognitive users about sensing and data transmission and individual sensing and decision making has to be performed by the cognitive users itself.

Each MAC protocol described below is working in the opportunistic spectrum access (OSA) mode that is working only over the spectrum white spaces which are free channels not currently utilized by the licensed users. The authors in [33] have proposed a centralized and distributed cognitive MAC protocol based on a packet scheduling scheme for multimedia voice traffic. The performance parameter considered by Ali and Zhuang in [33] is fairness while guaranteeing the required QoS for the voice traffic of multimedia data. In [33], the queue is maintained for data of both primary and cognitive users which is voice traffic in both cases and the access scheme of MAC employed is a fixed TDMA based access scheme. Two MAC protocols are proposed in [33] called centralized MAC I and distributed MAC I and have shown the improvement in the fairness in channel access in comparison to the existing MAC protocols. The steps which are followed in the proposed centralized MAC I [33] are given below:

1) Checking for the new packet generation of voice data by the cognitive user.
2) Ordering of cognitive users in terms of current dropping rate in descending order. The dropping rate is defined as the percentage of the voice data packets that may be dropped without affecting the quality of the data.

3) Ordering of the same packet dropping rate cognitive users as per the number of packets queued for the transmission in descending order.
4) Cognitive nodes sense the idle primary user timeslot from the TDMA frame.
5) Allocation of the idle timeslot to the top cognitive users and packet transmission is performed.
6) If delay bound of a packet, which is defined for the voice packet as that delay above that packet transmission has no significance, is crossed, that packet is dropped from the queue.

The steps which are followed in the proposed decentralized MAC I [33] are given below:

1) Checking for the new packet generation of voice data by the cognitive user.
2) Backoff duration is calculated for the cognitive user which is function of packet dropping rate and size of the queue.
3) In order to avoid the trouble of small differences in the backoff timings of cognitive users, the sensing duration of the TDMA frame is divided into mini slots. In order to reduce collisions among cognitive user, each cognitive user maps the backoff value to the minislot ID.
4) Sense the channel from the start of timeslot to the mini slot corresponding to the ID. If the channel is idle, transmit packet; if busy, wait for the next timeslot. However, if the collision occurs, increase the backoff time and recalculate ID, and wait for next timeslot.
5) If delay bound of a packet is crossed, that packet is dropped from the queue.

Further, Su and Zhang in [34] have proposed a cross-layer based opportunistic MAC protocol for the distributed network where the physical layer spectrum sensing policy is integrated with packet scheduling at the MAC layer. They have employed two sensing policies namely, the random sensing policy and negotiation-based sensing policy. The performance of the saturated and non-saturated network is presented where the saturated network always has a non-empty queue with packets always scheduled for transmission. The primary network is assumed to operate over the TDMA access scheme however, the cognitive network accesses the spectrum in a random manner or is employing random access scheme. The proposed MAC protocol considers two cognitive transceivers at each device, one transceiver called control transceiver is always tuned to the cognitive network dedicated control channel and another called SDR transceiver which switches to the different licensed channel for sensing and data transmission. In this proposed MAC protocol, the control channel is divided into cycles of reporting and negotiation phase where the number of slots in reporting phase is equal to the number of licensed channels available as shown in Fig. 4.5. The cognitive users sense the channel and report its sensing result over the control channel's reporting phase at the corresponding licensed channel's reporting sub-slots. Therefore, the sensing information of a licensed channel which is sensed by a particular cognitive user is available to all the other cognitive users over the control channel and hence the cooperation is provided. Afterwards, the negotiation phase

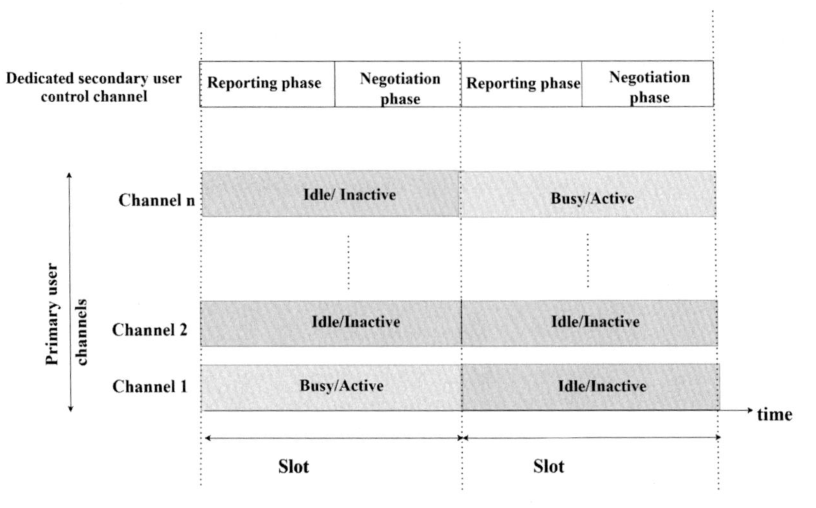

FIGURE 4.5

The proposed cross-layer based opportunistic MAC [34].

allows the cognitive users to negotiate and compete for the available idle channels by sending request-to-send (RTS) and clear-to-send (CTS) over the control channel. They have provided the analytical results and simulations which relate the packet arrival rate of the cognitive users with the transmission delay, primary channel utilization and the number of cognitive users.

In addition, the authors in [35] have proposed a CR-MAC protocol called TQCR-MAC with two levels of QoS support for real-time and non-real-time multimedia traffic. The proposed TQCR-MAC also consists of one dedicated control channel as that of the above discussed MAC for cognitive users to share their sensing information and coordination in spectrum access of available licensed channels. The frame structure of the proposed TQCR-MAC is shown in Fig. 4.6, which consists of three intervals called the sensing period, ATIM period, and communication period. Further, the division of ATIM period is performed into the beacon and control period. The TDMA access scheme is employed by the cognitive users to access the channel. In the proposed MAC protocol, it is assumed that each CR maintains a primary user's channel occupancy list which is a 2-D array of values related to the primary user channel number and its idle or busy status. This channel occupancy list is updated by the sensing information which is obtained by the sensing period where CR users sense the primary user channels. All CR users are said to tune to the control channel during ATIM period where they contend with each other and the real-time traffic packet is given priority in the contention in comparison to the non-real-time traffic packet. Since TQCR-MAC has given priority to real-time traffic in comparison to the non-real-time packet, therefore, the real-time packet may capture the channel access from the already scheduled non-real-time traffic on a slot basis. The proposed MAC protocol has been compared with already existing similar application proto-

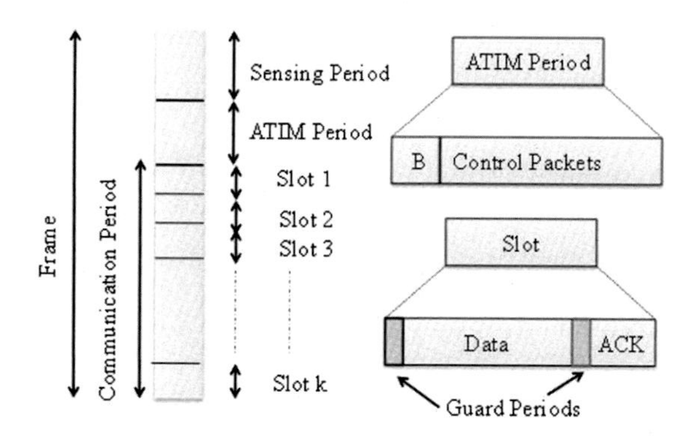

FIGURE 4.6

Frame structure of TQCR-MAC [35].

cols, however, TQCR-MAC has provided more throughput because of the provision of supporting both real-time and non-real-time data in a single frame on a slot basis. The additional advantage is also there in throughput due to the presence of a dedicated control channel.

Moreover, Cai et al. in [36] have proposed a distributed QoS aware MAC called QC-MAC for multichannel cognitive network. The authors' proposed MAC has provided support for heterogeneous multimedia traffic like voice, video and data by mapping the QoS demanded by these traffics and the primary users' channels which fulfill these demands. Therefore, based on the primary user's channel usage pattern, each cognitive user depending on its type of multimedia traffic to be carried out which can be either constant bit rate (CBR) or variable bit rate (VBR), find the channel for sensing and data transmission. A common control channel (CCC) in the UWB is considered for the distributed QC-MAC protocol. The QoS provisioning is enhanced in the QC-MAC by introducing service differentiation in the arbitrary sensing periods of different traffic flows by allowing smaller sensing window for real-time traffic in comparison to that of the non-real time traffic, so that the channel may be grabbed as soon as opportunity in the channel arrives. The flow of working of QC-MAC is shown in Fig. 4.7, where the cognitive user switches its sensing to the next channel if channel is busy or if the primary user resumes transmission. The greedy and ascending spectrum sensing policies are simulated in [36] and it is shown that the greedy scheme provides better delay performance in comparison to the ascending sensing policy for VBR traffic in low traffic load. However, when there is different type of multimedia traffic, like CBR and VBR traffic, ascending scheme outperforms the greedy sensing policy. A single half-duplex cognitive radio transceiver is employed in QC-MAC protocol for each cognitive user.

Further, Jha et al. in [37] have proposed a QoS provisioned distributed cognitive MAC called opportunistic multi-channel MAC (OMC-MAC) which provides efficient channel utilization, deals with multichannel hidden and exposed terminal problem and collision with the primary users. A single half-duplex transceiver is

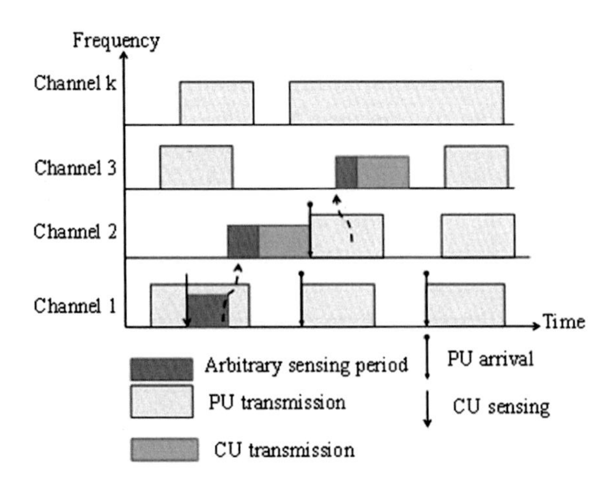

FIGURE 4.7

QoS-aware MAC (QC-MAC) [36].

considered to be present at each cognitive user and also a dedicated CCC is available with the cognitive network. The protocol is in the similar pattern as above considered protocols having sensing, contention and data transmission phase. However, the contention phase is further divided into two parts, one contention part is available for real-time or prioritized multimedia traffic cognitive user and other for the non-prioritized multimedia traffic cognitive user. The contention phase is employed to negotiate and reserve the idle licensed channel by the cognitive users and employ backoff algorithm [38]. The channel reservation and negotiation process in the OMC-MAC is shown in Fig. 4.8, where U is the cognitive receiver for Q transmitter and S is the receiver for R transmitter. In the sensing phase, each cognitive user has list of free licensed channels available with them as shown by X1, X3, X4, and X9 for Q transmitter. Similarly, other cognitive users list of available channels is also shown in Fig. 4.8. During the contention phase, the corresponding transmitter and receiver agree on the common available channel to both of them and then start transmitting their data on the selected data channel during data transmission phase. The network allocation vector (NAV) is updated accordingly which will tell the neighboring users about the usage of channel and avoid hidden and exposed terminal problem. Therefore, no collision in the data transmission phase is guaranteed by OMC-MAC protocol by the authors.

Jinag and Liu in [39] have proposed a QoS aware multichannel cognitive MAC named QA-MAC as represented in Fig. 4.9. In QA-MAC, different channel reporting and data transmission policies are adopted depending on the channel utilization patterns to satisfy the QoS of real-time and non-real time multimedia traffic. This protocol again has a dedicated common control channel (CCC) to the cognitive network and channels are divided into slots. Each slot of CCC has two phases called reporting and contention period. The reporting period has number of slots equal to the number of licensed channels and each slot corresponding to the individual li-

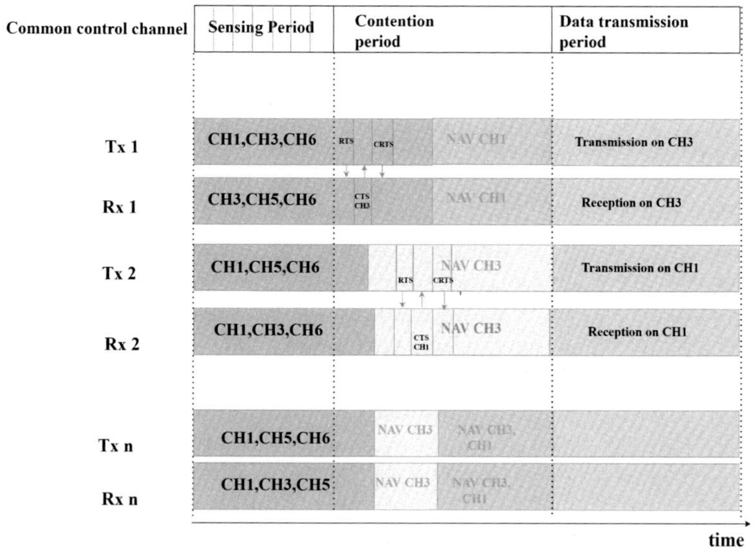

FIGURE 4.8

OMC-MAC for QoS provisioning [37].

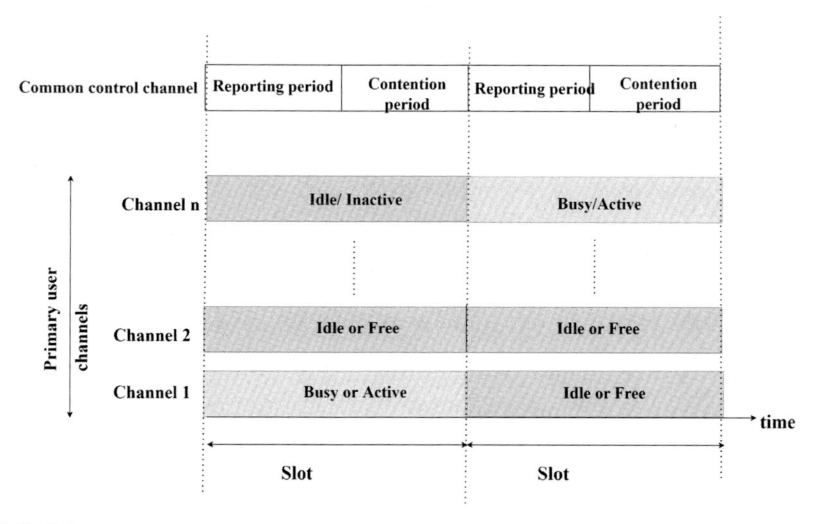

FIGURE 4.9

QoS-aware MAC (QA-MAC) [39].

censed channel. The sensing by the cognitive users is performed during the reporting phase, and spectrum information (SIP) of PU and cognitive user's service type (SST) data structures are then formed. The SIP field contains the information of primary

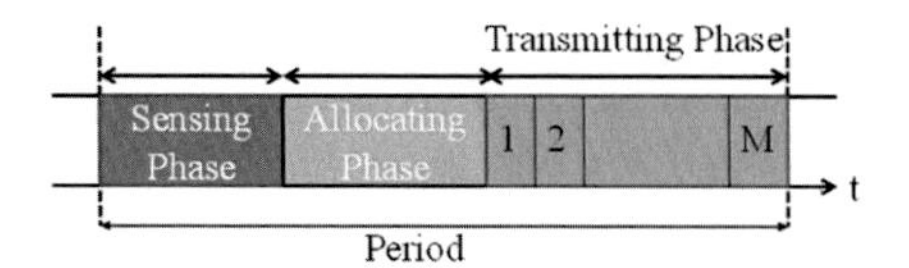

FIGURE 4.10

The principle of group-based MAC protocol [40].

user status i.e. idle or busy in a particular slot. However, the SST field contains the information of whether the cognitive user reporting the channel information wants to transmit real-time traffic or not. In case of real-time traffic, SST will be set to 1 and that cognitive user is set to higher priority and a licensed unreserved idle channel is reserved to the cognitive user. However, this field is set to 0 in case of non-real time traffic and cognitive users of non-real time traffic then compete during contention in the next contention period. In the proposed QA-MAC, channels are firstly reserved for the real-time traffic and then leftover unused channels are utilized by non-real-time traffic.

Another a group-based MAC protocol for QoS provisioning to the multimedia traffic is proposed by Song and Lin in [40]. This protocol has employed spectrum aggregation and dynamic spectrum access to provide QoS support for multimedia traffic. The fairness among the group of users is also guaranteed in [40]. Two transceivers are assumed at each cognitive user and one of which is always tuned to the CCC and other transceiver is exchanging data and sensing the licensed channels. In the proposed group-based MAC protocol, the cognitive users are divided into groups and each group has a leader user among those same users. The leader user is responsible for providing coordination and contacting with the manager. The manager is only a single entity in the whole network and is selected from a number of leaders. The manager is responsible for allocating channels to the various group users. The proposed MAC protocol principle is shown in Fig. 4.10, which consists of sensing phase, allocating phase, and transmitting phase. The functions performed in sensing, allocating and transmitting phase are described as:

Sensing phase: The manager allocates channels for sensing over the CCC through sense frame to each group leader. After sensing the channels listed in the sense frame, each group leader will take the sensed channels information from its member users and report it to the manager through CCC.

Allocating phase: Manager after getting the status information of licensed channels from the leaders, will take the bandwidth requirement of different groups through their leaders. After getting the required information, the manager finds a suitable channel allocation scheme accordingly and communicates it to the leaders. Afterwards, users of the respective groups will start using the allocated channel for its multimedia traffic.

Transmitting phase: Transmitting phase is divided into multiple time slots by the leader, which allows the member users to access the channels by contention scheme.

In [41], a completely different MAC is proposed in comparison to the above discussed systems, where instead of performing contention on a CCC, contention is performed at respective data channel and has avoided control channel saturation problem. In the proposed DC (distributed contention based)-MAC, coordination is provided by the rendezvous channel (RC) and each cognitive device is assumed to have single transceiver. The design utilizes a rendezvous channel as a media for distributing information about the nodes and channels in the network. The data channels which are the primary licensed channels are utilized for contention and data transmission. The idea or design of DC-MAC protocol is taken from existing C-MAC protocol which has almost similar scheme as that of the proposed DC-MAC in [40]. The home channel, home node, foreign channel, foreign nodes, representative node are the terms used to define the working of proposed MAC protocol. The home channel is the channel on which a node resides and all other channels with respect to it are termed as foreign channels. Therefore, a node residing on the home channel is called home node and if it resides on foreign channel, it is called foreign node. A representative node is defined for each channel which broadcasts the information of nodes residing on the home channel to RC. The criterion for selection of the representative node is dependent on the amount of traffic available at a particular node. Both the in-band and out-of-band sensing provision is available in DC-MAC protocol where in-band sensing is performed by home node on home channel while out-of-band sensing is conducted by home node on foreign channel. This is another major change available in this MAC protocol over the earlier ones. The channels are divided into time frames called super frames and, in the network, it is considered to have one RC channel and multiple data channels of primary users. The main components of RC channel and data channels are described below which are also shown in Fig. 4.11:

Rendezvous channel: The super frame of RC is composed of three intervals called Rendezvous channel beacon (RCB) interval, data transfer interval, and quiet interval. RCB interval allows the beacon from representative node of different data channels which contains the information of channel SNR, home nodes, and quiet interval distribution. During quiet interval, all cognitive users stop their data transmission so that sensing of incumbent or primary users can be performed properly, otherwise differentiating primary and cognitive users' transmission will be difficult. Moreover, data transfer interval performs the contention among home node and accordingly transfer data.

Data channels: The data channel's superframe is however divided into four intervals called home channel beacon (HCB) interval, home channel information beacon (HCI) interval, data transfer interval, and quiet interval. All home nodes including reference node exchange beacons during HCB. The reference node in this duration sends the information obtained from RCB, because of which all the home nodes of that particular channel have the idea of home channels of all other cognitive nodes. Further, HCI is used to choose the reference node from all the available homes nodes on home channel. Moreover, the function of data transfer and quiet interval is same as described above in rendezvous channel.

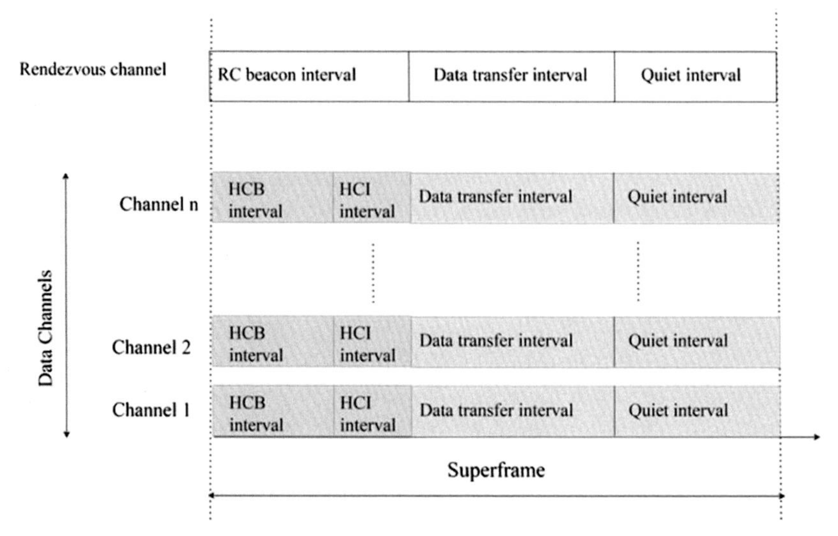

FIGURE 4.11

Distributed contention based-MAC that is DC-MAC for support to multimedia traffic [41].

For the real-time applications, the performance of DC-MAC is measured in terms of connection fail ratio which is the ratio of number of real-time connections failed to the total number of real-time connections originated. In addition, in the proposed DC-MAC protocol, four kinds of access category packets are prioritized which are: for voice, video, best effort, and background applications through packet marking scheme. Further, the performance improvement in the proposed DC-MAC is presented in comparison to the cognitive-MAC (C-MAC), and results have described that the proposed DC-MAC has outperformed C-MAC.

In addition, a whole spectrum decision framework is proposed for achieving the QoS of the different categories of the multimedia traffic which also has included MAC protocol unit embedded in it as shown in Fig. 4.12. The authors in [42] have categorized the data packets of multimedia traffic based on their type and accordingly mapped them to the available licensed channel where the QoS required by this traffic is fulfilled. The categorization of multimedia traffic is done by the application layer and traffic classification unit. There is one dedicated CCC available with the proposed MAC protocol for sharing the control information among cognitive users. The channels of primary user network are also categorized into eight different classes based on the channel bandwidth, its availability and primary user's fluctuation. As it is illustrated in Fig. 4.12, the left part of the proposed framework consists of QoS engine and right part indicates the wireless communication path. The application block provides the idea of QoS requirement of different applications generated by the user. The channel selection and management unit (CSMU) block get the information from the application block regarding the requirement of the QoS and accordingly provides

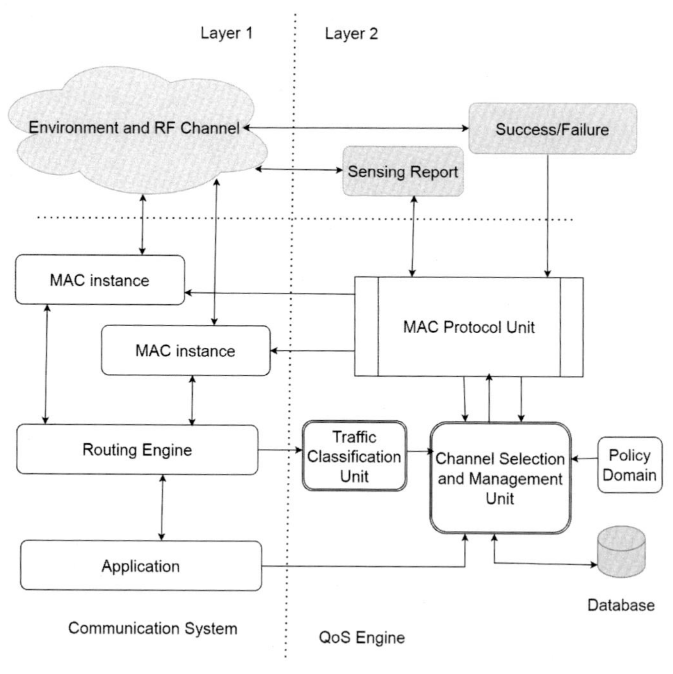

FIGURE 4.12

QoS based decision framework [42].

its spectrum decisions or accessing methods of the channels. The Policy domain, routing engine, and MAC protocol unit provides the input to the channel selection and management unit. In addition, the database block is used by the CSMU to store the channel availability information over time and space. One of the inputs to CSMU is from policy domain block which provides the local policy of transmission power, and suitable channels. Another input to CSMU is from the MAC protocol unit (MPU), which provides the appropriate MAC selection as per the QoS and information from the sensing report block. The sensing report block is being updated from the environment and RF channel block which is dealing with the environment directly. In addition, the success/failure report unit has the information about the success/failure rate of the packets transmitted over the channel and this information is also provided by the environment and RF channel unit. Moreover, the different blocks available at the right side of the framework that is in wireless communication path are dealing to provide communication between transmitter and receiver. A load balancing algorithm is given by the authors to balance the load of different categories of multimedia traffic on different available licensed channels.

Table 4.1 Comparison of various multimedia-based CR-MAC protocols.

Ref.	Multimedia CR-MAC	Type of multiple access scheme	Type of multimedia traffic considered	Central-ized	Dis-tributed	Green communica-tion/energy saving	Perfor-mance parameters evaluated
[33]	Centralized-MAC I and distributed MAC I	Fixed (TDMA)	voice	Yes	Yes	No	fairness
[34]	Cross-layer based opportunistic multichannel MAC	Random access	Not mentioned	No	Yes	No	traffic throughput, packet transmission delay
[35]	TQCR-MAC	Fixed (Contention and TDMA)	Real-time and non-real time	No	Yes	Yes	throughput, end-to-end delay
[36]	QC-MAC	Fixed (TDMA)	Voice, video, data	No	Yes	No	delay
[37]	OMC-MAC for QoS provisioning	Contention based	Delay sensitive and delay tolerant application	No	Yes	No	Throughput, collision probability, delay
[39]	QA-MAC	Employed both contention and non-contention scheme for different traffic	Real-time and non-real time	No	Yes	No	Throughput
[40]	Group based MAC protocol	Contention based	Not mentioned	No	Dis-tributed Adhoc	No	Fairness, throughput
[41]	DC-MAC	Contention based	Real time and non-real time traffic	No	Dis-tributed Adhoc	Yes	Aggregate throughput, real-time connection fail ratio

4.5 Challenges in cognitive radio based IoMT network

There are numerous potential challenges faced by internet of multimedia things network which are already mentioned in various literature. However, in this section IoMT challenges while implementing through cognitive radio are mentioned. As, the cognitive radio has to perform spectrum sensing, spectrum management/sharing and spectrum mobility tasks, and the major challenge of each operation has to be taken care of in IoMT network implemented with CR. Therefore, we have described the challenges and remedies or solutions available to deal with in the cognitive radio based IoMT network.

4.5.1 Spectrum sensing

Challenges: The spectrum sensing allows the cognitive users to detect the spectrum white space in the primary user network which then may be used for cognitive user's data/information transmission. In addition, the performance of spectrum sensing is measured by detection probability and false alarm probability values which should be maximized and minimized, respectively for getting better sensing results [43]. The high sensing accuracy of cognitive device helps in guaranteed delivery of multimedia content with low error in the data of the IoMT network when CR is providing service to the IoMT network. Therefore, to achieved high accuracy sensing results in the cognitive radio based IoMT network is a challenging task.

Solution:

1. **Sensing time:** One way to achieve significantly high sensing accuracy is to increase the sensing duration of the cognitive user, however it will reduce the multimedia traffic transfer time over the network and will decrease the data rate of the multimedia traffic. This might be tolerable for low data rate multimedia applications like email transfer or file transfer, however for delay sensitive multimedia content like video conferencing or interactive live interaction, which require high data rate, it will be cumbersome. The optimal duration for sensing needs to be devised in order to solve this sensing-data rate/throughput tradeoff problem and authors in [44] have found the optimal value of sensing time for achieving the target of detection and false alarm probability values.
2. **Sensing technique:** The spectrum sensing techniques employed for white space detection also play a vital role in sensing results accuracy. There are commonly three transmitter-based detection techniques namely, the energy detection, matched filter-based detection, and cyclostationary based feature detection. Among these, the cyclostationary based feature detection is highly accurate, however it is most complex. Moreover, the energy detection is least complex, highly employed spectrum sensing technique, however the accuracy is not very high. Therefore, in IoMT network, the distribution of the idle detected licensed channels will be based on the type of the multimedia traffic, its QoS and QoE requirement and the spectrum sensing technique employed to sense that channel. This will help in allocation of the highly accurate idle detected licensed channel to the most sensitive multimedia application and some delay tolerant and low data rate required multimedia application may be assigned with the channel which has somewhat lower sensing accuracy. However, since guaranteed delivery is essential, therefore some threshold for the sensing accuracy might be defined. For example, IEEE 802.22 standard has defined the acceptable range of miss detection probability to be 0.01 to 0.1 and detection probability range to be 0.9–0.99 [45].
3. **Cooperation techniques:** The cooperation among cognitive users helps to increase the efficiency of spectrum sensing results and have proved to provide high detection and low false alarm probability. There are various cooperative rules defined and employed like OR-rule, AND-rule, Majority-rule, k-out of M rule in the cognitive radio environment. However, the amount of overhead will increase

while employing cooperation in the network and might also increase the delay for transmission and reception of the data. This is because in the cooperation methods there is one central decision entity generally, called fusion center to which all the cognitive users forward their hard or soft sensing results to make the final decision about the availability or non-availability of the primary user sensed channel. This process is time consuming and may directly affect the QoS and QoE of delay sensitive multimedia applications.

4.5.2 Spectrum sharing/management

Challenges: The allocation of available primary user's channel to different cognitive users is the task of spectrum management. In the cognitive radio network based IoMT network, there are two different kinds of users that is primary licensed users having higher priority and secondary or cognitive users. The resource allocation among the users should be performed in such a way which avoids the interference at the primary user and along with that support the heterogeneous multimedia traffic over the cognitive network. Therefore, the allocation of such channels to the multimedia users is a challenging task. This is also a multichannel multiuser allocation problem, with higher priority setting for the primary users. The medium access control (MAC) protocol design for such heterogeneous multimedia traffic over the cognitive radio network is a crucial task. It is the MAC protocol which does all the scheduling and coordination among the cognitive users over the cognitive network. For heterogeneous multimedia traffic, the MAC protocol design in IoMT network should handle the data rate, latency and volume of the data and should provide flexibility in the content delivery.

Solution:

1. **Centralized and distributed resource allocation:** As described earlier in the cooperative spectrum sensing, the centralized controller or fusion center collect all the sensing information from different cognitive users and then make the decision about the presence and absence of the primary user on the channel. After the decision is made in the cooperative spectrum sensing, the resource or spectrum allocation is the next task in the line. In the centralized resource allocation policy, the central entity sometimes may be called base station or fusion center, will assign the spectrum to different cognitive users in the network. This will help a single coordination point of the network. However, the authors in [46] have presented to use decentralized resource allocation instead of centralized in order to reduce the infrastructure cost and to support heterogeneous multimedia traffic latency issue. It is also a much convenient solution in the wireless multimedia system. In the decentralized resource allocation policy, the cognitive users locally communicate with each other to make spectrum access decision. It is more convenient than the centralized resource allocation policy since infrastructure cost is not required. Further, the back-and-forth information exchange between each cognitive user and central entity is also minimized in distributed resource allocation.

In addition, a control channel may be set up in the MAC protocol of distributed spectrum allocation scheme which act as point of coordination without needing any extra infrastructure in distributed spectrum allocation.

2. **Spectrum overlay and interweave model of spectrum allocation:** The cognitive device has advantage of accessing the spectrum in three different modes that are opportunistic spectrum access (OSA)/spectrum interweave, spectrum underlay and spectrum overlay. In OSA mode, the cognitive users only utilize the spectrum white spaces i.e. the unutilized spectrum of the licensed users is employed. In this mode of operation, the availability of free spectrum is essential for cognitive device to transmit. However, the spectrum underlay mode of cognitive user will not require the free spectrum and can operate simultaneously with the transmission of the primary user over the same band. However, in the underlay mode of operation, the interference temperature limit at the primary receiver should not be crossed. Therefore, the cognitive transmitter has to limit it's transmit power in underlay mode so that it will not interfere the primary signal reception at the receiver. However, OSA mode gives the freedom to the cognitive device to operate without any constraint on the transmit power. In addition, the spectrum overlay mode allows the cognitive device to operate simultaneously, with the primary device in its spectrum band however certain coding techniques like dirty paper coding need to be employed for mitigating the interference of cognitive user at primary receiver. The received signal to noise plus interference ratio (SINR) at the primary receiver will be kept same as when the cognitive device is not in the picture by allowing cognitive device to use its transmit power to relay primary user data. Therefore, the loss in SINR at the primary receiver which is caused by the interference created by cognitive device is compensated by the cognitive device relaying primary user data by utilizing its own transmit power.
   These different modes of operation of cognitive device will help in IoMT network to carry heterogeneous multimedia traffic. Therefore, the data rate variation may be provided by the OSA, spectrum underlay, and spectrum overlay modes of operation of cognitive device in IoMT network. The high data rate and delay sensitive multimedia content may utilize OSA mode of operation. However, if guaranteed delivery of content is needed with less data rate requirement, spectrum underlay mode will be best. However, if guaranteed delivery, least delay and high data rate is required, spectrum overlay mode need to be employed which will increase the system complexity requiring extra coding blocks. Therefore, the adaptation in resource allocation and the efficient resource utilization like spectrum and power is possible in cognitive radio based IoMT network.
3. **Resource allocation management models** Authors in [1] have described different resource allocation models in cognitive network supporting multimedia traffic. These models are briefly described below:

   a) **Machine learning-based resource allocations techniques:** Although cognitive radio is a self-aware radio, however, machine learning algorithms help it to achieve high level of cognition, where it may itself train from the environment and make decision. Supervised, unsupervised, and reinforcement-based machine learning algorithms help in the design of cognitive radio based IoMT network.

b) **Game theory-based resource allocation:** Application of game theory in cognitive network for resource allocation is highly exploited field. It basically considers primary and cognitive users as players and resource allocation is the game which players play. These players want to win the game and some of them may behave as selfish players which want to maximize its own objective with no consideration of the global objective. In cognitive radio based IoMT network, game theory-based resource allocation will allow the cognitive players to optimize its channel allocation based on the incoming multimedia traffic.

c) **Cross-layer resource optimization:** Cross-layer based resource optimization will allow the interaction of different layers of OSI or TCP model to satisfy the QoS and QoE requirement of the user which is directly interacting with the application layer.

4.5.3 Spectrum mobility

Challenges: Since the cognitive radio allows dynamic spectrum mode of access in comparison to the fixed spectrum access, therefore the issue of network fluctuations may rise. This is because, once the primary user comes back into its channel where cognitive user is operating, cognitive user has to switch to another unutilized band. This switching will definitely introduce some delay and it is not always possible that the channel to which switching is performed will have same characteristics and performance as that of initially occupied channel of cognitive multimedia user. Therefore, this network fluctuation which caused by spectrum mobility of the cognitive user will affect the QoS and QoE of multimedia data.

Solution:

1. **Scalable video coding (SVC):** The most suffered and impacted traffic may be the one which is delay sensitive and require high data rate. Therefore, H.264 video coding method which is a scalable video coding offers a flexible traffic rate for media streaming to match the varied transmission conditions. In SVC, a scalable bit stream can be considered as a hierarchy of video layers, consisting of a mandatory base layer and a number of enhancement layers. As higher layer data is successfully received and decoded, the perceived quality of the video is improved [26]. Further, as per the cases described in [1], SVC can be done in three different modes based on spatial, temporal, and SNR. In the temporal mode, the number of frames is adjusted while in the spatial mode the size is changed, and in the SNR mode, the quality is modified in order to compress the video contents Further, Bocus et al. in [26] have considered SVC in the cognitive radio environment where the interference limit of primary users and two different networks need to be considered. They have performed subcarrier, bit, and power allocation in OFDMA based cognitive radio network for transmission of video content through SVC.

2. **Hybrid mode operation of cognitive radio:** Hybrid mode of spectrum underlay and OSA will allow the seamless transmission over the primary channel whether it is occupied by PU or not [47]. The OSA mode allows the multimedia traffic to be carried by the cognitive device over the channel when no primary user is there. However, underlay mode allows the operation when primary user has regained the access of the channel. Therefore, there is no need to switch the cognitive user from the current spectrum. However, maintaining data rate same is somewhat difficult in hybrid mode since underlay mode allows transmission with less power affecting data rate to be carried.
3. **Adaptive modulation and coding (AMC):** It allows the higher order modulation over the channel where fading is less severe to obtain higher data rate. However, the complex channel coding techniques will allow the cognitive user to fight with the channel impairments rising due to the severe fading however, it will increase the complexity of the system. In addition, the adaptive modulation is generally employed with advanced transmission schemes like multi-carrier code division multiple access (CDMA), multiple-input multiple-output (MIMO), and cooperative transmissions.
4. **Handoff management:** The switching of the channel to another licensed channel is called handoff which often arises in the cognitive network. This handoff delay minimization can be performed if instead of reactive handoff, proactive handoff scheme is employed. In the reactive handoff, the handoff is initiated only when the primary user comes back into the channel. However, in the proactive scheme, the spectrum prediction module allows the cognitive node to initiate the handoff well in advance when there is high probability of primary user coming back to the channel. There are many schemes that have been proposed to alleviate the issue of spectrum handoff for multimedia streaming over CRNS as described in [48].

4.5.4 Miscellaneous challenges

Other challenges which are not mentioned above and are common to the IoMT network are described below:

1. **Video compression:** The major challenge is posed by live streamed video content or live interactive session which is delay bound and employs large bandwidth. However, the compression techniques like MPEG4, H.264/AVC, H.265.HAVC, VP9 by Google, DAALA by Mozilla Corporation, PRISM [49,50], DISCOVER [51], are providing large amount of compression of video content to be carried out to reduce the bandwidth requirement. However, these results into large computation power as complexity is high of these compression techniques. However, there is tradeoff between the compression techniques complexity and error resilience. Moreover, the compressive sensing-based compression techniques provide both these advantages simultaneously by sparse signal structures to reduce the size of the transmitted or stored data [2]. The authors have also discussed several video encodings and compressing techniques in [5].

2. **Bulky data and software defined networking:** Unlike the scalar data acquisition by the sensor nodes of IoT, the multimedia data acquisition from the physical world is bulky in nature. Some of the applications which produce this heavy data are immersive 360° video, distributed gaming, free-viewpoint video, augmented reality (AR), virtual reality (VR), and extended reality (XR) [6]. Further, tiny multimedia devices have limited memory resources, thus the acquired multimedia data needs to be quickly processed and transmitted in the air to vacant the space in the memory for the incoming data, as most of the multimedia sources are continuous in nature. However, the data forwarded by these devices are captured by the routers and switches of the network which are hardware components and lack adaptability and flexibility. In addition, the data processing capacity of these routers and switches are limited in accordance with the amount of data produced by the multimedia devices, hence it drains the battery power, introduce latency and commence the memory restriction in the device. The solution to this problem is the implementation of software defined networking (SDN) which allows the reconfigurable and flexible network structure with control plane of the routers and switches to be reprogrammed in software. This allows the adaptability of these devices as per the amount of the traffic generated by the multimedia devices of the IoMT network and hence reduces the latency of traffic and memory restriction is also diminished.
3. **Multipath fading:** Deep fades occurring due to the multipath fading and corrupts a number of adjacent bits of the multimedia data. The interleaving and effective channel coding techniques allow the cognitive multimedia user to minimize the error at the receiver. At the transmitter, interleaving allows to spread the errors while doing de-interleaving at the receiver. While in the channel coding, the redundancy is added which further increase the data to be carried and it may again pose new challenge. In addition, the user may take advantage from multipath propagation by utilizing diversity and spatial multiplexing techniques, where the diversity methods increase the reliability of the system by reducing the bit error rate, however, the spatial multiplexing allows to increase the data rate of the system.
4. **Energy harvesting and green communication:** The IoMT devices are energy constrained or battery limited devices. As it has been mentioned earlier that in order to do the bandwidth saving in IoMT network, various video compression techniques are employed. However, it may lead to high energy consumption. Thus, we can say that there is a tradeoff between the achievable compression and the energy utilization for a specific level of user experience. Various researchers have explored the energy harvesting concept to fulfill the demand of energy constrained or battery limited IoMT devices. Different energy sources have been explored to harvest energy like wind energy, solar energy, piezoelectric, radio frequency (RF) source etc. Among these sources of energy harvesting, the RF-energy harvesting has been explored a lot now-a-day because of the easy and continuous RF-waves availability in the environment. The provision for energy harvesting at the MAC-protocol design help the system to improve the energy efficiency. In addition, since the cognitive radio has additional task of spectrum sensing and mo-

bility in comparison to the conventional user, therefore, the harvesting the energy will reduce the cost of energy consumption done in performing these tasks.

4.6 Summary

The rapid growth in multimedia traffic on the globally connected environment has imposed a huge challenge in the advancement and usage of multimedia-based applications and services. A connected paradigm globally available to the end-users in which the smart heterogeneous multimedia things can interact and cooperate with each other/things connected to the Internet to facilitate multimedia-based services and applications is known as internet-of-multimedia things (IoMT). In order to support the demand of multimedia traffic in the current era that is internet-of-things (IoT), the multimedia device needs to be bandwidth/spectral efficient and should have capability to support green communication. Further, an excellent higher processing capability requirement and ability to support different multimedia content through IoMT is a challenging task. Therefore, in this chapter, a review of the major challenges faced by the IoMT network and use of cognitive radio to handle those challenges has been presented. Further, the deployment of different methods to overcome these challenges and solutions are offered. Further, the major breakthrough in the design of protocols for IoMT devices in the cognitive network is comprehensively discussed. In addition, the MAC protocols which provided the solutions to satisfy the QoS/QoE constraints laid down by the network are studied and incorporated in this chapter.

References

[1] Md Jalil Piran, Quoc-Viet Pham, S.M. Riazul Islam, Sukhee Cho, Byungjun Bae, Doug Young Suh, Zhu Han, Multimedia communication over cognitive radio networks from QoS/QoE perspective: a comprehensive survey, Journal of Network and Computer Applications 172 (2020) 1–44.

[2] Sheeraz A. Alvi, Bilal Afzal, Ghalib A. Shah, Luigi Atzori, Waqar Mahmood, Internet of multimedia things: vision and challenges, Ad Hoc Networks 33 (2015) 87–111.

[3] H. Geng, Internet of Things and Data Analytics Handbook, Wiley, Hoboken, NJ, USA, Jan. 2017.

[4] M. Botterman, Internet of Things: an early reality of the future Internet, in: Proc. of Workshop Rep., Eur. Commission Inf. Soc. Media, 2009, pp. 1–30.

[5] A. Nauman, Y.A. Qadri, M. Amjad, Y.B. Zikria, M.K. Afzal, S.W. Kim, Multimedia internet of things: a comprehensive survey, IEEE Access 8 (2020) 8202–8250.

[6] F. Alvarez, et al., An edge-to-cloud virtualized multimedia service platform for 5G networks, IEEE Transactions on Broadcasting 65 (2) (Jan. 2019) 369–380.

[7] A. Rego, A. Canovas, J.M. Jimenez, J. Lloret, An intelligent system for video surveillance in IoT environments, IEEE Access 6 (2018) 31580–31598.

[8] K.P. Seng, L.-M. Ang, A big data layered architecture and functional units for the multimedia Internet of Things, IEEE Transactions on Multi-Scale Computing Systems 4 (4) (Oct. 2018) 500–512.
[9] Quality of Service Technical White Paper, Microsoft, Redmond, WA, USA, 1999.
[10] Series P: telephone transmission quality, telephone installations, local line networks recommendation ITU-T, document ITU-T P.10/G.100, Switzerland, Geneva, 2008.
[11] K. Brunnstrom, D. Hands, F. Speranza, A. Webster, VQEG validation and ITU standardization of objective perceptual video quality metrics, IEEE Signal Processing Magazine 26 (3) (2009) 96–101.
[12] M. Amjad, M.H. Rehmani, S. Mao, Wireless multimedia cognitive radio networks: a comprehensive survey, IEEE Communications Surveys and Tutorials 20 (2) (2018) 1056–1103.
[13] A.K. Moorthy, K. Seshadrinathan, A.C. Bovik, Image and video quality assessment: perception, psychophysical models, and algorithms, in: Percept. Digit. Imag.: Methods Appl., 2017, pp. 55–81.
[14] A. Khan, L. Sun, E. Jammeh, E. Ifeachor, Quality of experience driven adaptation scheme for video applications over wireless networks, IET Communications 4 (11) (2010) 1337–1347.
[15] Y.A. Rahama, M.S. Hassan, M.H. Ismail, A stochastic-based rate control approach for video streaming over cognitive radio networks, IEEE Transactions on Cognitive Communications and Networking 5 (1) (2019) 181–192.
[16] J. Mitola III, Cognitive radio: an integrated agent architecture for software defined radio, Ph.D. Thesis, Royal Institute of Technology (KTH) Sweden, May 2000.
[17] Dinesh Datla, Spectrum surveying for dynamic spectrum access networks, Master's of Science, Thesis, University of Madras, 2004.
[18] NTIA, U.S. frequency allocations [online], http://www.ntia.doc.gov/files/ntia/publications/2003-allochrt.pdf.
[19] E. Hoossain, D. Niyato, Z. Han, Dynamic Spectrum Access and Management in Cognitive Radio Networks, Cambridge University Press, New York, 2009.
[20] J. Mitola, Cognitive radio for flexible mobile multimedia communications, in: Proc. IEEE Int. Workshop Mobile Multimedia Commun., San Diego, CA, USA, 1999, pp. 3–10.
[21] C. Stevenson, G. Chouinard, Z.D. Lei, W.D. Hu, S. Shellhammer, W. Caldwell, IEEE 802.22: the first cognitive radio wireless regional area network standard, IEEE Communications Magazine 47 (1) (2009) 130–138.
[22] A.B. Flores, R.E. Guerra, E.W. Knightly, P. Ecclesine, S. Pandey, IEEE 802.11af: a standard for TV white space spectrum sharing, IEEE Communications Magazine 51 (10) (2013) 92–100.
[23] L.Z.W. Lee, K.K. Wee, T.H. Liew, S.H. Lau, K.K. Phang, An empirical study and the road ahead of IEEE 802.16, IAENG International Journal of Computer Science 43 (3) (2016) 1–12.
[24] S. Filin, T. Baykas, H. Harada, F. Kojima, H. Yano, IEEE Standard 802.19.1 for TV white space coexistence, IEEE Communications Magazine 54 (3) (2016) 22–26.
[25] O.A. Popescu, Y. Yao, M. Fiedler, P.A. Popescu, A management architecture for multimedia communication in cognitive radio networks, in: Fei Hu, Sunil Kumar (Eds.), Multimedia Over Cognitive Radio Networks, CRC Press, London, 2014, pp. 3–25.
[26] M.Z. Bocus, J.P. Coon, C.N. Canagarajah, J.P. McGeehan, S.M. Armour, A. Doufexi, Resource allocation for OFDMA-based cognitive radio networks with application to H. 264 scalable video transmission, EURASIP Journal on Wireless Communications and Networking 2011 (2011) 245673/1–245673/10.

[27] N. MohdAripin, R.A. Rashid, N. Fisal, A.C. Lo, S.H.S. Ariffin, S.K.S. Yusof, A cross-layer approach in sensing and resource allocation for multimedia transmission over cognitive UWB networks, EURASIP Journal on Wireless Communications and Networking (2010) 1–10.

[28] H. Kushwaha, Y. Xing, R. Chandramouli, H. Heffes, Reliable multimedia transmission over cognitive radio networks using Fountain codes, Proceedings of the IEEE 96 (1) (Jan. 2008) 155–165.

[29] L. Yang, H. Zhao, M. Jia, Cross-layer scheduling design for multimedia applications over cognitive ad hoc networks, China Communications 11 (7) (July 2014) 99–109.

[30] T. Jiang, H. Wang, A.V. Vasilakos, QoE-driven channel allocation schemes for multimedia transmission of priority-based secondary users over cognitive radio networks, IEEE Journal on Selected Areas in Communications 30 (7) (August 2012) 1215–1224.

[31] A. Bhattacharya, R. Ghosh, K. Sinha, D. Datta, B.P. Sinha, Noncontiguous channel allocation for multimedia communication in cognitive radio networks, IEEE Transactions on Cognitive Communications and Networking 1 (4) (Dec. 2015) 420–434.

[32] Shweta Pandit, Ghanshyam Singh, Spectrum Sharing in Cognitive Radio Networks: Medium Access Control Protocol Based Approach, Springer International Publishing, USA, 2017.

[33] K. Ali, W. Zhuang, Link-layer resource allocation for voice users in cognitive radio networks, in: Proc. of IEEE Int. Conf. Commun. (ICC'2011), Kyoto, Japan, 2011, pp. 1–5.

[34] H. Su, X. Zhang, Cross-layer based opportunistic MAC protocols for QoS provisionings over cognitive radio wireless networks, IEEE Journal on Selected Areas in Communications 26 (1) (Jan. 2008) 118–129.

[35] V. Mishra, L.C. Tong, S. Chan, J. Mathew, MAC protocol for two level QoS support in cognitive radio network, in: Proc. Int. Symp. Electron. Syst. Design (ISED'2011), Kochi, India, 19–21 Dec. 2011, pp. 296–301.

[36] L.X. Cai, Y. Liu, X. Shen, J.W. Mark, D. Zhao, Distributed QoS aware MAC for multimedia over cognitive radio networks, in: Proc. IEEE Glob. Telecommun. Conf, GLOBECOM, Miami, FL, USA, 2010, pp. 1–5.

[37] S.C. Jha, U. Phuyal, M.M. Rashid, V.K. Bhargava, Design of OMC-MAC: an opportunistic multi-channel MAC with QoS provisioning for distributed cognitive radio networks, IEEE Transactions on Wireless Communications 10 (10) (Oct. 2011) 3414–3425.

[38] Shweta Pandit, G. Singh, Backoff algorithm in cognitive radio MAC-protocol for throughput enhancement, IEEE Transactions on Vehicular Technology 64 (5) (May 2015) 1991–2000.

[39] F. Jiang, X. Liu, A QoS-aware MAC for multichannel cognitive radio networks, in: Proc. IET Int. Conf. Commun. Technol. Appl. (ICCTA'2011), Beijing, China, 2011, pp. 262–266.

[40] H. Song, X. Lin, A group-based MAC protocol for QoS provisioning in cognitive radio networks, in: Proc. 11th IEEE Singapore Int. Conf. Commun. Syst., Guangzhou, China, 19–21 Nov. 2008, pp. 1489–1493.

[41] M. Vishram, L.C. Tong, C. Syin, Distributed contention-based MAC protocol for cognitive radio networks with QoS provisioning, in: Proc. 19th IEEE Int. Conf. Netw. (ICON'2013), Singapore, 11–13 Dec. 2013, pp. 1–6.

[42] V. Mishra, L.C. Tong, C. Syin, QoS based spectrum decision framework for cognitive radio networks, in: Proc. 18th IEEE Int. Conf. Netw. (ICON), Singapore, 12–14 Dec. 2012, pp. 18–23.

[43] A. Kumar, P. Thakur, S. Pandit, G. Singh, Analysis of optimal threshold selection for spectrum sensing in a cognitive radio network: an energy detection approach, Wireless Networks 25 (7) (Oct. 2019) 3917–3931.
[44] Y.C. Liang, Y. Zeng, E.C. Peh, A.T. Hoang, Sensing-throughput tradeoff for cognitive radio networks, IEEE Transactions on Wireless Communications 7 (4) (2008) 1326–1337.
[45] IEEE 802 22-2011(TM) standard for cognitive wireless Regional Area Networks (RAN) for operation in TV bands was published as an official IEEE standard, http://www.ieee802.org/22/, 2011. (Accessed 6 August 2019).
[46] J. Huang, H. Wang, X. Bai, W. Wang, H. Liu, Scalable video transmission over cognitive radio networks using LDPC code, International Journal of Performability Engineering 8 (2) (Mar. 2012) 161–172.
[47] Prabhat Thakur, Alok Kumar, Shweta Pandit, Ghanshyam Singh, S.N. Satashia, Advanced frame structures for hybrid spectrum access strategy in cognitive radio communication systems, IEEE Communications Letters 21 (2) (Feb. 2017) 410–413.
[48] M. Jalil, N.H. Tran, D.Y. Suh, J.B. Song, C.S. Hong, Z. Han, QoE-driven channel allocation and handoff management for seamless multimedia in cognitive 5G cellular networks, IEEE Transactions on Vehicular Technology 66 (7) (Jul. 2017) 6569–6585.
[49] Anne Aaron, Rui Zhang, Bernd Girod, Wyner–Ziv coding of motion video, in: Proc. 36th IEEE Asilomar Conference on Signals, Systems and Computers, Pacific Grove, CA, USA, vol. 1, 3–6 Nov. 2002, pp. 240–244.
[50] Bernd Girod, Anne Margot Aaron, Shantanu Rane, David Rebollo-Monedero, Distributed video coding, Proceedings of the IEEE 93 (1) (2005) 71–83.
[51] Xavi Artigas, Joäo Ascenso, Marco Dalai, Sven Klomp, Denis Kubasov, Mourad Ouaret, The discover codec: architecture, techniques and evaluation, in: Proc. Picture Coding Symposium, Lisbon, Portugal, vol. 17, 2007, pp. 1103–1120.

CHAPTER

Multimedia nano communication for healthcare – noise analysis

5

Shyam Pratap Singh[a], Vivek K. Dwivedi[b], and Ghanshyam Singh[c]

[a]*Department of Electronics and Communication Engineering, Galgotia College of Engineering and Technology, Greater Noida, Uttar Pradesh, India*

[b]*Department of Electronics and Communication Engineering, Jaypee Institute of Information Technology, Noida, India*

[c]*Centre for Smart Information and Communication Systems, Department of Electrical and Electronics Engineering Science, University of Johannesburg, Johannesburg, South Africa*

5.1 Introduction

Currently, the nanotechnology provides a new set of tools to design and manufacture advanced devices to the engineering community which can create, process and transmit multimedia content at the le. The wireless interconnection ubiquitously deployed multimedia nano-devices with existing communication networks and ultimately the Internet defines a truly cyber-physical system which is further referred to as the Internet of Multimedia Nano-things (IoMNT). It is a specialized subset of internet of things (IoT) which enables the integration and cooperation among heterogeneous multimedia devices with different sensing, computational, and communication capabilities and resources. A multimedia-aware cloud combined with the multiagent systems can support impressive multimedia services and applications. The realization of IoMT rather has some additional challenges and stringent requirements as compared to IoT. One of the most important metrics that should be studied to assess the quality of any communication system is the noise. Basically, the noise degrades the signal transmission quality and affects the system throughput as it may impose re-transmission of data packets or extra coding to recover data in the presence of errors. However, the proper characterization of noise is one of the main challenges in the information theoretical analysis of intra-body electromagnetic nano communication. Recent development in the field of nanotechnology has evolved the nano communication (NC) as a new communication paradigm, which take place in extreme environments. Fundamentally, it is stem from nanotechnology which further based on production of devices and systems at the nanoscale level, termed as nanomachines. These nanomachines can achieve all the functionality of the contemporary

Internet of Multimedia Things (IoMT). https://doi.org/10.1016/B978-0-32-385845-8.00010-1

communication system. Specifically, the nanomachines can accomplish all the signal processing such as encoding, transmission, propagation, detection, and decoding [1]. However, the extreme environments include high temperature and radiation such as in adversarial chemical environments, lossy propagation environment like in the health monitoring and targeted drug delivery, anticipated defense environment for example, in Nuclear Biological and Chemical (NBC) challenges among the others. These extreme environments are susceptible to substantial challenges that impede monitoring, control, and communications. Moreover, the noise is primary and critical challenge with the fading and interference.

Further, the nano communication is classified in to two classes such as the molecular nano communication (MNC) and electromagnetic nano communication (EMNC). The MNC is inspired by the biological natural phenomenon of biology and uses natural/artificial Nanomachines [2] whereas, the EMNC is inspired by miniaturization of traditional communication devices and uses terahertz waves radiated by plasmonic nano-antennas [3,4]. However, the noise generation in MNC is intrinsically different than that in EMNC [5]. Till the date, additive inverse Gaussian (AIG) noise, random propagation delay noise, stable distribution noise, modified Nakagami distribution noise, radiation absorption noise (RAN), sampling and counting noise, molecular displacement noise, drug delivery noise, reactive obstacle noise, external noise, and system noise have been presented in open literatures of MNC system. On the other hand, the Johnson–Nyquist noise/thermal noise, black-body noise, Doppler-shift-induced noise, Molecular absorption noise, and body radiation noise have been introduced as noise in EMNC system. However, the noise is always represented and characterized by well-established physical or stochastic model. Therefore, this chapter develops a thorough understanding in addition to presenting physical and/or stochastic model for different types of noises in NC system. Furthermore, this Chapter discusses different open research challenges on the noises in NC system.

5.2 Noise in nano communication and statistical tools

Any unwanted signal, either man-made or natural, which interferes with the desired signal is known as noise, it may be random or deterministic. Man-made noise arises from electrical or electronic devices, such as different radiating sources, filtering and alias problems whereas natural signals arise from random thermal motion of electrons, atmospheric and cosmic sources. Specifically, in the MNC, which is inspired by the biological natural phenomenon, molecular concentration and their effects among the others are the key reason behind the noise. Though, efficient design may eliminate or minimize the man-made noise but there is no such direct way for natural noises. However, every noise includes physics that can be represented either mathematically in terms of physical variables or statistically using parameters of random processes correspondingly; it is known as a physical or stochastic model. Further, physical and stochastic models are based on block schemes and random processes, respectively.

Furthermore, the noise is one of the most critical impairments of any communication system among the others. It degrades the performance of the system and affects the overall system throughput. It may also impose re-transmission of information signal and/or requires some error correcting techniques to recover the information. Therefore, it should be carefully model to assure the quality of service (QOS). Specifically, in NC system noise is more prone due to miniaturization, as in case of MNC systems where Molecules are not only information, but these are transmitter and receiver (known as Transmitter- Bio Nano Machine (T-BNM) and Receiver-(R-BNM) respectively) as well [6]. However, noise model can be developed with in-depth understanding of medium characterization which further depends on the number of interrelated parameters. These parameters include but not limited to medium temperature, permittivity, and refractive index, dielectric of the medium, polarization mechanism and absorption cross section area. Again, it is noticeable that on the one hand, in biological matter, polarization mechanism in time domain is different at THz frequency band due to many relaxation times and nonsymmetric responses [7]. On the other hand, in the presence of high-water content, dielectric of tissues given by Debye Relaxation model [8].

However, the stochastic model and the physical model are two models to represent the noise in NC system, be it EMNC or be it MNC. In the case of physical model, the noise is expressed mathematically in terms of all the relevant physical parameters. Although, in case of the stochastic model, the noise is presented using statistical parameters. The physical model and the stochastic model are based on block schemes and random processes, respectively. The physical model offers a mathematical analysis of all the physical processes involved in the generation of noise. On the other hand, in the stochastic model all of the physical processes are included statistically.

It is important to mention that the noise study in MNC system is possible statistically using all stochastic parameters. The simulation of physical system generates sets of noise data which are used to test the ability of stochastic model. The stochastic model must characterize all the physical processes due to which the noise is generated [9]. In timing channel-based system, irrespective of different channel type, channel is reduced to an additive noise and modeled using the stable distribution family [10]. Specifically, for the channel A the noise follows the well-known Levy distribution. Though, the alpha-stable distribution is used to realize a more realistic noise model under the room acoustic scenario [11]. However, in a pulse modulation based MNC system, the noise is due to host of sources, but the prediction of dominant component is very difficult. Examples of such cases are as follows: inconsistent pulse emission, contamination by atmospheric or physical disturbances, and so on [9]. In general, a Gaussian distributed additive noise is considered [12], which is accurate for a high velocity turbulent flow with 3-dimensional channel [13]. At the same time, the noise model under random walk diffusion without drift, such as in low pressure and enclosed environment, is lacking.

In contrast, the number of literatures present study on the noise in EMNC system [14–16]. Though, all out of the presented study in the literatures includes one or another parameter but none of the single study includes all the physical parameters. The noise model presented in [14–16], only includes the medium radiation energy but ignores the molecular absorption due to the transmitted signal. However, the authors in [17] update the noise model of [14–16] by including molecular absorption noise due to the transmitted signal in addition to self-induced noise and atmospheric noise. Furthermore, the authors in [18] present noise for in-vivo communication which includes both the radiation of the medium and the molecular absorption due to the transmitted signal and radiation from medium a well.

Further, irrespective of types and sources of noise or type of communication, there are few common tools which are employed in the analysis of noise. In particular, the time-averaged representation, power spectral density (frequency domain), and the correlation function (time domain) are major tools to represent both random and deterministic noise. Therefore, the remaining of this section present these tools briefly in tabular form as given in Table 5.1. For detailed study [19] can be referred.

Note:

- For the ergodic random processes, a single time realization can be used to obtain the moments of the process.
- All the expressions in the table are only correct if the stochastic process is both stationary and ergodic.
- As the measuring time T is finite the quantity hence, only estimated values of the moments are possible.
- In practice, the only quantity accessible to measurements is $n(t)$, which forces us to assume a stationary, ergodic stochastic process.
- $T \to \infty$, shows that the observation time was finite and here sufficient time period is needed.
- Averaging any signal, be it random or deterministic, gives some insight information but detailed information is lost.

As mentioned previously, the NC can be either MNC or EMNC. Also, the noise in these two different types is intrinsically different. Therefore, the fundamentals on various types of noises in MNC and EMNC are presented in the preceding section.

5.3 Fundamentals on various noises in MNC

Various sources of noise in MNC system are first hitting time and difference of the first hitting, release timing of a single information molecule, unwanted effect in the molecular dynamics due to temperature fluctuations by virtue of EM radiation, sampling/counting effect due to discreteness of information, uncertainty in the position of received molecule due to Langevin force, unwanted effect in Particulate Drug Delivery Systems (PDDS) due to uncertainty in cardiovascular channel, the uncertainty in the number of received molecules because of reactive obstacle, uncertainty by reason of different sources, unintended transmitter or other sources and nonlinearity in the system owing to different impairments of transmitter, channel, and receiver. This subsection presents basics of different noises in MNC system.

Table 5.1 Parameter employed in the analysis of noise.

S. No.	Parameters	Description
1.	Mean value: $E\left(n(t)\right) = \lim_{T\to\infty}\frac{1}{T}\int_{-T/2}^{+T/2} n(t)dt$	• Gives dc or average value of $n(t)$. • Larger is the averaging time T smoother the fluctuations of $n(t)$.
2.	Mean-square value: $E\left(n(t)^2\right) = \lim_{T\to\infty}\frac{1}{T}\int_{-T/2}^{+T/2} \lvert n(t)\rvert^2 dt$	• Gives the time averaged power P of $n(t)$. • If $n(t)$ is the noise voltage or current, then the scaling factor will be equivalent to a resistance (generally, taken as 1 Ω) • Unit of root-mean-square (rms) value, $\sqrt{n(t)^2}$ is same as $n(t)$.
3.	AC component: $\sigma(t) = n(t) - \overline{n(t)}$	• Gives AC or fluctuating component of $n(t)$.
4.	Variance: $E\left(\sigma^2(t)\right) = \lim_{T\to\infty}\frac{1}{T}\int_{-T/2}^{+T/2} \left\lvert\sigma^2(t)\right\rvert^2 dt$	• Gives the measure of how strong the signal is fluctuation about the mean value.
5.	Time averaged power of $n(t)$: $E\left(n(t)^2\right) = \left\lvert E(n(t))\right\rvert^2 + \lim_{T\to\infty}\frac{1}{T}\int_{-T/2}^{+T/2} \left\lvert\sigma^2(t)\right\rvert^2 dt$	• Shows that time averaged power of $n(t)$ is equal to the sum of powers in the dc and ac components.
6.	Realization of time-based statistics: $\widehat{\eta(t_1)} = \frac{1}{N}\sum_i x(t_1, \xi_i)$	• $\widehat{\eta(t)}$, gives the realization of $\eta(t_1)$. Here, $\eta(t_1)$ is the mean value of random variable $x(t_1)$. Furthermore, $x(t_1)$ is obtained by sampling of stochastic process $x(t)$ at time t_1.
7.	Time average of sample: $\eta_T = \frac{1}{T}\int_{-T/2}^{+T/2} x(t)dt$ and $E(\eta_T) = \frac{1}{T}\int_{-T/2}^{+T/2} E(x(t))\,dt = \eta$	• η_T, valid only if $x(t)$ is stationary process, gives time average of the given sample if only a single sample of $x(t, \xi 1)$ is available, as $\widehat{\eta(t)}$ is applicable if large number of samples are available. • Here, η_T is random variable with mean $E(\eta_T)$. • Now, if $\eta_T \approx \eta$ for $T \to \infty$ then random process is known as mean ergodic.

(continued on next page)

Table 5.1 (*continued*)

S. No.	Parameters	Description
8.	The auto-correlation function: $E\{f^*(t)f(t+\tau)\} = R_f(\tau) = \lim_{T\to\infty}\frac{1}{T}\int_{-T/2}^{+T/2} f^*(t)f(t+\tau)dt$	• Gives a similarity of the signal $f(t)$ with itself versus τ. • For slowly varying signal $R_f(\tau)$ is flat. • For noise has $R_f(\tau)$ sharp peak for $\tau = 0$ (no time shift) and quickly falling to zero for increasing τ. • $R_f(\tau)$ is equal to inverse Fourier transform of $\lvert F_T(f)\rvert^2$, where $F_T(f)$ is the Fourier transform of $f(t)$.
9.	Cross-correlation function: $E\{f^*(t)g(t+\tau)\} = C_{fg}(\tau) = \lim_{T\to\infty}\frac{1}{T}\int_{-T/2}^{+T/2} f^*(t)g(t+\tau)dt$	• Gives a similarity of the signal $f(t)$ and $g(t)$ versus τ. • Detect the signals masked by additive noise.
10.	Power spectral density: $P = \int_{-\infty}^{+\infty} S(f)df = R_f(0)$ where $S(f) = \lim_{T\to\infty}\frac{1}{T}\lvert F_T(f)\rvert^2 = f\{R_f(\tau)\}$	• $S(f)$ gives the power over each frequency increment. • For noise $S(f) = R_f(\tau)$ should be used rather than FT. As FT of random quantity is also a random quantity. And does not give any insight of the information.

However, in particular case of diffusion based molecular nano communication (DMNC), it is worthy to mention that as the diffusion processes are random in nature and hence, DMNC is studied statistically in terms of stochastic parameters of the model. However, first of all, the sets of noise data are generated by simulation of the physical model. After that these generated data are used to test the ability of stochastic model to represent the behavior of the physical processes behind the noise generation.

5.3.1 Additive inverse Gaussian (IG) noise

Consider a molecular communication system in which information molecules are encoded onto carrier molecules by the transmitter and released into a fluidic medium. These encoded molecules are received by the receiver after propagation through a fluidic medium. The time, type, number, concentration, or the identities of the molecule themselves are used to encode the information. The receiver reacts (active receiver) with them and decodes the information. Further, it is assumed that inherent randomness due to Brownian motion, with negligible friction in the medium, is the only impairment that affects the communication link. This inherent randomness in turn creates an unwanted effect on time of first arrival and hence is modeled as noise and can be appropriately represented by the Wiener process [20]. However, the PDF of position is given by Gaussian distribution with mean $v.x$ and variance $\sigma^2.x$. Though, the PDF of first arrival time at receiver boundary is given by inverse Gaussian distribution more specifically, by additive white Gaussian noise (AIGN) with mean and variance as μ and μ^3/λ, where x is the time, v is the positive drift velocity, σ^2 is the variance of Wiener process, μ is the mean of AIGN distribution and $\lambda = l^2/\sigma^2$ with l denotes boundary position.

5.3.2 Normal inverse Gaussian noise

Consider diffusive molecular communication overflow-induced one-dimensional diffusive channels with perfectly absorbing (passive) receiver and encoding is based on difference in first hitting time of same molecule and/or that of two different molecules [21]. Again, the random propagation delay creates an unwanted effect but this time it is characterized over difference in the first hitting time of a single or two molecule and termed as noise. However, this time it is appropriately modeled by Normal Inverse Gaussian Noise (NIGN) as NIGN distribution has flexibility, heavy-tailed property, and skewness [22] in addition it has smaller approximation error than Gram–Charlier [23] and Edgeworth [23]approximation. As mentioned above, based on the types of encoding there are two scenarios of noise characterization but both of them are given by NIGN distribution. In the first scenario, the difference in first hitting time of a single molecule is modeled as NIGN. Though, in the second scenario, the difference in first hitting time of two different molecules is also modeled as NIGN. It is important to mention that in either of the above scenarios only a limited set of molecules are used in encoding.

5.3.3 Stable distribution noise

Consider a flow induced diffusive molecular communication system with active receiver, where the meaning of flow induced is that additional drift, which is exerted on the diffusion process, whereas the receiver reacts with the received encoded molecules to decode the information. Further, a general class of timing channel is considered where encoding is one of the three possible types such as: 1) the information molecule is encoded over release time of single information molecules (Channel-A), 2) time difference between the release time of two information molecules of same type (Channel-B), and 3) that of different types (Channel-C). In any of the above channel types, there exists a random propagation delay as an unwanted effect, termed as noise [11]. The noise distributions for different types of mentioned channel are given by different subtypes of stable process specifically, the Levy distribution, standardized stable-I and Standardized stable-II are the noise distributions for Channel-A [11, Eq. (3)], Channel-B [11, Eq. (19)], and Channel-C [11, Eq. (25)], respectively.

5.3.4 Modified Nakagami distribution noise

Consider a diffusive molecular communication within straight tube of 245 cm long with 4 cm diameter and 2 mm thickness. Further, consider the passive flow means drift velocity inside the channel is zero. Although, the humidity and temperature inside the tube is approximately zero. The on-off-keying (OOK) line coding is employed by system to convert alpha-numeric message into binary code, but the concentration (binary amplitude shift keying) based encoding is used to modulate the information signal. An electronic spray is used as transmitter which emits M molecules. A chemical sensor, which converts concentration into electrical signal placed at a distance d from transmitter, is used as receiver. With appropriate signal processing information is decoded as original message. Further, the noise arises from a host of sources for molecular communication system described above and it is quite tough to account all of them or to decide which one is dominant component. An inconsistent emission of mechanical pulses and atmospheric contamination is the example among the others. Though, the additive Gaussian distribution is an appropriate noise model for 3-dimensional channel under high turbulent. However, the experimental results, after applying Levenberg–Marquardt technique for curve fitting, show that the modified Nakagami distribution is suitable noise model for the channel without induced drift velocity [24].

5.3.5 Radiation absorption noise

Considering one of the real-world scenarios of molecular communication system where the external effects in addition to stochastic noise model due to molecular dynamics within the medium of propagation, cause noise [1]. One of such typical scenarios includes coexistence of MNC, EMNC, and traditional microwave communication in some typical multiple access technology [25]. In such a scenario, the EM radiation and its subsequent absorption by the medium molecule causes rise or fluctuation in temperature, which in turn affect the molecular dynamics and can be considered exogenous source of the noise and termed as radiation absorption noise (RAN). The characterization of such noise depends on the intrinsic impedance of different tissues of human body such as skin, brain, and blood and on frequency of incident EM wave [26–29]. In addition, the polarizations of incident EM radiation are another major factor influencing the RAN [30]. However, it is noteworthy to mention that RAN model hinges on law of energy conservation and is derived by equating Stefan–Boltzmann law with power density of EM wave (Poynting vector). Finally, an application of such noise ranges from cell signaling to chemical engineering is needed to explore.

5.3.6 Sampling and counting noise

Considering a diffusion based molecular communication system relying on emission, diffusion, and reception. The emission processes modulate the particle concentration rate, through the release/capture of particles into/from the emission gaps, using input signal ($T(t)$) at the transmitter. Although, the flow induced diffusion process helps in propagation of modulated signal to the receiver. Also, the reception is the process of sensing the variations in the particle concentration at receiver which in turn recovers the output signal ($R(t)$). In other words, the reception is the process of generation of output signal with help of chemical receptors. Symbolically, the modulated particle $r_T(t)$ is the output of transmitter and input to the propagation whereas, the propagated signal ($c_R(t)$) is the output of the propagation and input to receiver [9].

Now ideally, $r_T(t)$ should be function of $T(t)$ but due to unwanted perturbation on $r_T(t)$, input to the propagation is realized as $\widehat{r_T(t)}$. The unwanted perturbation is due to the discreteness of the particles caused by a particle flux between the transmitter and external space. Further, as the signal $\widehat{r_T(t)}$ is achieved by sampling of $r_T(t)$ itself and therefore the unwanted perturbation is termed as sampling noise $n_s(t)$. Now, $\widehat{r_T(t)}$ propagate via induced diffusion and expected to produce $c_R(t)$ at the receiver but due to unwanted perturbation on $c_R(t)$, the input to the receiver is realized as $\widehat{c_R(t)}$. The unwanted perturbation in this case is due to the variation in $c_R(t)$ by virtue of nonhomogeneous particle concentration, which is further, due to differences in particle concentration created by $\widehat{r_T(t)}$. Although, the difference in particle concentration is due to random movement and discreteness of the particles. As $\widehat{c_R(t)}$ is achieved by fluctuating and imprecise counting of $c_R(t)$ and therefore the unwanted perturbation is termed as counting noise $n_c(t)$.

5.3.7 Molecular displacement noise

Consider a pure diffusive molecular communication system with point source transmitter, time slotted channel of duration T each and passive receiver. The transmitter emits fixed number of molecules at the beginning of every time slot. The receiver perfectly absorbs the molecule as they cross the receiver for the first time and number of molecules received in the current slot convey the information to the receiver [31]. However, the propagation medium exerts two types of forces on the information molecule, by virtue of continuous collision with the molecule of fluid of bigger size, known as a viscous force and a random force (given by Langevin equation, $F(t)$), respectively. It is observed that mean square displacement $\langle x(t)^2 \rangle \approx 2Dt$ for $t \gg t_r$ whereas $\langle x(t)^2 \rangle \approx \left(\frac{\alpha}{m}\right) * Dt^2$ for $t \ll t_r$, where D is the diffusion coefficient, α is the drag coefficient, m is the molecular mass and t_r is relaxation time. However, for small time duration prediction of position of the molecule is very much uncertain. This uncertainty, by virtue of random force, introduces randomness in the instantaneous position or displacement of the molecules termed as molecular displacement noise and quantify by $M = U - U'$. Here, U and U' represent the number of molecules correspond to the molecular displacement $x(t)$ and $x'(t)$, respectively in the kth time slot, where, $x(t)$ is the molecular displacement in the kth time slot and $x'(t)$ is defined as $x'(t) = x(t) - \Delta t$. Here, Δt as displacement error at time t (which time of kth time slot) arises due to the action of $F(t)$ for a small duration t_1 around time t.

5.3.8 Drug delivery noise

Considering Molecular communication for particulate drug delivery systems (PDDS) where the transmitted message/drug nanoparticle, time varying and flow induced diffusion channel and received message are injected drug, cardiovascular system and delivered drug, respectively. The drug delivery noise (DDN) is consisting of all the noise effects that is injection, propagated or at reception. The basics each type of DDN is as follows.

Drug injection (DI) noise: is the first kind of DDN, prime reason behind the DI noise is the mechanical imperfection of drug injection device such as a computer-controlled infusion syringe pump. The drug injection rate at drug injection site, which is found to be inaccurate due to the imperfection the device, is the measure of DI noise. The mechanical friction and compression phenomenon in the pump apart from toxicity level limits injection rate. All of these effects collectively are responsible for DI noise. However, the average injection limit, maximum drug injection rate $A(t)$, leakage in the injection $\eta(t)$, and sampling duration T_s are used to model the DI noise.

Drug propagation (DP) noise: is the second kind of the DDN, prime reason behind the DP noise is the turbulent induced stochastic motion of drug nanoparticles, due to which those nanoparticles get lost randomly at the level of blood vessels bifurcations. As they get diverted towards the organ and tissues other than targeted one and the effect is termed as DP noise. It is important to mention that the agitation of drug nanoparticles is more at the higher concentration and therefore, the effect is more at higher concentration. However, cardiovascular topology and diffusion coefficient (D) are used to characterize the DP noise.

Drug reception (DR) noise: is the third kind of the DDN, the prime reason behind the DR noise is the stochastic behavior of ligand binding process by which drug nanoparticles bind to the surface of diseased cells. However, the weak chemical affinity between receptor and ligands, the impairment due to blood flow, cell-irregularities, and small interaction surface of receptor are major cause behind the stochastic behavior of the binding process. In addition, the thermal energy due to the Brownian motion in the blood medium and a chemical potential characterizes different energy interaction at the surface of the deceased cell such as, chemical energy produced by reaction between receptor and ligands and kinetic energy of blood. The overall effect of the above processes produces DR noise that impede the delivery of drug nanoparticle to the diseased cells. The major parameters used in the modeling of DR noise are the number of ligands (N_L) and receptors (N_R) of the drug nanoparticle, the decease cells, range of attraction (ξ), atmospheric temperature (T_P) bond equilibrium length (ζ), characteristic length (χ), average cross-sectional velocity during reception $(u_l(t))$, and spheroid dimensions of drug nanoparticle (radius (a) and aspect radio (γ)).

5.3.9 Reactive obstacle noise

Considering a diffusive molecular communication via diffusion based on free diffusion over the diffusive channel (blood vessels as an example) with reactive obstacle and passive and absorptive receivers. The transmitting nanomachine (TN) and receiving nanomachine (RN) are the point sources placed in an unbounded fluidic environment which ensures elastic collision of the molecules with the boundary.

However, the information molecules collide and react with a reactive obstacle with partial absorption present the channel and hence absorb a part of the information molecules. Also, a part of information molecules reflect back towards TN and never reach the RN. At RN, successfully reached molecules are counted without absorbing (in case of passive RN) or with absorption (in case of absorptive RN). It is noteworthy to mention that the passive RN does not affect the diffusion of information molecules and hence these molecules can enter into the boundary of passive RN without any chemical or physical interaction. Further, the presence of reactive obstacle severely affects received signal which is given as the number of molecules received at RN and termed as reactive obstacle noise. However, the reaction probability and radius of the reactive obstacle are the main parameters used in modeling of the reactive obstacle noise.

5.3.10 External noise

Considering a diffusive molecular communication with induced diffusion and passive receiver where the channel includes advection and molecule degradation, but the temperature and viscosity is constant throughout the channel. However, as the receiver is a passive receiver it neither obstruct diffusion nor initiate chemical reaction and therefore, focus lies on the impact of the propagation environment only. Assume that A be the information molecules emitted by the intended transmitter or by unintended sources. Also, the receiver follows a first-order degradation with constant diffusion of A molecules. Most important, the receiver only counts A molecules perfectly if they are within V_{obs} (volume of receiver) and any other molecules are ignored [34]. In addition, as the focus is on the impact of unintended information molecules on the receiver, hence the receiver is centered at the origin whereas, the intended transmitter is located at $\{-x1, 0, 0\}$. Further, the probable concentration of A molecules at the receiver is C_A and the diffusion coefficient of all A molecule is D_A. Now, as the diffusion of all A molecules is independent so the impacts of the individual sources can be represented by superposition, which can be stated as the accumulative impact of numerous noise sources can be given as the sum of the impacts of all the individual sources [34]. Assuming $U - 1$ sources of A molecules are the unintended transmitter, be it another transmitter or be it leaking source, then complete signal at the receiver can be given by: $N_{aobs}(t) = N_{a1}(t) + \sum_{u=2}^{U} N_{au}(t)$, where $N_{a1}(t)$ is the signal from the intended transmitter. Though $N_{au}(t)$ represents the unwanted signal at the receiver and termed as external noise by virtue of the superposition of accumulative impact of multiple noise sources including their advection. However, the above definition of noise is based on the fact that under certain condition, various interferences such as multiuser or intersymbol interference are approximated as a noise.

5.3.11 System noise

Considering an experimental diffusive molecular communication system over induced diffusion-based channel. The transmitter is consisting of a spray assisted by a microcontroller, an LCD display and a push buttons to control the spray. The isopropyl alcohol is used as carrier and a fan is used to induce the flow in the propagation channel which is an air medium. The input to the microcontroller is converted into a binary stream and then transmitted through different modulation schemes precisely achieve by controlling the spray. However, the receiver is consisting of an alcohol sensor and a microcontroller along with an analogue to digital converter to display the received data. Specifically, for the isopropyl alcohol as a carrier, MQ-3, semiconducting metal oxide gas sensor is employed to detect the signal at receiver. MQ-3 can be used to sense the concentration of different types of alcohol. The received data can be input to a computer through serial port for analysis purpose. The reception is based on simple detection without use of any error correcting techniques [13].

Further, considering the end-to-end system, a mathematical model for impulse response is presented based on experimental results. The experimental impulse response shows nonlinearity when compared with theoretically derived results [13]. Even though, precise reason behind the nonlinearity is unknown but transmitter imperfections, turbulent flow and receiver imperfections are most likely impairment behind the nonlinearity particularly, the inconsistent droplet size for different trial and/or response and recovery time of the sensor. Though, this nonlinearity is inherent but under certain assumption, such as 1-dimensional channel and absorbing receiver, it is modeled as nonstationary process and further reduced to a Gaussian distribution by simplifying the assumptions. In other words, the nonlinearity introduced by the system is modeled as the output of a linear system in presence of some unwanted effect/signal. This unwanted effect/signal is termed as system noise and effectively represents the nonlinearity in experimental model with respect to derived model.

5.4 Fundamentals on various noises in EMNC

Intra body noise sources in EMNC may include Doppler-shift-induced noise, Johnson–Nyquist noise, and Black-body noise. The main cause of these noises is particle fluctuation which may be either natural or triggered by EM radiation. However, the nano-antennas inside body radiates EM waves which is absorbed by biological cell and liquid in the medium, subsequently radiate it in terms of heat. This causes another source of noise in EMNC, is termed as molecular absorption noise. Finally, the local source of radiation gives rise to body radiation noise. This subsection presents basics of various noises in EMNC system.

5.4.1 Johnson–Nyquist noise/thermal noise

Consider an EMNC under intra-body which is full of both biological cells and nano-antenna (heat sources) of size more than tens of nanometers and surrounded by fluidic medium. However, the term "particle" is used for both biological cells and nano-antenna. The fluctuations of charged particles or in particular of electrons in a conducting media dissipate power which causes an increase in temperature, termed as thermal or Johnson's Nyquist noise and such noise, degrade the detection of the signal, is common at the front-end of all communication system [35]. Though, in general, the electronics device used such as mixer, detectors and amplifiers among the others are the basic sources of thermal noise. However, specifically, in an in-vivo EMNC, the thermal cycle of heat exchange in between the environment and the receiver is primary cause of thermal noise. The thermal noise follows the Gaussian distribution under the assumption of a very large number of receivers which are independent, whereas individual contribution is insignificant. In other words, the even though individual distributions of the receiver follow nonGaussian distribution but the convolution of sufficiently large number of the elementary distribution results in a Gaussian distribution. Finally, it is noteworthy to mention that detecting a nano-receiver at the receiver is responsible for the thermal noise [35].

5.4.2 Black-body noise

Consider the same scenario as in the case of thermal noise, each and every physical body radiate the EM wave continuously and spontaneously. A part of EM energy radiated by the nano-antennas is seized by the particle of human body. By virtue of which the particle starts vibrating and generate the heat. In turn an increase in temperature is observed. Now, as temperature of the medium is above the absolute zero, therefore, the human body particle behaves like black-body and corresponding noise is termed as black body noise [35]. Further, the radiation is given by Plank's law according to which, higher the temperature gradient, higher is the radiation at every wavelength. From above it is evident that in intra-body medium, even in absence of transmitting signal, black-body noise is always present. Hence, the black-body noise is counted as natural noise due to thermal EM radiation within or in vicinity of a body. However, the amount of radiated energy is given by spectral radiance ($E_f(f, T)$), which is defined as the radiated power emitted per unit, per unit solid angle over per unit frequency. Further, classically, black-body radiation is characterized by chaotic EM radiation. Even though, the average spatial distribution of the radiation is found to be homogeneous and isotropic but the amplitudes of spectral component of magnetic induction and the electric field strength, consisting of infinitely many independent and infinitesimal component, are completely random. However, the amplitudes of magnetic induction and the electric field strength at given frequency or mode follow central limit theorem and are given by Gaussian distribution.

5.4.3 Doppler-shift-induced noise

In contrast to the fluctuations of charged particles which results in the thermal noise, the internal fluctuations, due to the heat transfer mechanism of the intra-body system, result in random and irregular velocity of the cells [35]. This motion in the soft bio-nano machine, of micro and nano size, is governed under mesoscopic dynamics namely, by the Brownian motion [36]. The exerted force, known as Langevin force, varies very rapidly over any infinitesimal time interval. Therefore, it is characterized by some statistical parameters such as first and second moments. However, this relative velocity of the source (nano-biosensors implanted inside human body) with respect to the observer (nano-receiver implanted inside human body) results in Doppler-shifts to signal frequency transmitted by the source. In addition, due to the presence of scatterers in the medium, plane wave reaches to the receiver via multiple path results in random amplitudes, phases and angles of arrival of each of the waves. Taking it all together, the unwanted effect on the information signal is known Doppler-shift-induced noise.

5.4.4 Molecular absorption noise

The EM energy at specific frequencies of the terahertz band excites the molecule to vibrate internally. This means motion of the atoms within the molecules is periodic whereas the motion of the molecule as a whole is constant rotational and translational. Due to the internal vibration of molecules a part of the propagating energy converts into kinetic energy and in turn get lost, lead to an increase in temperature of the molecule [37]. By virtue of which the molecules re-radiate the EM energy at the same frequency at which the incident EM waves have triggered this motion. This unwanted energy is termed as molecular absorption noise (MAN). In simple words, the re-emission of a part of absorbed EM energy (characterized absorption coefficient) by the molecule to the EMNC channel is termed as MAN. As the vibration resonant frequencies highly depend on the internal structure of the molecule, therefore, MAN highly depends on the frequency which makes the spectral absorption unique frequency selective. Further, the MAN is consisting of atmospheric noise, the self-induced noise and MAN due to transmission signal. This classification is based on the primary source of the EM energy. Atmosphere is treated as a black body radiator, because of its temperature absorption property, produce EM energy which is absorbed and in turn re-emitted by the absorbing molecule, is known as atmospheric or background noise. Further, as atmospheric noise is generated by virtue of local radiation sources already present in the medium and radiation is due to the original energy state of the medium molecules only just before transmission occurs. Therefore, it is independent of the transmitted signal.

On the other hand, if the molecules of the channel itself, which are in vicinity of the absorbing molecules, produce the EM energy which is absorbed and in turn re-emitted by the absorbing molecule then corresponding noise is known as self-induced noise. It is noteworthy to mention that self-induced noise depends on the transmitted signal as it is after the transmission occurs (probably). Finally, it is most important to mention that if transmitted signal alone produces the EM energy which is absorbed and in turn re-emitted by the absorbing molecule then corresponding noise is known MAN due to transmission signal.

5.4.5 Body radiation noise

The background noise largely depends on the temperature in addition to medium composition. However, as tissue of human body are isothermal in nature. Therefore, atmosphere noise varies with medium of transmission in accordance with its refractive index. Further, the self-induced noise largely depends on transmitted signal [20]. Furthermore, the composition of the medium in human body and frequency in THz wave affect the speed of light in human body. In totality, total MAN in vivo is characterized differently and termed as body radiation noise.

5.5 Physical and/or stochastic models for noise in MNC

In conjunction with section three this section presents various representations of noise in MNC. These representations facilitate analysis of different noises in terms of respective physical/statistical parameters.

5.5.1 Additive inverse Gaussian noise (AIGN)

First arrival time (N), which is represented as continuous-time Wiener random process $W(x)$, is the key parameter of AIGN noise [20]. Let $W(x)$ represent the position of an information molecule following Brownian motion with v as the positive drift velocity, d as the diffusion coefficient, and σ as the variance of the Wiener process. Further, for perfectly absorbing receiver, only first arrival time at the receiver boundary is important. If location of the transmitter and receiver are on real positive axis ($l > 0$) and at the origin, then the first arrival time is given by $N = \min\{x : W(x) = l\}$. However, the PDF of $Nf(n)$, is given by inverse Gaussian noise as [20]:

$$f_N(n) = \begin{cases} \sqrt{\frac{\lambda}{2\pi n^3}} e^{-\frac{\lambda(n-\mu)^3}{2\mu^2 n}}, & n > 0 \\ 0, & n \leq 0 \end{cases} \tag{5.1}$$

where, $\mu = l/v$ and $\lambda = l^2/\sigma^2$. Also, μ and μ^3/λ are the mean and variance, respectively. It is noticeable that for $v = 0$ the distribution does not represent IG, whereas for $v < 0$ information molecule will never reach to the receiver. Therefore, $v > 0$ is valid condition and already taken as one of the assumptions.

5.5.2 Normal inverse Gaussian noise (NIGN)

The difference of the first hitting times of two molecules is very important in MNC if encoding is based on a limited set of the information molecules and the receiver is based on their arrival time and/or order. Then first hitting time difference of the two information molecules becomes very important for fruitful reception. However, as deviation from the predefined first hitting time difference is due to random propagation delay, therefore termed as random propagation delay noise. First hitting time or random propagation delay itself is given by inverse Gaussian (IG), whereas the difference of the first hitting times of two molecules is given by normal IG as [21]:

$$f_N(n) = \frac{\alpha\delta}{\pi} e^{\left(\delta\sqrt{\alpha^2-\beta^2}-\beta\sqrt{y-\mu}\right)} \times \frac{K_1\left(\alpha\sqrt{\delta^2+(y-\mu)^2}\right)}{\delta\sqrt{\alpha^2-\beta^2}} \tag{5.2}$$

where, $\alpha > 0$, $-\alpha < \beta < \alpha$, $\delta > 0$, and $\mu \in R$ are parameters and $K_1(\cdot)$ is the modified Bessel function of the third kind and type 1.

5.5.3 Stable distribution noise

Different traditional wireless system has already employed stable distributions to characterize the noise which are closed form PDF namely, Gaussian, Cauchy, and Levy distribution respectively [11]. On the other hand, for three types of timing channels in MNC, closed form PDF for stable noise distributions are proposed terms of complex error functions. It is noticeable that this class is the generalization of previously proposed, random propagation delay noise.

5.5.3.1 Levy distribution noise

The random propagation delay (T_n) of information molecule encoded in the release timing of a single information molecule is characterized by Levy distribution and hence termed as Levy distribution noise. It is remarkable that T_n includes the distance between the transmitter and the receiver and the diffusion coefficient of the information particle as parameter. However, the closed form PDF of above noise is given as [11]:

$$f_{T_N}(t_n) = \frac{d}{\sqrt{4\pi D t_n^3}} e^{-\left(\frac{d^2}{4Dt_n}\right)} \tag{5.3}$$

where, d is distance between the transmitter and the receiver, and D is diffusion coefficient of the information particle as parameter.

5.5.3.2 Standardized stable noise-I

Let us take, T_{x1} and T_{x2} be the release timing of first and second information molecules respectively. Also, T_{n1} and T_{n2} be the RPD of first and second information molecules respectively. Then random propagation delay, $L_n = T_{n2} - T_{n1}$ of information molecule encoded in $L_x = T_{x2} - T_{x1}$ (always positive) is characterized by Standardized stable noise-I (say) and hence termed as standardized stable noise-I. However, the closed form PDF of L_n is given as:

$$f(x; 1/2, \beta) = R\left\{\frac{z}{\pi x}\left[\sqrt{\pi}e^{-z^2} - 2jF(z)\right]\right\} \tag{5.4}$$

where

$$F(z) = e^{-z^2}\int_0^z e^{t^2}dt \tag{5.5}$$

Here, z is defined as:

$$z = \frac{1+\beta-j(1-\beta)}{2\sqrt{2x}} \tag{5.6}$$

where β is parameter of the distribution.

5.5.3.3 Standardized stable noise-II

Similar to the previous case, let us take, T_{xa} and T_{xb} be the release timing of information molecules of type-a and type-b respectively. Also, T_{na} and T_{nb} be the RPD of information molecules of type-a and type-b respectively. Then the random propagation delay, $Z_n = T_{nb} - T_{na}$ of information molecule encoded in $D_x = T_{xb} - T_{xa}$ (may be positive or negative depending on order types of information molecules-a and -b) is characterized by standardized stable noise-II (say) and hence termed as standardized stable noise-II. Finally, the closed form PDF of Z_n is given as:

$$f(z_n;\beta) = \begin{cases} \frac{1}{\sqrt{8\pi z_n^3}}\left[(1+\beta)K\left(-p_{z_n},q_{z_n,}\right)+(1-\beta)L\left(-p_{z_n},q_{z_n,}\right)\right], & z_n > 0 \\ \frac{2(1-\beta^2)}{\pi(1+\beta^2)}, & z_n = 0 \\ \frac{1}{\sqrt{8\pi |z_n|^3}}\left[(1-\beta)K\left(q_{z_n},p_{z_n}\right)-(1+\beta)L\left(q_{z_n},p_{z_n}\right)\right] & z_n < 0 \end{cases} \tag{5.7}$$

with,

$$K(a,b) = \frac{1}{\sqrt{\pi}}\int_0^{\infty} e^{-t^2/4}e^{-bt}\cos(at), \quad b > 0 \tag{5.8}$$

and

$$K(a,b) = \frac{1}{\sqrt{\pi}}\int_0^{\infty} e^{-t^2/4}e^{-bt}\sin(at), \quad b > 0 \tag{5.9}$$

where, $p_x = (1+\beta)/\sqrt{8}\,|x|\, q_x = (1-\beta)/\sqrt{8}\,|x|$, $\beta = (\sqrt{D_a} - \sqrt{D_b})/(\sqrt{D_a} + \sqrt{D_b})$ with D_a and D_b as diffusion coefficient of information molecules of type-a and type-b, respectively. Also, $z = p_x - jq_x$ and $z = p_x - jq_x$ for $x > 0$ and $x < 0$, respectively.

5.5.4 Modified Nakagami distribution noise

An experimental noise model for MNC system over pipe channel with no induced flow is characterized by two parameters namely, the shape and spread parameters of the distribution. Both the shape and spread parameters increases monotonically with the propagation distance. The most important fact of above noise is that in contrast to traditional Nakagami distribution, which is defined in the domain $[0, +\infty)$, modified Nakagami distribution noise is defined in the domain $(-\infty, +\infty)$. In other words, the noise extends all the way from negative to positive domain and can be negative mathematically in contrast to standard Nakagami distribution with strictly positive random variable. However, the scaling down by 2 is applied to the integral of the unmodified PDF in the range $(-\infty, +\infty)$ that will exceed 1. The PDF of the modified Nakagami distribution noise is given as [24]:

$$f_N(z; \mu; \omega) = \frac{1}{|z|\,\Gamma(\mu)} \left(\frac{\mu}{\omega} z^2\right)^{\mu} e^{\left(-\frac{\mu}{\omega} z^2\right)} \tag{5.10}$$

where, shape parameter μ and ω are the shape and spread parameters, respectively whereas z is random variable.

5.5.5 Radiation absorption noise

The absorption of EM radiation by the molecules of different human tissues causes temperature fluctuations which is an unwanted effect in molecular dynamics. This unwanted effect, severely degrade the MNC link, is as an exogenous noise source and termed radiation absorption noise (RAN). On the one hand, most of the literatures in molecular nano communication (MNC) focus on molecular dynamics within the medium to present different forms of stochastic noise. On the other hand, in various real-life deployments, external effects are equally influence molecular dynamics and thereby cause noise. Hence, the RAN finds applications in different important field ranging from cell signaling to chemical engineering. However, the PDF of RAN is given as:

$$f_T(T) = \sum_{n=1}^{3} 8\eta\varepsilon\sigma T^3 \frac{\beta_n}{\alpha_n} \exp(-\alpha_n \left(2\eta\varepsilon\sigma T^4\right)) \tag{5.11}$$

where, η is the intrinsic impedance of the body tissues [26–28], ε is the emissivity of the body [29], σ is the Stefan–Boltzmann constant, T is the body temperature in K, and α_n/β_n are coefficients along each of three-principal directions (n equal to 1, 2, and 3) [30].

5.5.6 Sampling and counting noise

The sampling/counting noises are the most relevant noises in DMNC. Continuous information signal $T(t)$ is modulated as $r_T(t)$ by emission process at transmitter. However, due to discreteness of information, $r_T(t)$ is actually realized as $\widehat{r_T(t)}$. In other words, $r_T(t)$ is sampled by information itself results in $\widehat{r_T(t)}$. This unwanted perturbation on emitted information $r_T(t)$, related to emission process, is due to the discreteness of information is termed as sampling noise (SN), expressed as $n_s(t)$.

$$n_s(t) = \widehat{r_T(t)} - \left\langle \widehat{r_T(t)} \right\rangle \tag{5.12}$$

However, under certain assumptions the effect of $n_s(t)$ is expressed stochastically in terms of $\widehat{c_T(t)}$ as:

$$\widehat{c_T(t)} = \begin{cases} Poiss(r_T(t)), & r_T(t) > 0 \\ -Poiss(-r_T(t)), & r_T(t) < 0 \end{cases} \tag{5.13}$$

With

$$\widehat{r_T(t_n)} = \frac{\widehat{c_T(t_n)} - \widehat{c_T(t_{n-1})}}{t_n - t_{n-1}} \tag{5.14}$$

where, '*Poiss*' is Poisson counting process. The discrete information signal $\widehat{r_T(t)}$ is expected to be realized as $c_R(t)$ at receiver. However, due to imprecision and fluctuations, by virtue of random movement and discreteness of information, $c_R(t)$ is actually realized as $\widehat{c_R(t)}$. This unwanted perturbation on received information $c_R(t)$, related to diffusion process, is due to the imprecision and fluctuations in information is termed as Information Counting Noise (CN), expressed as $n_c(t)$. However, under certain assumptions the effect of $n_c(t)$ is expressed stochastically in terms of $\widehat{N_P(t)}$ as

$$\widehat{N_P(t)} \sim Poiss(c_R(t)) \tag{5.15}$$

with

$$\langle \widehat{N_P(t)} \rangle = c_R(t) \frac{4}{3} \pi \rho^3 \tag{5.16}$$

where, '*Poiss*' and $\langle . \rangle$ represent Poisson counting process and average operation, respectively.

5.5.7 Molecular displacement noise

Molecular motion in fluid medium, due to its successive collision with other molecules in the medium, is governed by a continuous random Langevin force [31]. This random force causes uncertainty in the position of received molecule. Further, this uncertainty in the instantaneous position of received molecule because of random force is termed as Molecular displacement noise. Furthermore, symbolically it is given by $M \sim N(\mu_{M,\sigma 2M})$, which is Gaussian distribution [9]. Here, $M = U - U'$, where U and U' are the number of received molecules at displacement x and x', respectively. Also,

$$\mu_M = \sum_{j=0}^{k-1} N_{av} \left(q_j - q_j' \right) \quad \text{and} \quad \sigma_M^2 = \sum_{j=0}^{k-1} N_{av} \left(q_j \left(1 - q_j\right) + q_j' \left(1 - q_j'\right) \right) \tag{5.17}$$

5.5.8 Drug delivery noise

In view of uncertainty in cardiovascular channel, concentration of drug nanoparticles, targeted location of the drug delivery and exact timing are most important attribute of any Particulate Drug Delivery Systems (PDDS). In sequence, injection, propagation and reception noises respectively are the unwanted effects which introduces deviations in above attributes and degrade the performance of PDDS. The injection noise arises due to the pressure difference between blood flow and syringe, turbulences around the needle as well as due to the limitations of injection device. On the other hand, random dispersion of nanoparticles, by virtue of irregularly shaped blood vessels, and exhibiting Brownian motion is responsible for propagation noise. However, the random mechanical forces and stochastic behavior of the chemical reactions which complicate the penetration of drug nanoparticles to the tissues surrounding the blood vessels is responsible for reception noise [32]. Though explicit stochastic models are not available for above noises. However, this section presents an insight to above noises.

Drug injection noise: is due to the mechanical limitations of the injection device such as, mechanical friction of pump, compression phenomena, and drug injection rate. However, major component contributing to injection noise includes injection leakage, maximum injection rate, and sampling rate. The injection leakage is uncontrolled leakage of drug nanoparticles, due to pressure difference of needle and concentration gradient of the drug solution. It is important to avoid delivery of drug nanoparticle at the location of healthy tissues, but this limits the maximum injection rate. Whereas mechanical deficits of syringe such as, the friction of the rubber piston and/or the presence of compressible gaseous bubbles in the drug, limit the sampling rate. Collectively, above components give rise to drug injection noise. Though, no explicit model exists to represent the DI noise but parameters such as, average injection limit, maximum drug injection rate $A(t)$, leakage in the injection $\eta(t)$, and sampling duration T_s characterize the DI noise as follows.

Drug injection leakage is due to concentration gradient of the drug solution, and that pressure gradient between the needle and the blood flow. Primarily it is depending on drug leakage rate ($\eta(t)$). Where $\eta(t)$ is depending on drug syringe spill rate ($i(t)$), symbolically, we can write $\eta(t) = g(i(t))$. In addition, $\eta(t)$ depends on sampling rate $1/T_s$. Further, the drug concentration at the sight of injection should not create toxic effect, which is directly related to the heartbeat period and the blood velocity period and limits the maximum drug injection rate $A(t)$. Furthermore, maximum rate by which the drug can be injected depends on the friction in the rubber piston of the syringe and on the presence of compressible bubbles of gas and represented as $|f| \geq \frac{1}{T_s} \Rightarrow X(f) = 0$. All the above-mentioned effects limit the injection rate, which is an impairment in DDS and hence, treated as noise.

Drug propagation noise: is due to the stochastic nature of the motion of drug nanoparticles in a possibly turbulent blood flow introduces the stochastic behavior in motion of drug which in turn produces drug propagation noise. Drug propagation noise gives completed probabilistic description of the drug nanoparticle at the targeted site. Drug propagation noise depends on following factors; The model depends on the ligands in a drug nanoparticle (N_L), receptors in the diseased cells (N_R), the temperature (T_p), the spheroid dimensions drug nanoparticle (radius a) and an aspect-ratio (γ), the maximum attraction distance (ξ) among the others. However, the Poisson Binomial distribution and Bernoulli distribution are used as stochastic representations for drug propagation noise. The different models collectively, used to represent DP noise are as follows. Single Nanoparticle Propagation Noise is used to represent the random movement of one injected nanoparticle. The probability of delivery one single nanoparticle, injected at drug injection site a time τ, at drug delivery site at time t is given as:

$$\mathrm{P}[y_s(t)=k]=h^k(t,\tau)(1-h(t,\tau))^{1-k} \tag{5.18}$$

where $k \in \{0, 1\}$. Drug Propagation Poisson Binomial Noise is used to represent drug propagation rate $y_p(t)$ as:

$$\mathrm{P}\left[y_p(t)=k\right]=1/K\sum_{n=0}^{K-1} e^{-2ink\pi/K}\prod_{m=1}^{K-1}\left(1+\left(e^{\frac{2ik\pi}{K}}-1\right)p_s(t,mT_s)\right) \tag{5.19}$$

where k is the number of nanoparticles reached to delivery site after propagation. $p_s(t, mT_s)$ is the probability of drug reception at time t that was injected at time $\tau = mT_s$. Drug Propagation Poisson Noise is the approximated Poisson Binomial Noise using Le Cam's theorem to represent drug propagation rate $\mathrm{y_p(t)}$ according to which drug reception rate is given by:

$$y(t)=\mathrm{Pois}(\lambda(t)) \tag{5.20}$$

where, $\mathrm{Pois}(\lambda(t))$ denotes inhomogeneous Poisson process with rate $\lambda(t)$ as:

$$\lambda(t)=\int_{-\infty}^{+\infty} x(\tau)h(t,\tau)d\tau \tag{5.21}$$

Drug reception noise: is due to the stochastic nature of ligand binding interactions during binding of drug nanoparticles to the surface of unhealthy cells. However, different kind of interacting energies such as, thermal energy due to Brownian motion, the kinetic energy due to the blood flow, energy due to chemical potential, energy due to ligands and the receptor of reaction, are responsible for stochastic nature ligand binding. Also, interaction with small surface, the cells irregularities, weak affinity of legend receptor, negative flow of blood are major reason for drug reception noise. However, drug reception probability is given as:

$$\mathrm{p}_r(t)=\pi r_0^2 m_R m_L e^{-\frac{\chi a\beta_w(t)}{k_B T_p r_0 m_R}\left(\left(\frac{\alpha}{\gamma}+\zeta\right)F_s+\frac{a^2}{r_0}R_s\right)} \tag{5.22}$$

where, r_0 is the radius of nanoparticle located at a ligand-receptor, m_R and m_L are the surface density of receptor and ligands respectively, k_B is the Boltzmann constant, T_p is the blood temperature, F_s and R_s are the coefficient of drag and rotation, respectively due to blood flow, $\beta_w(t)$ is the wall shear stress, a is the nanoparticle size, ξ is maximum attraction length, ζ is the bond characteristic length, χ is the ligand-receptor bond characteristic length, and α is the nanoparticle characteristic size.

5.5.9 Reactive obstacle noise

The reactive obstacle severely affects the number of received molecules and is responsible for reactive obstacle noise. Though the literatures simulate number of received molecule at the receiver as a function of radius of obstacle and reaction probability of obstacle but probability distribution to illustrate the reactive obstacle noise is missing from literatures. However, channel impulse response under reactive obstacle noise is presented [33]. As the RN is passive and released molecules can enter anywhere throughout the boundary of passive RN. Thus, the channel impulse response at any arbitrary point (x, y, z) due to the transmitter located at (x_0, y_0, z_0) can be given as [33]:

$$h(x, y, z, t) = \frac{1}{(4\pi Dt)^{3/2}} \exp\left(-\frac{d^2}{4Dr}\right), \quad t \geq 0 \tag{5.23}$$

where D is the diffusion coefficient, d is the distance between TN and RN. Also, the corresponding number of the received molecule at RN is given as:

$$N_P = \alpha V_{rx} h(x, y, z, t) \tag{5.24}$$

where α is the total number of molecules released by TN at time $t = 0$, V_{rx} is the volume of spherical RN given as $V_{rx} = \frac{4\pi}{3R_{rx}^3}$ with R_{rx} as the receiver radius.

5.5.10 External noise

All the signals from unintended sources be it other transmitter or be it leaking source are accounted for external noise. Let consider a total $U - 1$ sources of A molecules are the unintended transmitter, be it another transmitter or be it leaking source, thus a total signal at the receiver is: $N_{aobs}(t) = N_{a1}(t) + \sum_{u=2}^{U} N_{au}(t)$, where $N_{a1}(t)$ is the signal from the intended transmitter [34]. However, $N_{au}(t)$ are signal component from unwanted signal at the receiver, superposition of accumulative impact of $N_{au}(t)$ gives rise the overall noise at the receiver. Moreover, above definition of noise is based on the fact that under certain condition, various interferences such as multiuser or Intersymbol interference are approximated as a noise. However, channel impulse response due to the noise source, the expected concentration of molecules at the receiver points due to an emission of one molecule by the noise source at t^*. Where, * denotes dimensionless representation for all quantities. The dimensionless channel impulse response at (x^*, y^*, z^*) due to presence of noise source at $(x_n^*, 0, 0)$

is given by [31,34]:

$$C_a^* = \frac{1}{(4\pi t^*)^{3/2}} \exp\left(\frac{-|r^*|^2}{4t^*} - k^* t^*\right) \tag{5.25}$$

where r^* is the time-varying effective distance between (x^*, y^*, z^*) and $(x_n^*, 0, 0)$. Further, $N_{au}(t)$ is given by:

$$\overline{N_{an}^*(t^*)} = \iiint_{-\infty,0,0}^{t^*, r_{obs}^*, 2\pi} \int_0^{\pi} r_i^* \overline{N_{agen}^*(\tau)} C_a^* \sin\theta d\theta d\varphi dr_i^* d\tau \tag{5.26}$$

Eq. (5.26), under uniform concentration assumption (UCA), becomes:

$$\overline{N_{an}^*(t^*)} = V_{obs}^*(t^*) \int_0^{t^*} C_a^*(r_a^*, \tau) d\tau \tag{5.27}$$

Eq. (5.27) yields:

$$C_a^*\left(r_a^*, t^*\right) = \frac{1}{(4\pi t^*)^{3/2}} \exp\left(\frac{-\left|r_{eff}^*\right|^2}{4t^*} - k^* t^*\right) \tag{5.28}$$

Furthermore, the asymptotic solution of Eq. (5.26) in absence of flow and in absence of flow and degradation is given by [34, Eq. (17)] and [34, Eq. (19)], respectively.

5.5.11 System noise

A nonlinearity is observed in experimental impulse response of a diffusive molecular communication system over induced diffusion-based channel. Precise reason behind the nonlinearity is unknown but transmitter imperfections, turbulent flow and receiver imperfections are most likely impairment behind the nonlinearity. In particular, an inconsistent droplet size for different trial and/or response and recovery time of the sensor are the major causes. Under certain assumption, such as 1-dimensional channel and absorbing receiver, impulse response of the noise model simplifies to Gaussian distribution as [13]:

$$h_2 = \frac{Md}{\sqrt{(4\pi D t^3)}} \exp\left(\frac{(vt-d)^2}{4Dt}\right) \tag{5.29}$$

where M is the number of released molecules in short burst, D is the diffusion constant, v is average speed of the flow, d is the Tx-Rx distance, and t is the time. However, Eq. (5.29), after applying corrections, yields:

$$M_2 = \frac{a}{\sqrt{(t^3)}} \exp\left(\frac{-b(ct-d)^2}{t}\right) \tag{5.30}$$

where a is the correction factor due to the scaling factor α (related to sensor resume/respond times) and D, (related to turbulent flow) b is the correction factor due to D, (related to turbulent flow) and scaling factor α. Whereas c is the correction factor due to average speed of the flow and scaling factor α.

5.6 Physical and/or stochastic models for noise in EMNC

In conjunction with section four this section presents various representations of noise in EMNC. These representations facilitate analysis of different noises in terms of respective physical/statistical parameters.

5.6.1 Johnson–Nyquist noise/thermal noise

Johnson–Nyquist noise/thermal noise: is due to fluctuations of charged particles and thereby power dissipation by electronic elements at Nano-receiver [35]. This power dissipation follows thermal cycle between environment and Nano-receiver. It is imperative to mention that temperature T develops noise voltage $V(t)$ at open terminal of a resistor R, which follow a Gaussian distribution. The main assumption behind this distribution is that detector consisting of sufficiently large number receivers which are independent and each of them contribute to an infinitesimal amplitude. However, PDF of the noise is:

$$p(x) = \frac{1}{V_{rms}\sqrt{2\pi}} e^{-\frac{x^2}{2V_{rms}^2}} \tag{5.31}$$

whereas the available power spectral density at the load resistance, using maximum power transfer theorem, can be given as:

$$G_a(f) = \frac{G_v(f)}{4R} \tag{5.32}$$

With $G_v(f)$, the mean square spectral density of thermal noise, as:

$$G_v(f) = 2Rhfe^{-\left(\frac{hf}{k_bT}-1\right)} \tag{5.33}$$

5.6.2 Black-body noise

The black-body noise is due to subsequent radiation of EM energy seized by biological cell by virtue of EM radiation from nano-antennas. As the biological cell behaves like black-body and hence, corresponding noise is known as black-body noise [35]. This noise is always present, even in absence of signal transmission. The PDF of a_n can be given as [31]:

$$p(a_n) = \frac{1}{\langle a_n \rangle} \sqrt{2\pi}\, e^{\left[-\frac{a_n^2}{2\langle a_n^2 \rangle}\right]} \tag{5.34}$$

where n is a particular mode of the radiation field.

5.6.3 Doppler-shift-induced noise

Is due the random and irregular velocity of Nano-antennas (source) circulating with blood inside the body. However, the motion is induced by internal fluctuations of cells by virtue of heat transfer by EM radiation and govern by the Brownian motion [36]. As this varying velocity of Nano-antennas inside the body induces Doppler-shifts to the transmitted frequency and therefore termed as Doppler shift induced noise. Let, $R(f,t)$ be the received signal due to the Doppler shift, then the PDF of in-phase component of $R(f,t)$ can be denoted by x and can be given as:

$$f(x) = \frac{1}{\sigma\sqrt{2\pi}} e^{-\frac{x^2}{2\sigma^2}} \tag{5.35}$$

where, σ is the RMS value [35].

5.6.4 Molecular absorption noise

In the terahertz, total path loss in human tissues is consisting of spreading loss and molecular absorption loss, caused by expansion of wave during propagation and molecular absorption respectively. The modified Friis transmission equation [38] and transmittance of the medium [15] respectively are used to represent the spreading loss and molecular absorption loss respectively. Molecular absorption loss, which causes molecular absorption noise (MAN), is approximately twice of spreading loss at a given frequency of operation and distance. However, MAN is due to radiation of the EM energy by virtue of vibrating molecules that has previously absorbed it [16]. Further, MAN is contributed by both background/ sky/ atmospheric noise and the self-induced noise. Self-induced noise and background noise respectively are dependent and independent on the transmitted signal. It is important to mention that as use of graphene material and its derivatives suppress the thermal noise [37]. Whereas effect of MAN is negligible for a transmission distance ten of millimeters. Hence, MAN is the dominant noise in terahertz nano communication. Furthermore, the probability density function (PDF) of the MAN, conditioned on x_m is represented by [39]:

$$f_N(n \mid X = x_m) = \frac{1}{\sqrt{2\pi N_m}} e^{-\frac{n^2}{2N_m}} \tag{5.36}$$

where, n is noise with nose power N_m for mth signal.

5.6.5 Body radiation noise

Body radiation noise: is due to the radiation from the tissue of human body, which is local source of radiation. As this radiation is completely due to the original energy value of the molecules in the medium before transmission start. Therefore, it is independent of the transmitted information signal. However, similar to the sky noise it is based radiation from tissues of human and is termed as body radiation noise (BRN) [20]. Though there is no separate PDF to characterize the BRN but Planck's function is used to describe it.

$$B(T_0, f) = \frac{2h\pi (nf)^3}{c^2} \left(e^{\frac{hf}{k_B T_0}} - 1 \right)^{-1} \tag{5.37}$$

with k_b and h as Boltzmann's and Planck's constant, respectively.

5.7 Simulation results of different noises under nano communication

This section presents the graphical representation of different noise under NC system. It is noteworthy to mention that PDF for all the presented noises is not available in the literatures. However, this section presents the graphical representation of all the available PDF. See Figs. 5.1–5.6.

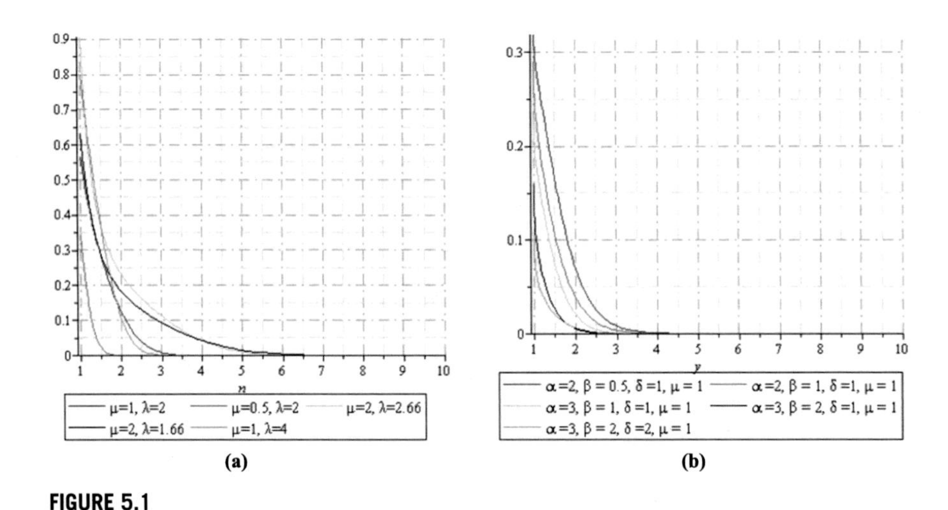

FIGURE 5.1

(a) Additive Inverse Gaussian (IG) noise PDF, (b) Random propagation delay noise PDF.

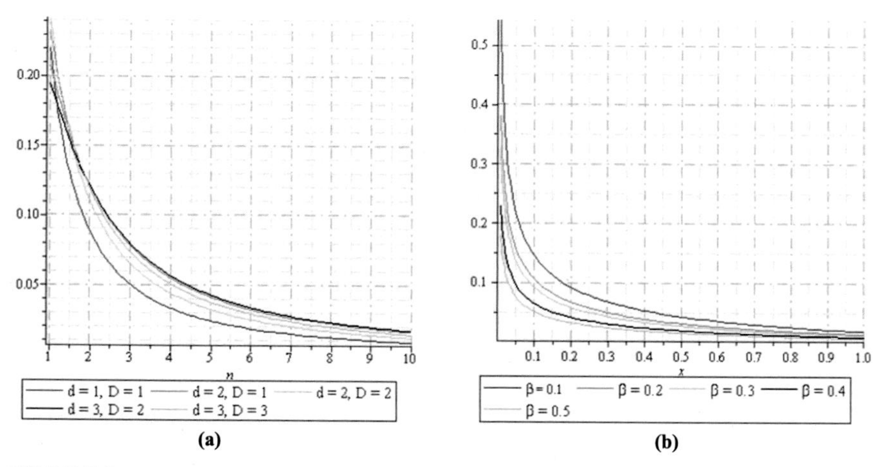

FIGURE 5.2

(a) Levy distribution noise PDF, (b) Standardized stable noise PDF.

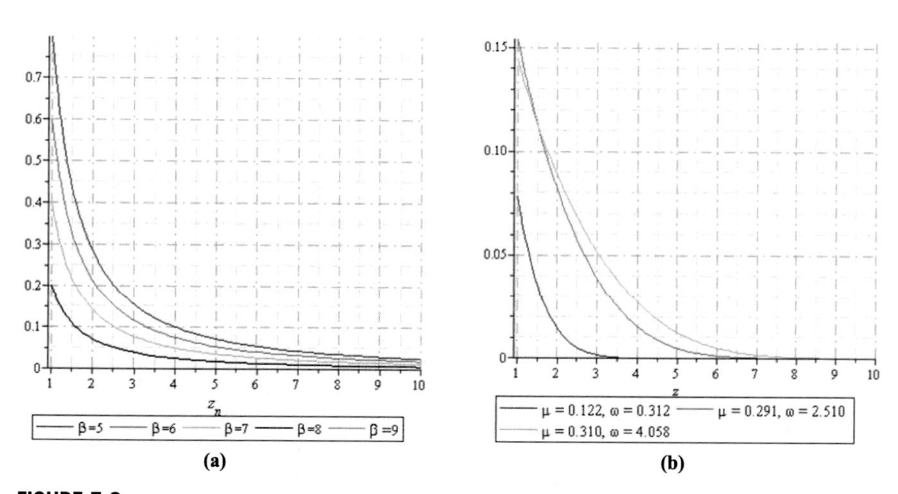

FIGURE 5.3

(a) Standardized stable noise-II PDF, (b) Modified Nakagami distribution noise PDF.

5.8 Open research challenges on noises in nano communication

In this section different open research challenges of both types of NC system be it MNC or be it EMNC has been discussed.

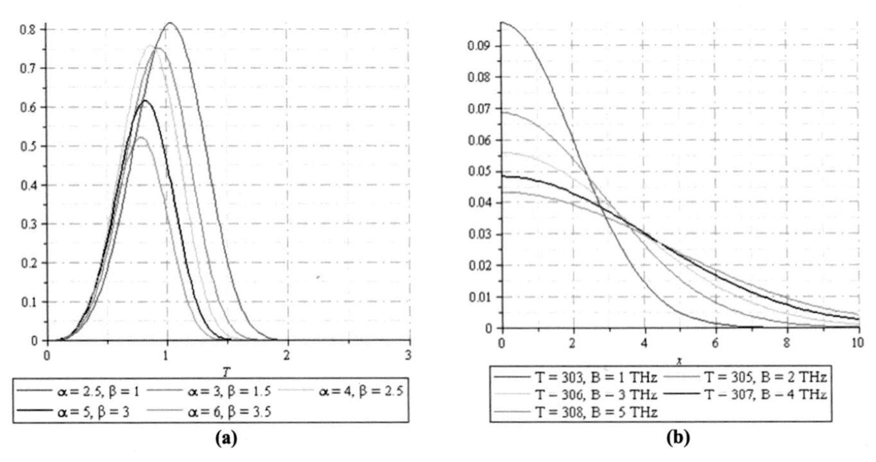

FIGURE 5.4

(a) Radiation absorption noise PDF, (b) Johnson–Nyquist noise PDF.

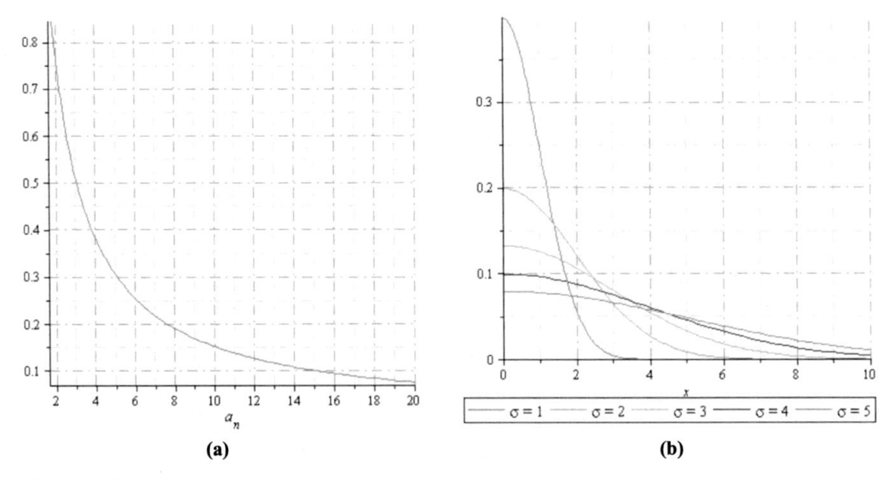

FIGURE 5.5

(a) Black-body noise PDF, (b) Doppler-shift-induced noise PDF.

5.8.1 Challenges on the noises in MNC

Based on the types of sources, there are a total eleven noises in the MNC system. First four noises namely, AIGN, NIGN and stable noise belong to a common noise source referred to as random propagation delay. Though, Generalized Gaussian Distribution (GGD) is presented as a generalization of AIGN distribution in [40]. However, to unify the noise model in MNC, which may include all of the random propagation delay-based noise models, is one of the open research challenges. Also, modified Nakagami noise models can be included in this unified model. Neglecting friction,

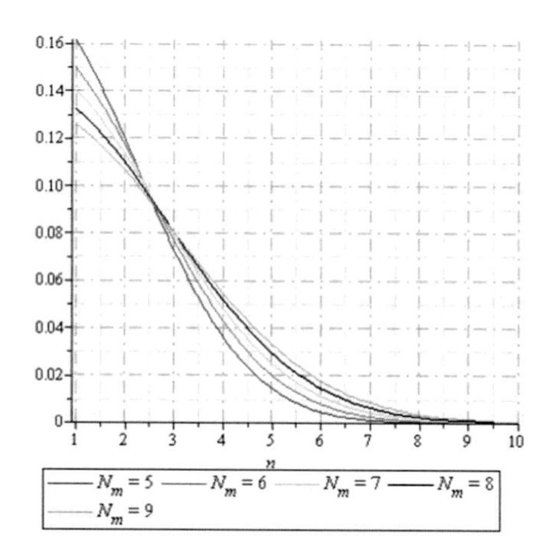

FIGURE 5.6

Molecular absorption noise PDF.

limited set of molecules used in encoding, decision on dominant component, in case of AIGN, NIGN, and modified Nakagami noise model respectively, need further attention in order to add the accuracy in the noise models. While presenting Radiation Absorption Noise in [1] the sources of EM absorption, which results in an increase in temperature, are traditional microwave communication in the frequency range of MHz to GHz and EMNC in THz frequency range. Inclusion of increase in temperature due to sky effect and/or black body radiation as a noise source in MNC is another open research challenge. In sampling and counting noise particle flux between the transmitter and the external space is responsible for discreteness of the particles. This particle flux is the key factor behind the sampling noise. On the other hand, fluctuating and imprecise counting of $c_R(t)$ at the receiver is the major cause of counting noise. Efficient techniques of particle flux control and efficient counting approaches at the transmitter and receiver respectively, needs further attention of both, the researchers and the technocrat. The uncertainty in the prediction of the position for a small duration is presented as displacement noise. However, in MNC the overall process of communication is very much slow. Therefore, assumption of small-time duration is appropriate. In addition, the effect of some other random forces, which introduces randomness in the instantaneous displacement, should be included. In drug delivery noise, a maximum drug injection rate $A(t)$, leakage in the injection $\eta(t)$ and concentration which is responsible for blood vessels bifurcations has to be optimized. In addition, some contemporary topology should be employed for cardiovascular topology. Finally, different energy interaction at the surface of the deceased cell, especially thermal energy due to the Brownian motion, needs to be revisited. Reaction obstacle in the medium is responsible for reaction obstacle noise, which has been analyzed with respect to noise reaction probability and radius of the reactive obstacle only. Therefore, different absorption properties of the obstacle need further exploration. The external noise has been defined for passive receivers over the channel with advection and molecule degradation under assumption of constant temperature and viscosity throughout the channel. The assumption is not practical. In addition, suitability of first-order degradation reaction at the receiver is an open research challenge. Finally, assuming multiuser or Inter-symbol interference as a noise is not a valid assumption. Nonlinearity in the impulse response characterizes the system noise, proposed on the basis of experimental data, is most accurate under the given scenario. Under certain assumptions it is approximated as a Gaussian distribution. However, inherent physical parameters, really responsible for the noise, are largely unknown. Therefore, insight of the involved parameters is the real challenge for researchers and technocrats involved in the noise modeling.

As a concluding remark, it is observed that on the one hand, statistical models (CHF, CDF, and CHF) for some of the noise models have been proposed with the missing simulation model. On the other hand, for others simulation models are available with the missing statistical models. To propose a missing model of a particular noise is a research plethora.

5.8.2 Challenges on the noises in EMNC

In general fluctuations of electrons in a conducting media dissipate power causing an increase in temperature which is modeled as Johnson's Nyquist noise. Whereas, in an intra-body EMNC, where biological cells and nano-antenna (heat sources) are surrounded by fluidic medium, the thermal cycle of heat exchange in between the environment and the receiver is the primary cause of Johnson's–Nyquist noise. So, in principle, this cycle of energy must be incorporated to add accuracy in the noise model. Further, black-body noise in case of intra-body EMNC is mainly due to the re-emission from particle black-body of human tissue. But the structure of human tissues can be any shape out of cylindrical and spherical. Therefore, assumption of homogeneous and isotropic radiation does not replicate the practical scenario. Furthermore, internal vibration of molecules due to absorption of EM energy from different sources results in re-radiation of EM waves which in turn results in Molecular absorption noise. As the vibration resonant frequencies highly depend on the internal structure of the molecule and hence, MAN highly depends on the frequency. However, statistical modeling of such frequency dependent noise is the open research challenge of EMNC. Again, in the case of MAN in EMNC, different correlated phenomena comprise the overall effect. But the proportions of different phenomena are largely unknown. In addition, adequate consideration of medium complexity (such as Stokes shift) and temporal characterization of absorption is missing from the literature.

Similar to the noise model in MNC, to propose various missing statistical (CHF, CDF, and CHF) and/or simulation noise models is a concluding research opening of EMNC.

5.9 Summary

Recently, the internet of bio-nanothings (IoBNT) has been envisioned as a paradigm shift, which is going to engineer the nanomachines, program them and interconnect with outside world. However, its realization includes some extreme environments, for example, high temperature and radiation such as in adversarial chemical environments, lossy propagation environment like health monitoring and targeted drug delivery, anticipated defense environment as in the nuclear biological and chemical challenges among the others. The miniaturized communicating devices under the extreme environments are susceptible to substantial challenges that impede monitoring, control, and communications. The noise is indomitable effect of extreme environments. Further, as the communication at nanoscale is classified as MNC and EMNC. Therefore, there are two different classes of noise at nanoscale with various modes in each class. This chapter presents different types of noises in each of the class in details. Further, due to the randomness involved in the noise representation, irrespective of types and sources of noise or type of communication, there are few common tools which are employed in the analysis of noise. In particular, time-averaged representation, power spectral density (frequency domain) and the correlation function (time domain) are major tools to represent both random and deterministic noise. Therefore, the chapter starts with the statistical tools used in the noise analysis along with the introductory level on the noise in the nano communication which may be MNC or EMNC.

However, it is noteworthy to mention that the noise and its modeling is the fundamental of Technological Readiness Level (TRL) which assures its success like technology development and technology demonstration therefore the fundamentals of noises in both MNC and EMNC are well elaborated. To represent the noise in NC system, two potential models like stochastic- and physical-model are discussed. The physical model offers a mathematical analysis of all the physical processes involved in generation of the noise. However, in the stochastic model all the physical processes are included statistically. In both the cases, the mathematics is commonly exploited tool of representation of different types of noises. Further, the noise model can be developed with in-depth understanding of medium characteristics which in turn depends on the number of interrelated main parameters such as temperature, permittivity/refractive index of the medium, polarization mechanism and absorption cross section area which are used to characterize the medium.

Finally, several open challenges on different classes/types of noise are explored. To unify the noise model in MNC, which may include all of the random propagation delay-based noise models, is one of the open research challenges. Further, the inclusion of increase in temperature of molecules as a noise source in MNC is another open research challenge. Assumption of small-time duration in MC is not appropriate, as overall process of communication is very much slow therefore, some contemporary topologies should be employed for cardiovascular topology. The external noise has been defined for passive receivers over the channel with advection and molecule degradation under assumption of constant temperature and viscosity throughout the channel however, it is not practical. Further, in case of external noise, the assumption of multiuser or inter-symbol interference as a noise is also not valid. Several inherent physical parameters, really responsible for the noise, are largely unknown. In addition, an adequate consideration of medium complexity (such as Stokes shift) and temporal characterization of absorption is also missing from the literature.

References

[1] S. Pratap Singh, Shekhar Singh, Weisi Guo, Saket Mishra, Sanjay Kumar, Radiation absorption noise for molecular information transfer, IEEE Access 8 (2020) 6379–6387.
[2] Dogu Arifler, Capacity analysis of a diffusion-based short-range molecular nano-communication channel, Computer Networks 55 (6) (2011) 1426–1434.
[3] Ian F. Akyildiz, Josep Miquel Jornet, Electromagnetic wireless nanosensor networks, Nano Communication Networks 1 (1) (2010) 3–19.
[4] Hadeel Elayan, Pedram Johari, Raed M. Shubair, Josep Miquel Jornet, Photothermal modeling and analysis of intrabody terahertz nanoscale communication, IEEE Transactions on Nanobioscience 16 (8) (2017) 755–763.
[5] Youssef Chahibi, Molecular communication for drug delivery systems: a survey, Nano Communication Networks 11 (2017) 90–102.
[6] Fei Li, Weiting Liu, Cesare Stefanini, Xin Fu, Paolo Dario, A novel bioinspired PVDF micro/nano hair receptor for a robot sensing system, Sensors 10 (1) (2010) 994–1011.
[7] Yu Feldman, Yu A. Gusev, M.A. Vasilyeva, Dielectric relaxation phenomena in complex systems, Tutorial, Kazan Federal University, Institute of Physics, 2012.
[8] K.M. Yaws, D.G. Mixon, W.P. Roach, Electromagnetic properties of tissue in the optical region, in: Optical Interactions with Tissue and Cells XVIII, vol. 6435, International Society for Optics and Photonics, 2007, p. 643507.
[9] Massimiliano Pierobon, Ian F. Akyildiz, Diffusion-based noise analysis for molecular communication in nanonetworks, IEEE Transactions on Signal Processing 59 (6) (2011) 2532–2547.
[10] Peter J. Thomas, Donald J. Spencer, Sierra K. Hampton, Peter Park, Joseph P. Zurkus, The diffusion-limited biochemical signal-relay channel, in: Advances in Neural Information Processing Systems, 2004, pp. 1263–1270.
[11] Nariman Farsad, Weisi Guo, Chan-Byoung Chae, Andrew Eckford, Stable distributions as noise models for molecular communication, in: Proc. IEEE Global Communications Conference (GLOBECOM), 2015, pp. 1–6.
[12] Ling-San Meng, Ping-Cheng Yeh, Kwang-Cheng Chen, Ian F. Akyildiz, MIMO communications based on molecular diffusion, in: Proc. IEEE Global Communications Conference (GLOBECOM), 2012, pp. 5380–5385.
[13] Nariman Farsad, Na-Rae Kim, Andrew W. Eckford, Chan-Byoung Chae, Channel and noise models for nonlinear molecular communication systems, IEEE Journal on Selected Areas in Communications 32 (12) (2014) 2392–2401.
[14] Josep Miquel Jornet, Ian F. Akyildiz, Channel capacity of electromagnetic nanonetworks in the terahertz band, in: Proc. IEEE International Conference on Communications, 2010, pp. 1–6.
[15] Josep Miquel Jornet, Ian F. Akyildiz, Channel modeling and capacity analysis for electromagnetic wireless nanonetworks in the terahertz band, IEEE Transactions on Wireless Communications 10 (10) (2011) 3211–3221.

[16] Ke Yang, Alice Pellegrini, Max O. Munoz, Alessio Brizzi, Akram Alomainy, Yang Hao, Numerical analysis and characterization of THz propagation channel for body-centric nano-communications, IEEE Transactions on Terahertz Science and Technology 5 (3) (2015) 419–426.

[17] Pu Wang, Josep Miquel Jornet, M.G. Abbas Malik, Nadine Akkari, Ian F. Akyildiz, Energy and spectrum-aware MAC protocol for perpetual wireless nanosensor networks in the Terahertz Band, Ad Hoc Networks 11 (8) (2013) 2541–2555.

[18] Rui Zhang, Ke Yang AkramAlomainy, Akram Alomainy, Qammer H. Abbasi, Khalid Qaraqe, Raed M. Shubair, Modelling of the terahertz communication channel for in-vivo nano-networks in the presence of noise, in: Proc. 16th Mediterranean Microwave Symposium (MMS), 2016, pp. 1–4.

[19] Bhagwandas Pannalal Lathi, Modern Digital and Analog Communications Systems, Oxford University Press, 1998.

[20] Kothapalli V. Srinivas, Andrew W. Eckford, Raviraj S. Adve, Molecular communication in fluid media: the additive inverse Gaussian noise channel, IEEE Transactions on Information Theory 58 (7) (2012) 4678–4692.

[21] Werner Haselmayr, Dmitry Efrosinin, Weisi Guo, Normal inverse Gaussian approximation for arrival time difference in flow-induced molecular communications, IEEE Transactions on Molecular, Biological and Multi-Scale Communications 3 (4) (2017) 259–264.

[22] Anders Eriksson, Eric Ghysels, Fangfang Wang, The normal inverse Gaussian distribution and the pricing of derivatives, The Journal of Derivatives 16 (3) (2009) 23–37.

[23] Norman R. Draper, David E. Tierney, Regions of positive and unimodal series expansion of the Edgeworth and Gram-Charlier approximations, Biometrika 59 (2) (1972) 463–465.

[24] Song Qiu, Siyi Wang, Weisi Guo, Experimental Nakagami distributed noise model for molecular communication channels with no drift, Electronics Letters 51 (8) (2015) 611–613.

[25] Ke Yang, Dadi Bi, Yansha Deng, Rui Zhang, Muhammad Mahboob Ur Rahman, Najah Abu Ali, Muhammad Ali Imran, Josep Miquel Jornet, Qammer H. Abbasi, Akram Alomainy, A comprehensive survey on hybrid communication in context of molecular communication and terahertz communication for body-centric Nanonetworks, IEEE Transactions on Molecular, Biological and Multi-Scale Communications 6 (2) (2020) 107–133.

[26] P.A. Hasgall, E. Neufeld, M.C. Gosselin, A. Klingenbck, N. Kuster, It is database for thermal and electromagnetic parameters of biological tissues, Version 3.0, 2015.

[27] Luca Zilberti, Damien Voyer OrianoBottauscio, Oriano Bottauscio, Mario Chiampi, Riccardo Scorretti, Effect of tissue parameters on skin heating due to millimeter EM waves, IEEE Transactions on Magnetics 51 (3) (2015) 1–4.

[28] Richard B. Schulz, V.C. Plantz, D.R. Brush, Shielding theory and practice, IEEE Transactions on Electromagnetic Compatibility 30 (3) (1988) 187–201.

[29] Bryan F. Jones, Peter Plassmann, Digital infrared thermal imaging of human skin, IEEE Engineering in Medicine and Biology Magazine 21 (6) (2002) 41–48.

[30] R. Arnaut Luk, Compound exponential distributions for under-mode reverberation chambers, IEEE Transactions on Electromagnetic Compatibility 44 (3) (2002) 442–457.

[31] Amit Singhal, Ranjan K. Mallik, Brejesh Lall, Effect of molecular noise in diffusion-based molecular communication, IEEE Wireless Communications Letters 3 (5) (2014) 489–492.

[32] Youssef Chahibi, Ian F. Akyildiz, Molecular communication noise and capacity analysis for particulate drug delivery systems, IEEE Transactions on Communications 62 (11) (2014) 3891–3903.
[33] Muneer M. Al-Zu'bi, Ananda S. Mohan, Steve S.H. Ling, Impact of reactive obstacle on molecular communication between nanomachines, in: Proc. 40th Annual International Conference of the IEEE Engineering in Medicine and Biology Society (EMBC), 2018, pp. 4468–4471.
[34] Adam Noel, Karen C. Cheung, Robert Schober, A unifying model for external noise sources and ISI in diffusive molecular communication, IEEE Journal on Selected Areas in Communications 32 (12) (2014) 2330–2343.
[35] Hadeel Elayan, Cesare Stefanini, Raed M. Shubair, Josep Miquel Jornet, End-to-end noise model for intra-body terahertz nanoscale communication, IEEE Transactions on Nanobioscience 17 (4) (2018) 464–473.
[36] Clifford V. Heer, Statistical Mechanics, Kinetic Theory, and Stochastic Processes, Elsevier, 2012.
[37] Rui Zhang, Ke Yang, Qammer H. Abbasi, Khalid A. Qaraqe, Akram Alomainy, Analytical modelling of the effect of noise on the terahertz in-vivo communication channel for body-centric nano-networks, Nano Communication Networks 15 (2018) 59–68.
[38] Frederick Wooten, Optical Properties of Solids, Academic Press, 2013.
[39] Josep Miquel Jornet, Low-weight error-prevention codes for electromagnetic Nanonetworks in the terahertz band, Nano Communication Networks 5 (1–2) (2014) 35–44.
[40] S. Pratap Singh, Sanjay Kumar, Closed form expressions for ABER and capacity over EGK fading channel in presence of CCI, International Journal of Electronics 104 (3) (2017) 513–527.

CHAPTER

The use of deep learning in image analysis for the study of oncology

6

Bailey Janeczko and Gautam Srivastava
Department of Math and Computer Science, Brandon University, Brandon, MB, Canada

6.1 The difficulties in meeting demand

The study of oncology, or the study of cancer or tumors if you prefer, is one in which has made groundbreaking discoveries over the years. However, there are many areas to improve. It is estimated that in 2020 on average 617 Canadians will be diagnosed with cancer every day, and 228 Canadians will die from it every day [1]. Effectively making cancer the leading cause of death in Canada as it has been for several years [2]. According to the WHO [3], early detection of cancer in any form can greatly improve your treatment and chance of survival. Therefore, early screening and detection are vital, followed by a good prognosis and treatment plan. Doctors often use computer-aided diagnosis or CAD, to aid in the diagnosis stage. They use machine learning to help highlight potential issues or anomalies in clinical images such as computerized tomography (CT) scans, magnetic resonance imaging (MRI), positron emission tomography (PET) scans, X-rays, ultrasounds and especially in the case of mammograms. In this chapter we will go over the current uses of CAD more specifically, the deep learning algorithms and practices currently used as well as the future and potential deep learning have in this specific field.

Canada does not have an abundance of Cancer healthcare centers. Four provinces only have 2 cancer treatment centers, two only have 1, and for the approximately 125,000 people living in the 3 territories, Nunavut, Northwest Territories, and Yukon, there currently exists no cancer treatment centers [16]. Worldwide, chemotherapy demands are expected to far exceed the physician workforce required [17]. More physicians are needed, but that is not a plausible solution, at least not in the short term. This is due to the amount of time required to become an oncologist. It often takes over a decade to fully train any sort of physician specialist and a Medical Oncologist is no exception to this rule. To become a medical oncologist, one must first complete their undergraduate studies, which technically can be completed in 3 years, but many take 4-year programs as the field for medicine is very competitive. Once those 3–4 years of Undergraduate studies are done, the prospective physician must go to Medical School, which is an additional 4 years of study. Once they have now completed 7–8 years of schooling, they must become a part of a residency program where they will

Internet of Multimedia Things (IoMT). https://doi.org/10.1016/B978-0-32-385845-8.00011-3

typically spend again around 4 years or so. Finally, they will go for their license and become a licensed physician. This is a lot of work for students to go through and due to the approximate 12-year time period it takes to complete any changes in demand for Oncologists in Canada cannot be satisfied relatively quickly. As a result of this, a massive delay in our medical system exists to make our supply of physicians meet our needs.

Canada has made clear we need more med students as most of those students who can endure the first 8 years and make it past medical school can expect to be accepted into a residency program right away. Residency programs for years now have accommodated nearly every student that comes out of medical school, in 2020, 2855 of the 3072 (92.94%) graduates that applied for residency programs got accepted to one. In 2019 an even more impressive 2819 of 3020 (93.34%) were accepted [19]. This of course also applies to those studying Radiational Oncology, where not a single one of the 67 applicants over the last 3 years has not been accepted to a residency program through the first iteration of searching.

6.1.1 The medical imaging equipment issue

The issue of physician shortages may not even be the biggest issue with analyzing, diagnosing, and treating patients. Top-of-the-line medical imaging software would go a long way to helping physicians not only speed up some of the processes but also get better resolution images to work from and diagnose. According to the European Coordination Committee of the Radiological Electromedical and Healthcare IT industry, there are some golden rules which should be followed to ensure that a countries medical imaging machines are up to date enough for the role they play in healthcare [28]. The rules are as follows:

(1) At least 60% of the installed equipment base should be younger than 5 years.
(2) Not more than 30% should be between 6–10 years old.
(3) Not more than 10% of the age profile should be older than 10 years.

So how does Canada stack up to these measures? Well, we can look to the Canadian Medical Imaging Inventory 2017 report to find the answers to that as shown in Table 6.1. For CT machines, only 33% of the units were 5 years old or newer, 40% were 6–10 years, 23% were 11–15 years, and 4% were 16–20 years old. That is a far cry from the 60%, 30%, 10%, and 0% respectively recommended on these machines. MRI machines are worse, as only 36% of MRI machines were 5 years old or newer, 33% were 6–10 years old, 26% were 11–15 years old, and 5% were 16–20 years old. On the chart provided below the different medical imaging machines in Canada are displayed along with their growth between 2007 and 2017 and the breakdown of age.

As you can see, despite Canada growing its inventory of medical imaging machines immensely since 2007, Canada still underperforms the recommendations put forth by the COCIR for having enough new equipment. This can be observed entirely across the board and it is easy to understand why. Two things can be causing Canadian hospitals and clinics to have older equipment. One of these reasons is they are

Table 6.1 Medical imaging inventory.

Machine	Units in 2017	Units in 2007	% Growth	% of units ≤ 5 years	% of units 6–10 years	% of units 11+ years
CT	561	419	34%	33%	40%	27%
MRI	366	222	65%	36%	33%	26%
PET-CT or PET	51	21	143%	32%	53%	16%
SPECT	330	603	−45%	13%	29%	57%
SPECT-CT	261	5	5120%	40%	48%	12%

Table 6.2 CT scanners cost in Canadian dollars [29].

CT scanner	New price range	Refurbished price range
16-slice CT scanner	$285,000 – $360,000	$90,000 – $205,000
64-slice CT scanner	$500,000 – $700,000	$175,000 – $390,000
128-slice CT scanner	$675,000 – $1,000,000	$225,000 – $650,000
256+ slice CT scanner	$1,350,000 – $2,100,000	Small market right now

either using refurbished or used machines or another reason, they are holding on to their machines for longer than suggested. The breakdown of pricing for CT scanners is shown in Table 6.2.

As you can see by the chart, refurbished CT scanners give quite a discount, and with how expensive these machines are, it is understandable that hospitals and clinics not only want to buy refurbished but want to try and keep their current machines running for as long as possible before replacing.

MRI machines are not any cheaper, and the refurbished market acts quite the same as the CT scanner. New MRI machines range between $1,000,000 and $2,100,000 for the 1.5 and 3.0 Tesla 70 cm Wide Bore machines. In comparison, you can find cheaper MRI machines refurbished for as low as $200,000 and ranging up to $700,000 for premium refurbished machines [32].

6.2 What is deep learning?

Deep learning is a subset of machine learning, that was created to simulate how our brain works. The similarities are loose here but our brains have a system of neurons that allow us to think and feel an emotion and intuitively, make decisions for us. These neurons in our brain receive massive amounts of input at times and in turn, those neurons essentially calculate all the inputs. If it passes a certain threshold the neuron will become active through a process called synaptic integration [6]. Deep Learning models can have many different forms of thresholds, some may have no thresholds at all. All of these different forms are referred to as activation functions. These activation functions are all within artificial neurons which is what an artificial

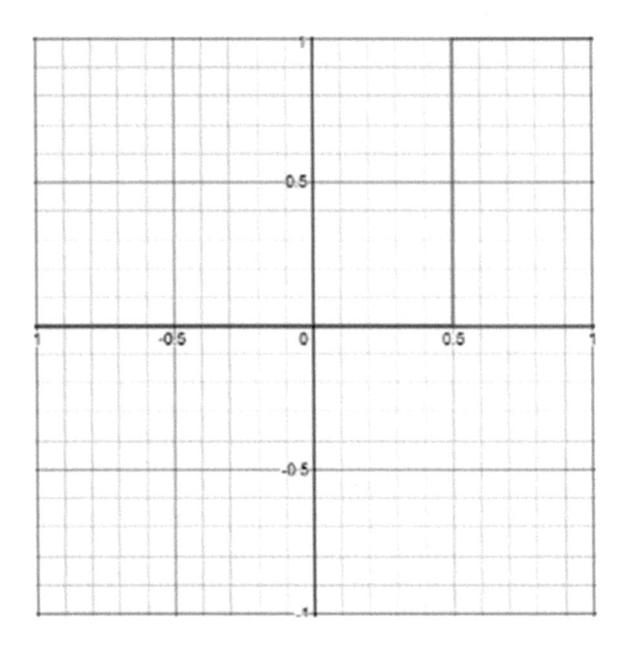

FIGURE 6.1

Binary step function.

neural network is composed of. Below you can find different and common activation functions within artificial neurons.

6.2.1 Neuron's and activation functions

Similar to the brain, deep learning in the form of artificial neural networks has its take on synaptic integration. This occurs in the form of activation functions. As mentioned before Deep Learning models can have many different forms of thresholds, some no thresholds at all. We will look at four of the most common activation functions in artificial neurons below; Binary Step Function, linear functions, Rectified Linear Unit (RelU), and Sigmoid functions.

6.2.1.1 Binary step function

Binary Step Function is an activation function in which the inputs are calculated linearly with weights and a bias (Fig. 6.1). This is done to create an output by which a threshold is created for activation. The graphical representation for this activation function can be found to the right.

6.2.1.2 Linear activation functions

Linear Functions are similar to Binary Step Function in the sense that inputs are combined linearly with weights and a bias (Fig. 6.2). However, unlike Binary Step Function activation, there is no threshold by which the output passes to be active.

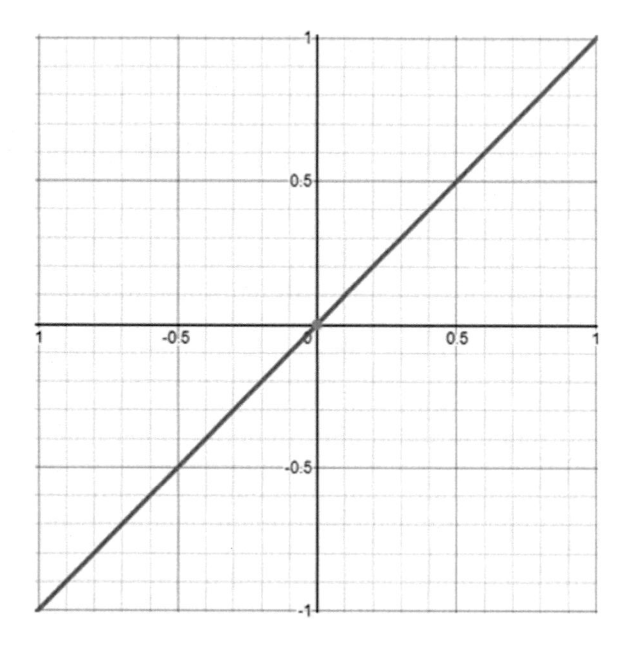

FIGURE 6.2

Linear activation functions.

The output for the neurons is not binary, but rather it is the linear output itself. The graphical representation of this activation function can be found to the right.

6.2.1.3 Rectified linear unit

Rectified Linear Unit is perhaps the most computationally efficient activation function and it is also one of the most used and simplest activation functions (Fig. 6.3). It is a linear activation function without an allowance of negative numbers. The function simply goes $y = \max(0, x)$ where x is the linear equation created by the model ensuring that the neurons will go to 0 for any output less than 0. The graphical representation of this activation function can be found to the right.

6.2.1.4 Sigmoid functions

Sigmoid functions are simply mathematical equations that equate all values of x to an output ranging between 0 and 1 thus creating an S-shaped graph. This was a great step in neural networks as it allowed the outputs of these functions to be more in line with standard probability situations. The most common form of sigmoid functions that are typically found in artificial neural networks is: $\frac{e^x}{e^x+1}$ or $\frac{1}{e^{-x}+1}$.

The graphical representation of this activation function can be found in Fig. 6.4.

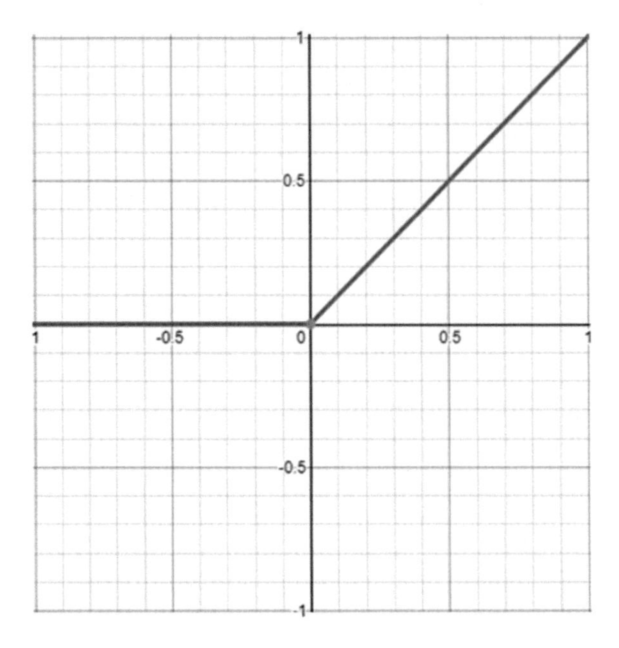

FIGURE 6.3

Rectified linear unit.

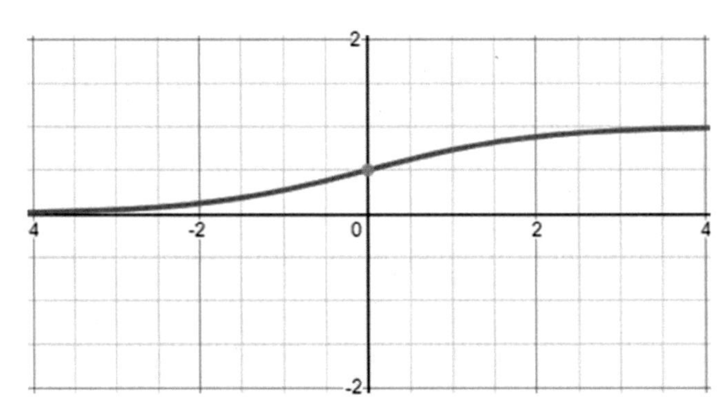

FIGURE 6.4

Sigmoid functions.

6.2.2 Feedforward neural networks

Now that we understand how artificial neurons work in an artificial neural network or ANN, let us break down how exactly these neural networks are mapped and how they work. Let us start where all machine learning models start, the input. In the ANN, these inputs will be placed in the input layer. Then comes the hidden layer(s) of neural networks, these layers contain neurons that depending on the structure of the neural network can take in varying amounts of input neurons, or previously hidden layers outputs to calculate a new set of outputs. These are dependent on the algorithms it finds most satisfying of the goal of the network, and the activation functions within the neurons. Layer by layer, changes are made through the model to try and optimize the output to the goal.

6.2.2.1 The limitations of FFNs

The ability to transform the data in so many ways, transformations upon transformations, allows these artificial neural networks to solve very complex issues. One of the biggest issues however with Feedforward Neural Networks, is the restriction they have on inputs, more specifically, FFNs must have a fixed set of inputs and outputs. The great thing about future models that we will get into is the ability to take new data all the time, oftentimes in real-world applications or real-time, to adjust the model accordingly. Besides this limitation though, FFNs have plenty of issues with reading in images and processing them, we will discuss this further later in the chapter. FFNs did not have much of a chance to contribute to this field, but it was the building block for which many models followed, and thus it is an important topic to study here.

Image analysis for machine learning models is vastly more complicated than most machine learning problems out there. The most difficult issue to get around when analyzing images, is how exactly to take in the input from them and turn it into meaningful features that we can extract information out of. The individual pixel information is the only way to directly take that image input, but individual pixel data is extremely difficult to work with. To work with images, we must input every pixel of that image, and a single-pixel alone can contain a vector of three colors, red, blue, and green, all ranging from 0 to 255 which accounts for the color of the image.

Now how can we get from pixel data to detecting something? how can a computer analyze these individual pixels to determine edges and patterns to start understanding how to see? The answer to this depends on the model or technique used to do such as this is not a one solution fits all problem. Let us look at some examples of how deep learning can detect these patterns and edges, as well as what can be done with this information.

6.2.3 Recurrent neural networks

Following Feedforward Neural Networks, Recurrent Neural Networks came. These networks worked by taking step by step data as inputs and producing step by step outputs. Each new step of data associated weights with previous steps, and thus helped form what the newest step of output would be. What this meant was in RNNs, unlike FNN, they have the ability for neurons in the same layer to connect and create maps that are not linearly forward like the feedforward neural networks were. This allows the models to have dynamic states, that prior, were not possible. This as a result adds a vast more complexity to models and as you would imagine from such, the potential from these models became immense. RNNs are best used for sequential data, or time-dependent data, where you could input step-by-step data into it and it would weigh differently as time, or steps, went on. This made them a huge hit in things such as language translators, taking in words in a sentence one word at a time to translate languages, and stock market predictors, taking in historical data, as little as by minute, to analyze when to buy and sell, particularly in day trading.

6.2.3.1 Vanishing gradient problem

The vanishing gradient problem occurs in many recurrent neural networks and can be defined as the earlier edges between layers having less and less weight are being adjusted, to the point of becoming obsolete. The problem occurs within backpropagation, as during training, adjustments are made to the weights of edges to get the result, and in many RNNs, the early layers become so small that even large changes to the weights of edges attached to them see little impact on the output. Thus, this is far from an optimal model that cannot make much for change in the early layers of the model. It is worth noting while less common, the problem that occurs here in RNNs that causes the vanishing gradient problem can also cause an exploding gradient problem that is the exact opposite of this.

6.2.4 Long-short term memory model

After discovering the vanishing gradient problem, the creation of Long-Short Term Memory models, or LSTMs, solved it and was a massive success. To prevent the vanishing gradient problem (or exploding gradient problem), it uses a different setup, that works as a state machine, with forget gates that have their activation of what to forget when new input is entered. This forget gate allows better control over the gradient's values and thus leading to the removal of the vanishing gradient problem.

LSTM is a type of RNN, and thus is a great deep learning model for sequence data. It would and has worked especially well in terms of not only diagnosing cancers, especially when paired with a CNN model but also in giving a prognosis of cancers through tissue sampling [21]. LSTMs have also been shown to be working very well in pairing with CNN models (which we will get to right away here) in classifying breast lesions for breast cancer through dynamic contrast-enhanced MRI scans [20] and if we step outside image analysis for a second, even make accurate predictions of cancer patient fluctuations in hospitals [22].

6.2.5 Convolutional neural networks

Convolutional neural networks, or CNNs, are neural network structures primed for pattern detection, making them especially useful for image analysis. CNNs have specialized layers, called convolutional layers that exist within the model. These layers do convolution operations, or if you prefer, these neurons within the convolutional layer, convolve the input. These convolution operations are what make CNNs so good at detecting patterns, so the question you may be asking is, how do they detect these patterns?

This may be easier to understand starting from the idea of having a predetermined idea of what patterns we are trying to detect and to start simply, we will consider an example of looking for a hard edge. What a convolutional operation does, is use a

Table 6.3 Convolutional matrix operator.

0	1	1
0	1	1
0	1	1

filter, typically a 3 × 3 matrix filter shown in Table 6.3, that contains values in which can be compared, section by section of an image, and after applying the filter, by taking the dot product of the filter and the input image section (the pixels within the section), we can see how similar the two are, and how likely it fits the profile of the pattern. So, for an edge, we have a filter that looks something like the table to the right. Imagine a black and white image in this case, and 0 is white, and 1 is black, and any value in between is a form of gray, where the larger the number the darker the pixel. Now if we run this filter over a section that looks something like this (see Table 6.4):

Table 6.4 Filter results.

0.1	0.7	0.9
0.2	0.8	1
0.1	0.9	1

Since this is very similar to the filter that was being run over it, we would get a very high convolutional output, that would then put placed to a new table, creating an image that would recognize this pattern, thus an image detecting all the left edges of an image. As you can see, if you do this through many neurons, detecting all sorts of different patterns and edges, you can get results that can accurately detect features in images, often used for things like facial recognition, but in our case, that can be used to detect tumors. These convolutional operations are unmatched by anything else in machine learning at detecting patterns and edges, and regardless of other techniques used, almost all image analyses, including and outside of the field of Oncology, use convolutional operations in some form or manner. They have been shown to diagnose all forms of cancers using all different types of clinical images [22,25,26], and are much of the credit for the potential deep learning has in the medical imaging field for Oncology.

6.3 Deep learning techniques and processes

6.3.1 Backpropagation

The key to changing the weights and biases within neurons to optimize the model is something called backpropagation. This is key to nearly all artificial neural networks and is a process that starts as soon as the first iteration of the model is created. Once you have the output layer, this is the layer with our predictions from the model, we can test it against the labeled output. Say this neural network is simply outputting probabilities for a classification, and one example, has the probability of the input being A at 0.5, probability of it being B at 0.3 and a probability of it being C at 0.2, and the real output should be B. What backpropagation then does, is it tries to adjust the neurons and edge weights to lower the error rate here, following all the edges back, any neuron or edges in this case that assigned more value to A or C would be lowered in weight, and any that assigned value to B, clearly didn't do enough (as B only had a probability of 0.3), so they would be raised in weight, all proportional to their influence in the model. This is a very time expensive process, but it is what allows often very complex neural networks to be optimized and improved upon and is what we must know when discussing deep learning models.

6.3.2 Image segmentation

Image Segmentation is a process used to separate, or segment, an image to create numerous sections, or segments. An easy way to think of this is as if you gave someone an image and told them to draw borders around objects in the photo, this is segmentation. The key to this idea is that by segmenting images using deep learning, we can focus on individual objects in the image, thus localizing it, in a process referred to as object localization. This allows you to focus on objects without background noise muddling with your algorithms.

6.3.3 Object localization in the study of oncology

The way a physician checks to see if someone has cancer is by running some scans, whether that be MRI scans, CT scans, mammography or otherwise, and attempting to find tumors within the scans. Many models, including CNNs, can have a technique that shows where it finds tumors. If they contain this feature this is called object localization.

Image segmentation, as discussed above, allows these deep learning models to pick one or numerous points of emphasis to classify. In the case of cancer detection, localizing specific areas on scans that may contain a tumor, narrows down the area in which these models will have to focus their efforts. Then not only can we scan whole clinical images to classify cancers or not, but we can get the model to detect where exactly it is detecting the tumor is forming.

Much of the concern around deep learning as a field, has been the notion that deep learning is much like a black box. You cannot see what is going on, and you do not know why the model makes the choices it does, which can cause issues when underlying issues lie in the data. Object localization would give more power to those reading the results from these models, as it is forced to not only make classifications on whole clinical images but is almost told why it believes so. As a result, it is forced to point out the tumor itself in question. (Note: That is not to say it explains why it believes certain areas in scans may or may not contain tumors.)

For most cases, this is not going to be wildly different for results, or in the case of tumor formation and cancers, often you will know where the suspected tumor is and do not need a model to point out such, but it does improve the model's capabilities, and at times it will provide very useful information that can be looked into further if need be, and as such it must be mentioned here before we move forward.

6.4 Data difficulties

Data is not easy to obtain. This is especially true when it comes to image analysis, and again even more so with clinical images. Clinical images are not only hard to gather in big enough numbers but much of it is unlabeled, meaning, there is no output attached to the data, it only provides the input image. This means to use something like a traditional CNN, we must first label all our data, typically thousands of images. This is far from ideal, and most times simply not plausible to do so, we do however have some workable solutions for this.

6.4.1 Stepping away from supervised-learning

To understand a way around this problem, we must first realize that the way deep learning, and machine learning models for that matter, have been talked about thus far through the chapter; taking in inputs, having labeled data that then gets tested against in the end to find the optimal predictor model, is not the only way. It is best to break down the learning process of machine learning models into 4 separate categories.

(1) Supervised-learning
You give the model entirely labeled data, all data points have inputs and outputs to test against. Examples of this include FNNs, RNNs, CNNs, and many standard classifications, and regression models.

(2) Semisupervised learning
You give the model some labeled data, and some unlabeled data, you then can test against the partial labeled data or test using an unsupervised-learning method and boost its performance using the partially labeled data

(3) Unsupervised learning
You give the model just inputs and typically get the model to in some way create the input and test it against. We will get more into this in the below sections. Examples of this include principal component analysis, clustering, the autoencoder, the restricted Boltzmann machine and deep belief networks.

(4) Reinforcement learning
This form of learning is a form of trial and error, the model runs trials, trying different outputs in sequence and taking in new input on the changes that occur, and will get rewarded when goals are being reached.

6.4.2 Dimension reduction

One of the best things to come from unsupervised learning is dimension reduction, that is, reducing the plethora of inputs into a lower dimension, to extract more useful and predictive features from such. There are two types of dimension reduction models that are worth close examination. These are Principal Component Analysis (PCA), which were the beginnings of dimension reduction as we know it, and Autoencoders, as well as the various forms that come in.

6.4.2.1 Principal component analysis

Principal Component Analysis is a simple form of dimension reduction, that plots the correlations between samples onto a 2D graph (can also do a 3D graph, but that is not often the case). I will not get into too much detail on how this is done, but the result is two (again sometimes three) features, PC1 and PC2, that are plotted as an axis on the 2D graph. These principal components are ranked from most to least influential in explaining the variance in the samples, so PC1 has the highest variance and should be considered the most important when clustering groups together on this graph.

6.4.2.2 Autoencoder

The Autoencoder was the deep learning-based improvement needed on the rudimentary PCA charts. Autoencoders are neural networks that are the main goal, like PCAs, is to extract useful features from data, by compressing the data, and reducing the dimensions.

The traditional autoencoder has an input layer, one or more encoder layers, a latent dimension that sits in the middle, often called the bottleneck, followed by one or more decoder layers, and our output layer. The input is compressed down, and dimensions are reduced in the encoder layers, which leads to a bottleneck. Then the image is recreated using the decoder. The goal of the autoencoder is to successfully encode the data to a lower dimension, then decode the data from that lower dimension to recreate the image as best it can, training the model to produce as low a reconstruction error as possible.

Once training is down, the autoencoder will have optimized how it compresses and decompresses the data, creating a vector of compressed down, particularly powerful features in the bottleneck of the model. The bottleneck is what holds all the information of value in this model, it is trained to recreate its input to optimize the use of that bottleneck, to optimize the process of reducing the dimensions, to give us a compressed version of the image, within the bottleneck.

There have been plenty of variations on the traditional autoencoder, I have listed below some of the most common ones.

6.4.2.2.1 Stacked autoencoder

Stacked Autoencoders are a variation of autoencoders that, instead of having one hidden layer that encodes, and one layer that decodes, has many layers that can encode and decode, you can think of this as stacking layers, hence, stacked autoencoder.

6.4.2.2.2 Sparse autoencoder

Spare Autoencoders are a variation of autoencoders that try to keep as many neurons as possible, inactive, that is, close to 0. They also do not have a bottleneck like most autoencoders, in fact, often they have more neurons in the middle of the network than in the input layer, but they can do this because they have something called a "sparsity penalty," which penalizes neuron activations. This is a form of regularization built into the model, meant to fight off overfitting that would occur without it due to it is high neuron count. Due to the higher neuron count and the added complexity with the sparsity penalty, this autoencoder while often very effective, is one of the more time extensive to train.

6.4.2.2.3 Denoising autoencoder

Denoising Autoencoders are a variation of autoencoders that were made to fight the inherent bias towards overfitting data that autoencoders can often face. Like sparse autoencoders, the denoising autoencoder tries to fight the possibility of overfitting, but instead of having a form of regularization built-in, they do this by adding noise into the input. The noise added to the data makes the encoder/decoder less able to memorize tiny details in training data that can cause overfitting, thus being a more optimized model for new data.

6.4.2.2.4 Variational autoencoder

Variational Autoencoders are a variation of autoencoders that were made to act similarly to a Generative Adversarial Network, by adding in a generation process, with a sampler that can be placed along with the bottleneck for decoding. This feature to the autoencoder allows the autoencoder to generate new data, using the input data and sampling inputs to generate realistic new outputs.

6.4.2.2.5 Convolution autoencoder

Convolutional Autoencoders are less common than the ones mentioned above, but it has been making some headway recently, especially in the case of image analysis, due to its use of course of convolving, the same process of detecting patterns and edges that CNNs have.

6.4.3 Synthetic data

Another solution to a lack of data for studying is the use of synthetic data. It takes a lot of data to best utilize the tools and models we now have available to us, and this could not be truer than with deep learning models specifically. The idea of creating original synthetic data is not new, but what is new is a method for doing so-called Generative Adversarial Network or GAN.

This model was created in 2014, by Ian Goodfellow and his colleagues [7]. GAN was purposed as a model that could generate new, original, synthetic data, by pitting two deep learning models against each other. Each of these deep learning models operated as their network, one was a generator, this network's job was to sample from the dataset to generate new and original synthetic data, while the other network, the discriminator's goal, was to take in the real and synthetic data and try and determine what was real, and what was fake. The goal of the generator model is to create data that increases the error rate among the discriminator, and of course, the goal of the discriminator is to lower said error rate, effectively putting the two models in competition with one another [18,23].

In images, often you can use autoencoders with these Generative Adversarial Networks, this allows you to compress these images down to a vector which can much more easily be worked on to create new realistic synthetic data. GANs have been a massive success since the idea was introduced just 6 years ago, it's been used to create art [8], edit facial features on images [9] and most importantly to the topic at hand, the ability to create clinical images.

This includes the creation of MRI scans [10], 3D MRI scans [11], PET scans [12], mammograms [13], and even the creation of CT scans through MRI scans alone, with radiation concerns surrounding CT/X-ray scans, this could be a big part of the future because of it [14].

6.4.3.1 Super resolution

GANs offered more than just image creation though, as the ability for this model to create what did not exist before has also been used to enhance images. Image enhancement, or superresolution imaging (SR), uses generative deep learning models to try and fill in the missing data from photos. In most cases that missing data ends up being the difference between an HD image, and one taken at a lower resolution. A big issue with high-resolution scans, such as 3D MRI scans, is that they are very time extensive, this is due to hardware limitations that then result in small spatial coverage and low signal to noise [15]. High-resolution CT scans or HRCT is another method used to obtain high-resolution CT scans. These traditionally use typical 64 slice CT scanners, these machines are optimized for resolution over speed, maximizing the spatial resolution and using especially narrow widths for slices. However, this technique comes with a downside. The main one is that because this technique takes much longer to run, the areas being scanned must be far for focused and localized. On top of that, HRCT images have high levels of noise, which is a huge downside to being able to diagnose more broadly [27]. Proper image enhancement can greatly improve classification results among neural networks, while furthering allowing physicians to get a clearer picture, in a quite literal sense [24].

6.5 The current uses of deep learning image analysis

Deep learning in image analysis in the study of oncology is a new research topic and as we discussed, the healthcare industry is not the most adaptive to making changes and perhaps for good reason. This does, however, leave this field of study in a bizarre place. The research aspect of deep learning image analysis on clinical images has yielded fantastic results in numerous forms of images and all sorts of tumor diagnostics as well as prognosis and treatment plans from said image analysis. There is no reason to believe deep learning image analysis will not take over the healthcare industry, including the field of Oncology. Deep learning algorithms are already very prevalent in the healthcare industry, including chatbots and smart health records, thus leaving room for so much more to be done in the form of diagnostic help.

The academic research side of this field has grown extensively, and we are just starting to see this potential being unleashed in the healthcare industry itself. In February of 2020, the largest hospital in Canada (by the number of beds) announced it was launching an advanced IT group called “CREATE” that would contain experts in data science, AI, and software engineering [31].

Across the border, John Hopkins University has done a lot of work using deep learning, including a key discovery. It was found by researcher Elliott K. Fishman that approximately 30% of those with pancreatic cancer, had screenings before that showed the tumor present, simply not distinguishable enough at first glance to notice. We know that early detection is key to fighting any form of cancer, and early detection maybe even more important with pancreatic cancer as it is the deadliest cancer of all [4,5]. This kind of research is what can save lives in the healthcare industry, and in comparison to the machines that are run to scan these images and the cost of the physicians getting second opinions, deep learning offers a cheaper and more reliable alternative that as Fishman argues, can not only perform on par with a radiologist but perhaps, outperform them [30].

6.6 The future of deep learning image analysis in the study of oncology

Deep Learning as a field is not going to stop growing as it is currently a powerhouse in AI and machine learning. As a result, we can fully expect it to continue dominating sectors like never imagined. There are plenty of discussions and concerns around what this future would be. The main concerns revolve around how much we should entrust deep learning models and algorithms, the protection that would be required for proper data privacy, especially in terms of the healthcare industry and data security. There is a lot of work to be done to ensure this data does not fall into the wrong hands. But there is a lot of benefits too.

Even just stepping away from pure image analysis for a moment, deep learning has already shown the potential to develop new drugs, make treatment plans, and track and predict hospital patient inflow to best optimize the industry. As for deep learning image analysis, it has not only shown potential, but results, in predicting all different forms of cancers and giving proper prognosis. Models such as Autoencoders, or CNNs when paired specifically with LSTM or other recurrent neural network models, have shown a real path for a new form of CAD tools in neural network form.

The future of deep learning in the industry is not entirely clear, but it likely includes more standard use of deep learning models in diagnostic assistance and more research using neural networks to find out new information. This is because deep learning does not just have the potential to assist physicians but rather it has the potential to help data engineers and doctors make discoveries and push the study of Oncology to new limits.

6.7 Conclusion

There is no way around the fact that cancer treatment will be in shorter supply for those that need it. Thus, we need to do a better and faster job of getting the information necessary to diagnose and treat these patients while being as accurate as possible. With the help of deep learning, this can be made easier for physicians as deep learning models can specialize in diagnosing all sorts of tumors with numerous different types of scans. Furthermore, we have the methods and techniques to do so without having extensive amounts of labeled data. Not only can deep learning image analysis algorithms and techniques help physicians in a form of Computer-Aided Diagnosis, but they can also help make discoveries. Thus, furthering research in training algorithms to detect tumors, give a prognosis, and create treatment plans will inevitably lead to more discoveries and in turn lead to improvement in the industry as a whole. While the concerns around the black box of deep learning will need to be addressed and being able to fully trust a deep learning algorithm entirely to diagnose or give treatment plans, at least in the field of Oncology may not be the case anytime soon. However, this does not mean that deep machine learning and deep learning powers should not and will not be used to contribute to the field in all sorts of innovative ways.

References

[1] Statistics Canada, Table 102-0561 – Leading Causes of Death, Total Population, by Age Group and Sex, Canada, Annual, Statistics Canada, Ottawa, CANSIM database, 2012, July 25.

[2] Statistics Canada, Table 13-10-0394-01 Leading causes of death, total population, by age group.

[3] Early detection of cancer, World Health Organization, https://www.who.int/activities/promoting-cancer-early-diagnosis, 2016.

[4] Cancer screening tests, https://www.cdc.gov/cancer/dcpc/prevention/screening.htm, 2019.

[5] Alessandro Cucchetti, Giorgio Ercolani, Matteo Cescon, Giovanni Brandi, Giovanni Taffurelli, Lorenzo Maroni, Matteo Ravaioli, Raffaele Pezzilli, Antonio Daniele Pinna, Estimation of the survival benefit obtainable from screening for the early detection of pancreatic cancer, Pancreas 45 (5) (2016) 714–719, https://doi.org/10.1097/mpa.0000000000000523.

[6] How do neurons work? Uq. Edu.Au. November 9, 2017, https://qbi.uq.edu.au/brain-basics/brain/brain-physiology/how-do-neurons-work, 2017.

[7] Ian J. Goodfellow, Jean Pouget-Abadie, Mehdi Mirza, Bing Xu, David Warde-Farley, Sherjil Ozair, Aaron Courville, Yoshua Bengio, Generative adversarial networks, arXiv, https://arxiv.org/abs/1406.2661, 2014.

[8] Ahmed Elgammal, Bingchen Liu, Mohamed Elhoseiny, Marian Mazzone, CAN: creative adversarial networks, generating 'art' by learning about styles and deviating from style norms, https://arxiv.org/abs/1706.07068 [Cs], June 2017.

[9] Gang Zhang, Meina Kan, Shiguang Shan, Xilin Chen, Generative adversarial network with spatial attention for face attribute editing, Openaccess. Thecvf. Com., https://openaccess.thecvf.com/content_ECCV_2018/html/Gang_Zhang_Generative_Adversarial_Network_ECCV_2018_paper.html, 2018.

[10] Generative adversarial networks for the creation of realistic artificial brain magnetic resonance images, Tomography 4 (4) (2018) 159–163, https://doi.org/10.18383/j.tom.2018.00042.

[11] Gihyun Kwon, Chihye Han, Dae-shik Kim, Generation of 3D brain MRI using auto-encoding generative adversarial networks, https://arxiv.org/abs/1908.02498 [Cs, Eess], August 2019.

[12] Jyoti Islam, Yanqing Zhang, GAN-based synthetic brain PET image generation, Brain Informatics 7 (1) (2020), https://doi.org/10.1186/s40708-020-00104-2.

[13] Shuyue Guan, Breast cancer detection using synthetic mammograms from generative adversarial networks in convolutional neural networks, Journal of Medical Imaging 6 (03) (2019) 1, https://doi.org/10.1117/1.jmi.6.3.031411.

[14] Yang Lei, Joseph Harms, Tonghe Wang, Yingzi Liu, Hui-Kuo Shu, Ashesh B. Jani, Walter J. Curran, Hui Mao, Tian Liu, Xiaofeng Yang, MRI-only based synthetic CT generation using dense cycle consistent generative adversarial networks, Medical Physics 46 (8) (2019) 3565–3581, https://doi.org/10.1002/mp.13617.

[15] Esben Plenge, Dirk H.J. Poot, Monique Bernsen, Gyula Kotek, Gavin Houston, Piotr Wielopolski, Louise van der Weerd, Wiro J. Niessen, Erik Meijering, Super-resolution methods in MRI: can they improve the trade-off between resolution, signal-to-noise ratio, and acquisition time?, Magnetic Resonance in Medicine 68 (6) (2012) 1983–1993, https://doi.org/10.1002/mrm.24187.

[16] Canadian Radiation Oncology Centres, Canadian Association of Radiation Oncology, http://www.caro-acro.ca/patients/canadian-radiation-oncology-centres/. (Accessed 10 October 2020).

[17] Brooke E. Wilson, Susannah Jacob, Mei Ling Yap, Jacques Ferlay, Freddie Bray, Michael B. Barton, Estimates of global chemotherapy demands and corresponding physician workforce requirements for 2018 and 2040: a population-based study, The Lancet Oncology 20 (6) (2019) 769–780, https://doi.org/10.1016/s1470-2045(19)30163-9.

[18] R-1 match interactive data, https://www.carms.ca/data-reports/r1-data-reports/r-1-match-interactive-data/. (Accessed 10 October 2020).

[19] Wenqing Sun, Tzu-Liang (Bill) Tseng, Jianying Zhang, Wei Qian, Enhancing deep convolutional neural network scheme for breast cancer diagnosis with unlabeled data, Com-

puterized Medical Imaging and Graphics 57 (April) (2017) 4–9, https://doi.org/10.1016/j.compmedimag.2016.07.004.

[20] Natalia Antropova, Benjamin Huynh, Hui Li, Maryellen L. Giger, Breast lesion classification based on dynamic contrast-enhanced magnetic resonance images sequences with long short-term memory networks, Journal of Medical Imaging 6 (1) (2019), https://doi.org/10.1117/1.JMI.6.1.011002.

[21] Dmitrii Bychkov, Nina Linder, Riku Turkki, Stig Nordling, Panu E. Kovanen, Clare Verrill, Margarita Walliander, Mikael Lundin, Caj Haglund, Johan Lundin, Deep learning based tissue analysis predicts outcome in colorectal cancer, Scientific Reports 8 (1) (2018), https://doi.org/10.1038/s41598-018-21758-3.

[22] Amani Alrobai, Musfira Jilani, Cancer incidence prediction using a hybrid model of wavelet transform and LSTM networks, Communications in Computer and Information Science (2019) 224–235, https://doi.org/10.1007/978-3-030-36365-9_19.

[23] Chanki Yu, Sejung Yang, Wonoh Kim, Jinwoong Jung, Kee-Yang Chung, Sang Wook Lee, Byung-ho Oh, Acral melanoma detection using a convolutional neural network for dermoscopy images, in: Nikolas K. Haass (Ed.), PLoS ONE 13 (3) (2018) e0193321, https://doi.org/10.1371/journal.pone.0193321.

[24] Honggang Chen, Xiaohai He, Cheolhong An, Truong Q. Nguyen, Adaptive image coding efficiency enhancement using deep convolutional neural networks, Information Sciences 524 (July 2020) 298–317, https://doi.org/10.1016/j.ins.2020.03.042.

[25] Wafaa Alakwaa, Mohammad Nassef, Amr Badr, Lung cancer detection and classification with 3D convolutional neural network (3D-CNN), International Journal of Advanced Computer Science and Applications 8 (8) (2017), https://doi.org/10.14569/ijacsa.2017.080853.

[26] Sunghwan Yoo, Isha Gujrathi, Masoom A. Haider, Farzad Khalvati, Prostate cancer detection using deep convolutional neural networks, Scientific Reports 9 (1) (2019), https://doi.org/10.1038/s41598-019-55972-4.

[27] Practice parameter 1 HRCT lungs, https://www.acr.org/-/media/ACR/Files/Practice-Parameters/HRCT-Lungs.pdf. (Accessed 10 October 2020).

[28] COCIR, Sustainable competence in advancing healthcare medical imaging equipment age profile&density, https://www.cocir.org/uploads/media/14008_COC_Age_Profile_web_01.pdf. (Accessed 10 October 2020).

[29] Meridian Leasing, CT scanner buyers guide: slice counts and pricing, Meridian Leasing, https://www.meridianleasing.com/blog/medical-equipment-blog/ct-scanner-buyers-guide, March 7, 2020.

[30] Johns Hopkins researchers use deep learning to combat pancreatic cancer, Healthcare IT News, https://www.healthcareitnews.com/news/johns-hopkins-researchers-use-deep-learning-combat-pancreatic-cancer, August 16, 2018.

[31] Hamilton Health Sciences Launches Advanced IT Group | Canadian Healthcare Technology, https://www.canhealth.com/2020/02/27/hamilton-health-sciences-launches-advanced-it-group/. (Accessed 10 October 2020).

[32] Meridian Leasing, MRI machine buyers guide: options and pricing, https://www.meridianleasing.com/blog/medical-equipment-blog/mri-machine-buyers-guide-options-and-pricing, October 10, 2020.

CHAPTER

7 Automatic analysis of the heart sound signal to build smart healthcare system

Puneet Kumar Jain[a] and Om Prakash Mahela[b]

[a]*Department of Computer Science and Engineering, National Institute of Technology Rourkela, Odisha, India*

[b]*Power System Planning Division, Rajasthan Rajya Vidyut Prasaran Nigam Ltd., Jaipur, India*

7.1 Introduction

Automatic analysis of the heart sound signal is typically performed in the following four steps; denoising, segmentation, feature extraction, and classification. Phonocardiogram (PCG) is the recording of the heart sound signal acquired using the digital stethoscope and widely used to analyze the heart sound signal. Since it uses a microphone to convert the audio signal to an electrical signal, it is highly prone to the noise generated due to various components including movement and speech of the subject or movement of the stethoscope while recording, and noise generated due to the ambient sources. Moreover, the physiological sounds generated due to lungs also contaminate the heart sounds in time and frequency domain both [1]. Therefore, in the first step, denoising algorithms are used to suppress the in-band and out-of-band noise from the PCG signal. Denoising is a process where the unwanted components are suppressed from the acquired signal to obtain the desired signal.

Once the signal is denoised, it is segmented into the systole (time duration from S1 to S2) and diastole (time duration from S2 to S1) periods by localizing the fundamental heart sounds (FHS), S1 and S2 [2–4]. Segmentation of the heart sound signal is important since the identification of these components and duration helps in the extraction of diagnostically important features. The presence of the murmurs in a particular period correlates with a class of heart valve diseases. Moreover, the frequency characteristics of each of the individual components can be extracted for the consideration of classification [5]. Once the segmentation is performed, features are extracted from the time, frequency, and time-frequency domain. Finally, using the extracted features and a suitable machine learning based classification technique, the PCG signal is classified either as a normal or abnormal case associated with a specific disease.

Internet of Multimedia Things (IoMT). https://doi.org/10.1016/B978-0-32-385845-8.00012-5

7.1.1 Motivation

Worldwide, CVDs results in more than 17.7 million deaths every year, which makes it the leading cause of mortality [6]. Moreover, it is predicted that it will increase to approximately 23 million deaths every year by 2030 [7]. The major impact of CVDs is in low and middle-income countries where it accounts 80% of the total mortality caused by CVDs worldwide. The reason behind it is the insufficiency of medical facilities specifically in the rural areas, due to which the disease could not be diagnosed at an early stage. Lack of timely intervention and proper medication at an early stage causes heart diseases to an extent at a level where it is difficult to cure [8]. Thus, it is of paramount importance to diagnose the CVDs before the situation gets worsen which will improve the expectancy of life. Among the CVDs, Aortic stenosis (AS), Mitral stenosis (MS), Mitral regurgitation (MR), and Mitral valve prolapse (MVP) are the heart valve-related diseases whose impact is increasing every year [9]. According to studies [10,11], a serious concern about heart valve disease is that in most of cases symptoms occur at a later stage when the condition gets worsen. Since the heart sound signals are generated due to the movement of the heart valves, the valvular defect will impact the characteristics of the heart sound components. Hence, analysis of these characteristics of the heart sound will help to diagnose the valvular diseases [12]. Moreover, heart valve diseases which are related to the structural defect of the valves, its signature may not be present in electrocardiography (ECG) which measures the electrical activity of the heart [13]. Therefore, the analysis of the heart sound signal is of paramount importance to diagnose the diseases at an early stage.

7.1.2 Contributions

This work focuses on the automatic analysis of the heart sound signal so that various CVDs can be detected at an early stage and proper medication can be started timely. It will reduce the impact of diseases as well as the cost to cure it. In this direction, following contributions have been presented.

7.1.2.1 TQWT based denoising algorithm for PCG signal

Time-frequency domain techniques localize the components of a signal in time and frequency both and hence very helpful to emphasize the FHS [14]. In the literature, various wavelet transform (WT) based methods have been proposed to emphasize the FHS [3,15,16]. WT based method can suppress the out-of-band noise by discarding the high frequency bands related to noise and also in-band noise by performing thresholding of the coefficients. However, due to the constant quality factor (ratio of center frequency to bandwidth) of the wavelet [17], it is difficult to adapt to the signal's oscillating characteristics and hence impacts the separation of murmur with FHS in some pathological cases [18]. On the other hand, in TQWT based transformation method, the quality factor can be tuned according to oscillations of the FHS with the help of scaling parameters used in the method.

The proposed work uses the TQWT method to decompose the signal up to 18 levels. Among these decomposed levels, a particular level having emphasized FHS is adaptively selected using the Fano factor as a quality parameter which was calculated on the absolute values of the signal. As observed in [18], the Fano factor will be relatively lower for the signal with the murmur as compared to the FHS because the variance of the murmur component will be larger compared to the FHS. Further, to address the issue of in-band noise in the selected level, an adaptive thresholding method proposed in [19] is used. The denoised signal is used to perform the segmentation by applying the HSMM based Springer's algorithm [20].

7.1.2.2 Analysis of multidomain features for PCG signal

The time and frequency-domain features from the segmented periods are extracted to incorporate the duration, energy, distribution, and oscillation characteristics of the signal. In addition to it, features are extracted from the decomposed signal using TQWT to incorporate the time localized frequency characteristics of the heart sound signal components. For this purpose, the individual entropy of systole and diastole periods of each detailed levels is calculated. Moreover, the ratio of the energy and entropy of high frequency bands to low frequency band is also calculated. It is because, in case of the presence of different murmurs, the frequency band of the significant components would very [5].

7.1.2.3 Classification of PCG signal

In the presented work, various classical machine learning algorithms including the SVM, KNN, and ensemble method have been presented. Moreover, to analyze the impact of hyper-parameters of these methods, variants of these methods have been tested to classify the PCG signal. The experiments with all the variants of the model on the complete dataset are carried out to show the efficacy of the selected features and to obtain the best model for the classification. Considering the limited number of recordings per class, a 10-fold cross validation technique is used to establish the robustness of the obtained results.

The rest of the paper is organized as follows: the literature survey of the denoising algorithms, segmentation algorithms, and feature extraction and classification algorithms is provided in Section 7.2. Section 7.3 presents a brief theoretical background of the TQWT and classification methods used in the work. The proposed heart sound signal classification method is described in section 7.4. Section 7.5 provides the obtained experimental results using the proposed method, and its comparison to various state-of-the-art methods. At last, concluding remarks of the work are mentioned in section 7.6.

7.2 Literature survey

In the literature, various methods for the automatic analysis of the heart sound signal have been proposed to address the complex nature of the signal and its susceptibility to various noise sources. A brief review of the work done towards the denoising, segmentation, feature extraction, and classification steps is described as follows.

7.2.1 Denoising algorithms

Considering the susceptibility to the noise of the PCG signal, various denoising algorithms to suppress the level of noise from the PCG signal have been proposed. A detailed review of these algorithms can be obtained in [2,5,21]. These algorithms can be broadly categorized as time-domain and frequency-domain algorithms. Generally, time-domain denoising algorithms are used to perform the preprocessing of the signal to remove the out-of-band noise. Conventional filters (Chebyshev IIR filter) [22], autocorrelation and adaptive noise cancellation (ANC) [23] are few examples of the time-domain filtering techniques. However, the time-domain representation of the signal does not provide spectral characteristics of the components presented in the signal. To exploit the spectral characteristics of FHS and other components, the signal is generally transformed into either frequency or time-frequency domain using a transform technique including discrete Fourier transform (DFT), short-time Fourier transform (STFT), discrete wavelet transform (DWT) etc. then the transformed signal is processed to suppress the level of noise. For example, Sanei et al. [24] used singular spectrum analysis to separate FHS and murmurs. In the time-frequency domain, DWT based algorithms are extensively used for the PCG signal [1,15,25–29]. This is because of the variable length of time window with respect to scale in DWT as compared to a fixed length of the window in STFT. Due to this feature, band-limited FHS will be distinctly present in some of the frequency bands of DWT and the murmur and noise components will spread throughout the levels and hence will be small in amplitude [30]. In DWT based denoising algorithms, choice of the mother wavelet and way of thresholding are crucial parameters [19,31]. In the literature, 'sqtwolog' [32], 'minimaxi' [26], 'rigrsure' [15,25], and 'heursure' [1] methods are used to estimate the threshold value. However, these threshold estimation methods do not incorporate the domain knowledge of the PCG signal. To address this issue, an adaptive threshold estimation method was proposed in [19] which uses statistical parameters including the mean, variance, and a new parameter called med_{75}. These parameters were calculated on the absolute values of the DWT coefficients. The parameter med_{75} is the 75th percentile value of the coefficient vector after sorting in ascending order. Further, the performance of the method was improvised by applying the estimated threshold value using a new threshold function called mid-function and its parameter were optimized using the genetic algorithm.

DWT based methods satisfactorily perform the denoising of the PCG signal and therefore widely used. However, the constant Q-factor imposes the restriction on its behavior to adapt according to the dynamics of the signal. To resolve this, Patidar and Pachori proposed to use the TQWT to separate the FHS and murmurs in the PCG signal [33]. However, the method used to optimize the parameters results in multiple time decomposition of the signal using TQWT and hence the algorithms require high computational time.

7.2.2 Segmentation algorithms

The purpose of the segmentation of heart sound signal is to localize the FHS, S1 and S2 and hence the systolic and diastolic periods. It helps in the extraction of relevant features from each component and duration. However, segmentation is a challenging task due to interbeat variation in length of the cardiac cycle [34], intra-beat amplitude variation of the components [35], and overlapping of noise in time and frequency domain [3,36]. One more challenge in the segmentation is that the frequency range of the FHS also varies from one pathological case to another. Considering these challenges, several segmentation algorithms have been proposed in the literature which are comprehensively reviewed in [5,19,37]. In literature, various algorithms performed the segmentation by first extracting the envelope (also called envelogram) of the PCG signal and then uses a peak finding algorithm to localize the FHS. In the time-domain, envelogram have been extracted using the squared energy [38], normalized average Shannon energy [39], cardiac sound characteristic waveform [35], and Shannon energy [40]. While in the frequency domain, envelopes are extracted using the S-transform [4], Hilbert transform (HT) [41], Wigner-Ville distribution [42], empirical mode decomposition (EMD) [43], and WT [15,44]. However, the performance of these envelope based methods is limited in the presence of external noise. Moreover, the requirement of efficient peak-finding and identification algorithms limits their use for a dataset comprised of the signals recorded in a different setup.

In another approach, features based classification of identified peaks as S1, S2, or other is performed to segment the heart sound signal. These approaches extracted the features from multidomain including time-domain [45], frequency-domain [46], and wavelet-domain [47]. Then the extracted features are applied to the classifier to identify specific peak as S1 or S2. In [48], features are extracted from the correlation coefficients matrix, while in [49], structural characteristics of the FHS are extracted based on shape context and an adaptive waveform filtering based on blanket fractal dimension. M. Shervegar and G. Bhats [50] first converted the spectrogram of the heart sound signal into a Bark spectrogram by converting the spectrogram to the bark scale and then the loudness index for all the frequency band is used to identify the occurrence of the FHS. In [51], narrow band-limited components (NBCs) are obtained using the variational mode decomposition and then the HT is applied on these NBCs to obtain the analytic signal representation (ASR) of it. Then the features extracted from the instantaneous spectral attributes and convex hull area measured from the ASRs are applied to an SVM model. The proposed method in [52] performs the segmentation in two steps: Detection and Selection. In the detection step, the sound events are identified based on the spectral flux, which indicates the rate of change in the power spectrum of the signal. In the selection step, the localization of FHS is performed based on the maxima position in the spectral flux. M. Mishra et al. [53] proposed a method to classify the FHS using a deep neural network technique, stack auto-encoder based on the combinatory feature extracted from the higher-order moments and cepstral-domain. In [54], a simple k-means clustering algorithm is used to classify the FHS.

More recently, Hidden Markov Model (HMM) and HSMM have been proposed to segment the heart sound signal [55,56]. These probabilistic models can naturally model the sequential and periodic nature of the heart sound components. In the HMM, the heart sound components are represented as hidden states. Transitions among these states are explicitly modeled using the transition probability matrix. It helps to model the constraints on the transition of one component to the next one. For example, a transition should occur from S1 to systole state and hence a high probability will be assigned to this transition, while the transition from the S1 to directly to the S2 state will be assigned low probability. These models predict the hidden state based on the observations produced by the particular state. For this, features extracted from the time, frequency, and time-frequency domain of the heart sound components are used as observation characteristics. Then, the recorded observations are modeled via statistical distributions of the particular state. In these models, it is important to incorporate the information about the sojourn time of the particular state i.e. the time spent by the model in a particular state. Schmidt et al. [57,58] proposed a generalization of HMM called HSMM which improved the performance of the model compared to the HMM-based method. While in [56], F. Renna presented an effective implementation of the HMM and HSMM in conjunction with 1-dimensional CNN architecture. To further improve the performance of these models, adaptation to the sojourn time of the different PCG signals or adoption for emission distributions have been proposed [59,60]. For the adaptation of emission distribution, SVM based approach [61] and maximum likelihood approach [60] have been proposed. The state-of-the-art segmentation algorithm uses the logistic regression based approach for emission distribution adaptation and a modified Viterbi algorithm [20].

7.2.3 Feature extraction and classification algorithms

The complex and nonstationary nature of the heart sound signal makes the classification of the PCG signal a challenging task. In addition to it, the signature of the FHS and murmurs overlaps in the frequency domain and varies from person to person and interperson as well. Selection of the features to analyze all these diagnostically important factors is the key in the classification step. Therefore, an extensive range of features from different domains have been studied in the literature [5]. In the time-domain, the statistical parameters including mean, standard deviation, skewness, kurtosis of the amplitude and duration of the S1, S2, systole, and diastole periods are extracted [5]. While most of the classification methods first segment the heart sound signal and then extract the features, there are few attempts to extract the features without segmentation of the signal [62,63]. S. Deng and J. Hans [62] classify the signal using the autocorrelation feature and diffusion map without segmentation of the signal. In [63], M. Zabihi et al. extracted the linear predictive coefficient and entropy in time-domain. Although time-domain features are easy to extract, they do not exploit the frequency characteristics of the heart sound components and therefore generally accompanied with the time-frequency domain features.

In the literature, several transformation techniques for the heart sound signal have been presented including S-transform [4], singular spectrum analysis [24], pseudo-affine Wigner-Ville distribution [39,42], STFT [64], EMD [65], Chirplet transform [66], Gammatone filter bank [67], Mel-frequency cepstral coefficients (MFCC) [68,69], wavelet synchrosqueezing transform [70], and TQWT [71]. Among all these, the most popular transformation technique is the WT due to its feature of multiresolution analysis [62,72–74]. In [65], M. Altuve et al. first decomposed the heart sound signal in the intrinsic mode functions (IMF) domain using two techniques including the EMD and the improved complete ensemble EMD (ICEEMD). Then Shannon's entropy and variance are obtained from the various time windows among different IMF related to different frequencies ranges. W. Zhang et al. in [75] first obtain the spectrograms of the signal and then the tensor decomposition is used to reduce the dimension of the feature vector. In [62], features are extracted using a diffusion map of the autocorrelation of the detailed and approximation level coefficients obtained using the DWT. M. Mohanty et al. [76] extracted thirteen features including filter leakage measure, spectral analysis, covariance measure, skewness, kurtosis, standard exponential algorithm and modified exponential algorithm.

Classification of the heart sound signals is generally performed using machine learning techniques including the classical and recently developed deep learning techniques. A comprehensive review of the proposed classification technique for the heart sound signal can be obtained in [2,5]. Deep learning based techniques are very handy to classify the heart sound signal due to its ability to extract the relevant features automatically [69]. In [77,78], a recurrent neural network technique, long-short term memory (LSTM), have been proposed to classify the PCG signal as normal or abnormal. Recently, S.L. Oh et al. [79] used the wavelet based deep learning method called deep wavenet to classify the signals into five categories. However, considering the limited number of samples per heart valve diseases, application of deep learning methods for the PCG signal is very limited. Mostly, neural network based algorithms have been used to classify the FHS to segment the heart sound signal instead of the classification of the signal [53,56,80]. On the other hand, classical machine learning techniques have been used extensively including SVM [62,68,75,76], KNN [68], Decision tree [81], random forest [70], and discriminant analysis [82]. Among these techniques, the SVM is the most popular technique to classify the heart sound signal due to its capability to classify the data by transforming the feature domain to a nonlinear sparse domain with the help of optimum nonlinear kernel function. Among the various nonlinear kernel functions, the most popular kernel functions for the heart sound signals are polynomial, Gaussian, radial basis function, and Sigmoid. Moreover, the SVM technique produces satisfactory results for high dimensional classification problem even in the case of a small dataset as in the case of heart signals. However, SVM is capable to classify the signal into two classes at a time, for a multiclass classification separate SVM models based on one-vs-all or one-vs-one classification approach is used.

Another popular approach for the heart sound signal classification is KNN due to its simplicity, ease of implementation and low training time compared to other complex machine learning techniques [68,83]. KNN classify the signal by measuring the distance between the features of the testing sample to the features of the training samples for which the label of the class is known. The performance of the KNN significantly depends on the k parameter and the selection of function to measure the distance. A suitable distance function among Euclidean, Manhattan, Minkowski, and Mahalanobis can be selected based on the nature of the feature. KNN and SVM also have been used in combination as a part of ensemble classification method [84]. S.K. Ghosh et al. in [66] performed the classification using a multiclass composite classifier combined with the sparse representation of the test sample and the distance metric from the nearest neighbor training samples.

In addition to uni-model classification techniques, the ensemble of multiple classification techniques have been proposed for the heart sound signal, where multiple classifiers participate to predict the class of the test sample. In this direction, M. Zabihi et al. in [63], trained 20 feed-forward neural networks according to the bootstrap approach and then predicted the final result based on majority voting of all the classifiers. While C. Potes et al. in [85] ensemble two classifiers including Adaboost and convolutional neural network (CNN) and the final prediction is obtained using a decision rule based on threshold values.

7.3 Methods and materials

7.3.1 Theoretical background about TQWT

In the TQWT developed by I.W. Selesnivk [17], the Q-factor of the wavelet can be tuned according to the oscillating characteristics of the signal. The following subsections briefly describes the decomposition and reconstruction steps of the TQWT, as described in [17].

Decomposition of the signal

Fig. 7.2 shows the decomposition steps using TQWT along with the DWT to provide a clear difference between these two methods. The two steps, filtering and scaling of the coefficients are the same in TQWT and DWT. In both the methods, the signal is decomposed by applying the analysis filters, one high-pass and one low pass, and then the decimation of the obtained coefficients. From the figure, it can be observed that the difference between these two techniques is that: in WT, the bandwidth of a particular level is typical half of the bandwidth of the previous level. For example, the bandwidth of the detailed level-2 coefficients is typically $\pi/4$ to $\pi/2$ as compared to the bandwidth $\pi/2$ to π of the detailed coefficients at level-1. On the other hand, in TQWT, the bandwidth ($(1-\beta)\alpha^j\pi$ to $\alpha^j\pi$) of particular level (j) can be tuned by tuning the alpha and beta parameters. The construction of analysis filters and scaling of the coefficient can be performed by the following steps.

STEP 1: First the scaling parameters, α and β are initialized as follows:

$$\beta = \frac{2}{Q+1}; \qquad \alpha = 1 - \frac{\beta}{r} \tag{7.1}$$

where r is the redundancy factor and Q is the quality factor. Although redundancy makes the transform nearly shift-invariant, the values of α and β should be less than one to avoid high redundancy, and the summation of α and β should be greater than one to achieve the perfect reconstruction.

STEP 2: Next step is to obtain the unitary DFT of the signal $x(n)$ with length N as the following equation [17]:

$$X(k) = \frac{1}{\sqrt{N}} \sum_{n=0}^{N-1} x(n) \exp\left(-j\frac{2\pi}{N}nk\right) \tag{7.2}$$

for $0 \leq k \leq N-1$.

STEP 3: In the next step, to construct the analysis filters, three frequency bands: Pass-band (P), Stop-band (S), and Transition-band (T) are obtained. These bands for the low-pass filter can be obtained as follows:

1. Pass band (low-pass filter): 0 to N_1
2. Stop band (low-pass filter): N_0 to N
3. Transition band : N_0 to N_1

where N_0 and N_1 are obtained as follows:

$$N_0 = \alpha \times N; \qquad N_1 = \beta \times N \tag{7.3}$$

Similarly, these three bands can be obtained for the high-pass filter where the pass-band and stop-band will get exchange, as shown in Fig. 7.1. For the transition band (N_0 to N_1), a 2π-periodic power-complementary function can be used in such a way that the summation of the output responses of both the filters becomes one [17].

STEP 4: In the last step, the output response of the high pass and low pass filters are decimated with a factor α and β, respectively, as shown in Fig. 7.2.

When a signal with sampling frequency F_s will be decomposed using these steps, the center frequency (F_c) of a particular level j can be obtained as:

$$F_c = \frac{1}{4}\alpha^{j-1}(2-\beta)F_s \tag{7.4}$$

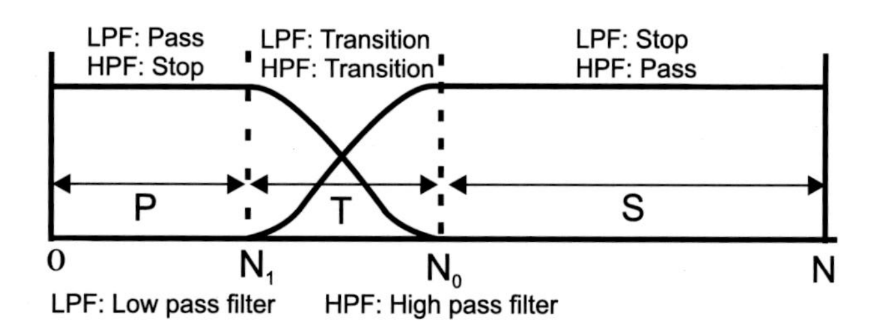

FIGURE 7.1

Construction of the analysis filters for the TQWT decomposition.

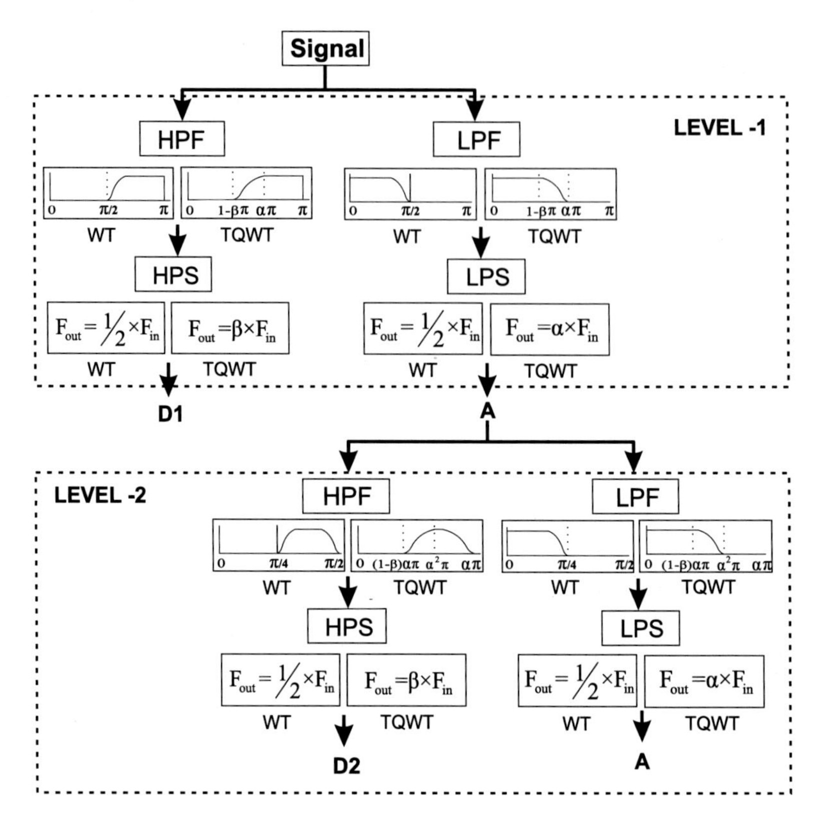

FIGURE 7.2

Decomposition of the signal using WT and TQWT.

The level j will cover the bandwidth (BW) as:

$$BW = \frac{1}{4}\alpha^{j-1}\beta F_s \tag{7.5}$$

Reconstruction using TQWT

The reconstruction of the signal can be obtained by applying the scaling and synthesis filters to the detailed and approximation level coefficients as shown in Fig. 7.3. First, the approximation coefficients and the last level detailed coefficients are rescaled by the factor $1/\alpha$ and $1/\beta$, respectively [17]. Then the synthesis filters are applied to the rescaled coefficients. The obtained responses of the two filters are added to obtain the approximation coefficient of the next level. Thus, applying these two steps successively till level-1, the signal can be reconstructed.

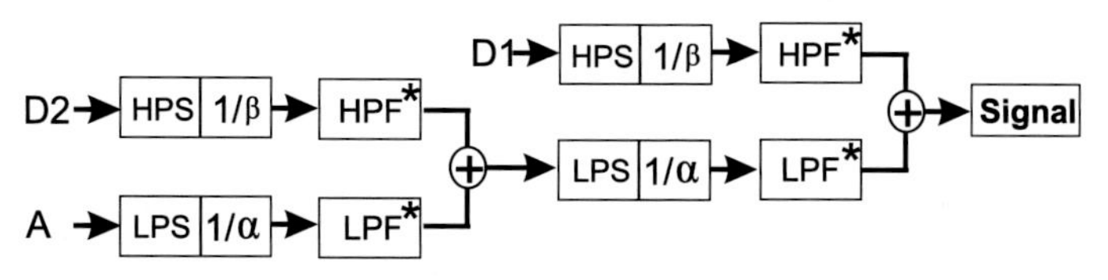

FIGURE 7.3

Reconstruction steps using TQWT.

A detailed description of the TQWT based decomposition and reconstruction steps can be obtained in [17] and the Matlab® (Mathworks, USA) implementation of the TQWT can be obtained in [86].

7.3.2 HSMM based heart sound signal segmentation

The segmentation of the heart sound signal in various duration by localizing the FHS is a very important step in the analysis of the signal. In normal and noise-free signal, the localization of S1 and S2 is quite easy since these two components prominently present in the signal. However, the presence of the noise and murmur makes the segmentation task challenging. As mentioned in the literature survey section, a lot of approaches to segment the heart sound signal have been presented. Recently, a HMM based algorithm is proposed, which is currently a state-of-the-art method for this purpose. The major contribution of the work in [20], is the inclusion of duration dependency and calculation of logistic regression (LR)-based emission probabilities to enhance the HMM. Moreover, use of an extended Viterbi algorithm has shown a significant improvement in the reported results. Following is a brief description of the working principle of the algorithm.

HMM makes inferences about the likelihood of being in a certain "hidden state", based on an observation generated by the particular state. It is governed by the Markov property, according to which the next state at time $t+1$ depends only on the current state at time t [87]. The HSMM proposed in [20] is described as follows:

$$\lambda = (A, B, \pi, p) \tag{7.6}$$

where,

$A = \{a_{ij}\}$ is the transmission matrix that defines the probability of moving from state S_i to state S_j.

$B = \{b_j(O_t)\}, 1 \leq j \leq N$ is the emission probability that describes the probability of an observation O_i by state S_i.

π is the initial state distribution.

$p = \{p_i(d)\}$ is the probability of sojourn time in state S_i for duration d. It is introduced by D.B. Springers et al. to incorporate the information about the expected duration of heart sound components.

The likelihood of the most probable sequence $\delta_t(j)$, is calculated using a modified version of the Viterbi algorithm as follows [20]

$$\delta_t(j) = \max_d \left[\max_{i \neq j} \left[\delta_{t-d}(i) \cdot a_{ij} \right] \cdot p_j(d) \cdot \prod_{s=0}^{d-1} b_j(O_{t-s}) \right] \tag{7.7}$$

Here, the observation density $\prod_{s=0}^{d-1} b_j(O_{t-s})$ describes the probability of observing all the observations from time $t-d$ to time t in state S_j. As an observation for a component, authors extracted the feature including the Homomorphic envelope, Hilbert envelope, wavelet envelope, and power spectral density envelope.

A detailed description of the algorithm can be obtained in [20] and the implementation using Matlab (Mathworks, USA) software can be obtained in [88].

7.3.3 Machine learning based classification algorithms

SVM, KNN, and ensemble methods are the popular machine learning methods for the classification of heart sound signal. These methods are described briefly as follows.

7.3.3.1 Support vector machine (SVM)

Since its development by Vapnik et al. in 1995 [89], SVM method has been used extensively for linear and nonlinear classification. The basic principle of the SVM is based on the representation of feature vector in high-dimension feature space using kernel functions and then an optimal hyperplane is obtained to separate the different classes. The hyperplane which is also known as the maximal margin hyperplane plays a decision boundary to maximize the limit of separation between two classes. Although, SVM is build to classify the input feature vector $\{x_i\}_{i=1}^N$ into two classes $y_i\,[-1, 1]$, it can be applied on a multiclass classification application using either one-vs-one approach or one-vs-all approach. In the one-vs-one approach, the separate SVM models are trained for every combination of the class and then a majority voting is used to classify the input as one of the class. In this approach, n(n-1)/2 SVM models will be required. While, in the one-vs-all approach, SVM models equal to the number of classes are trained to classify the signal as belongs to a particular class or not. A two-class SVM can be model as follows.

For an input feature vector $\{x_i\}_{i=1}^N$ and output classes $y_i\,[-1, 1]$, the hyperplane separating the input data in the state space can be expressed by

$$\omega^T x + b = 0 \tag{7.8}$$

where ω represents a weight vector and b is a bias. The output of SVM is constructed in such a way that the optimal hyperplane satisfies the following constraint.

$$y_i \left[\omega^T \psi(x_i) + b \right] \geq 1 - \xi_i, \quad i = 1, 2, 3...., N \tag{7.9}$$

where $\psi(.)$ represents a kernel function that maps the input feature vector to a high-dimensional space, ξ_i is a nonnegative slack variable that is used to minimize the training error. The classification boundary is obtained in such a way that the margin among all the hyperplanes gets maximized. Thus, the objective function of the SVM classifier is to minimize the following equation:

$$\min_{\omega,b,\xi} \Phi(\omega, b, \xi) = \frac{1}{2}\omega^T \omega + \frac{1}{2}\gamma \sum_{i=1}^{N} \xi_i^2 \tag{7.10}$$

$$\text{subject to } y_i \left[\omega^T \psi(x_i) + b\right] = 1 - \xi_i,\ i = 1, 2..., N \tag{7.11}$$

where γ is a penalty parameter that controls the trade-off of training errors. Thus, training an SVM model corresponds to solving a quadratic optimization problem to identify a hyperplane that minimizes the margins between the classes.

Kernel trick is generally used to solve a nonlinear classification problem with a linear classifier. These kernel functions transform the input feature vector into a high-dimensional space. In the literature, some widely used Kernel functions are linear, quadratic polynomial, cubic polynomial, and Gaussian functions which can be realized as follows:

Linear kernel:

$$\psi(x_1, x_2) = x_1^T x_2$$

Polynomial kernel:

$$\psi(x_1, x_2) = \left(x_1{}^T x_2 + 1\right)^p$$

where p represents the degree of polynomial such as quadratic polynomial ($p = 2$) and cubic polynomial ($p = 3$).

Gaussian kernel:

$$\psi(x_1, x_2) = \exp\left(-\frac{\|x_1 - x_2\|^2}{2\sigma^2}\right)$$

where σ represents the width of the kernel.

7.3.3.2 K-nearest neighbor (KNN)

KNN is a nonparametric supervised learning method that measures the distance between the input feature vector and labeled training samples and then classify the input vector based on the class label of k-nearest neighbors. For example, if $k = 1$, the input vector will be predicted to the closest neighbor based on distance measurement between the feature vector of unlabeled input and labeled training samples. For $k \geq 1$, the method measures the distance between the input feature vector and the training sample's feature vectors and then predicts the label for the input vector based on the majority voting among the k nearest neighbors. Following are the popular metrics to measure the distance:

Euclidean distance:

$$Dist_{Euclidean}(X, Y) = \sqrt{\sum_{i=1}^{n} (x_i - y_i)^2} \tag{7.12}$$

Cosine distance:

$$Dist_{cosine}(X, Y) = 1 - \frac{\sum_{i=1}^{n} x_i y_i}{\sqrt{\sum_{i=1}^{n} x_i^2}\sqrt{\sum_{i=1}^{n} y_i^2}} \tag{7.13}$$

where, X and Y are training and testing sample feature vectors of length n, respectively.

7.3.3.3 Ensemble of multiple classifiers

The ensemble method is a technique where multiple classifiers are used to solve the same problem and then combined their results to obtain improved results. Following are the popular approaches to perform the ensemble of multiple classifier methods.

Bagging: The name bagging is the abbreviation used for Bootstrap Aggregating which summarizes the key elements of this strategy [90]. The first step, bootstrap sampling involves creating multiple datasets by randomly drawing subsamples from the original dataset. These multiple datasets are created by different samples in a dataset, although the same sample may be present multiple times in multiple datasets or may be absent at all and thus allowing the replacement of samples. Each generated dataset is used to train a separate model. Since the datasets are independent of each other, separate models can run in parallel which makes this implementation a faster approach. The second step is aggregating the result of all the models trained using different datasets generated in the first step. The final result of the classification is obtained by majority voting among the results obtained from each model.

Boosting: In contrast to Bagging, Boosting improves the accuracy of the model by iteratively adjusting the weight of the sample. Thus, it is an incremental approach where the feedback of the samples which are misclassified are provided to the same model in the next iteration. In each iteration, the model is trained with the same dataset however, the weight of samples are adjusted according to the prediction in the previous iteration. The sample which is misclassified in the previous iteration will be given higher weights in the next iteration. In this way, the model is forced to learn specifically for the misclassified samples. The final model (strong model) is constructed by calculating a weighted mean of all the models (weak models) generated in all the iterations. Thus, a strong learning model is constructed using several weak-learning models. This approach is specifically effective in the application where few samples are hard to classify because the weak models are trained for specific samples in each iteration. Thus, the weak models may not perform satisfactorily on the complete dataset, they perform well for a subset of the dataset and hence improves the classification performance of the overall strong model. Among the boosting methods, Adaboost is the most popular method which uses different 'adjusting dataset' process and also a different process of combining all the weak models. To obtain the final classification result, instead of equal weights to each model, a weighted voting function is used. The boosting method has been reported with better accuracy compared to the bagging method, however, weak models are trained specifically for a subset of the dataset, it is more likely that the models get overfit over the training dataset.

Random subspace: This is an extension to the bagging approach where the multiple datasets are generated in the same fashion as in the case of bagging, however, instead of the samples, the features are randomly sampled with the replacement [91]. Therefore, this method is also called feature bagging. The random selection of features helps to reduce the correlation between the different models since each of them is trained on random samples of features instead of the entire feature set. Consequentially, it avoids overfitting of the models focusing on certain highly predictive features, which may not be good predictors for the unseen samples in the testing set. Due to this reason, the random subspaces method is an effective approach in the application having a large number of features compared to the number of samples in the training set.

7.4 Proposed methodology

The proposed method performs the classification of heart sound signal into five categories AS, MS, MR, MVP, and N as described in Fig. 7.4. The method performs the five tasks, preprocessing of the PCG signal, TQWT based denoising, segmentation using the Springer's algorithm, feature extraction, and classification of the signal, as described in the following subsections.

7.4.1 Data preprocessing

The dataset used in this study consists of five classes of heart sound signal with the sampling frequency of each signal is 8 kHz. Following are the data preprocessing steps applied on each signal:

Re-sampling: Since the frequency range of the FHS and various pathological sound lie below 500 Hz [5], the signal is downsampled from 8 kHz to 1 kHz sampling frequency.

Out-of-band noise suppression: Each file is filtered using the low-pass and high-pass Butterworth filter of order having the cut-off frequency of 20 Hz and 400 Hz respectively to suppress the out-of-band noise.

Normalization: To suppress the amplitude variation due to interclass variation on the amplitude of the heart signal, the filtered signal is normalized as follows:

$$x_{norm}(n) = \frac{x\,(n)}{\max\,(|x|)} \tag{7.14}$$

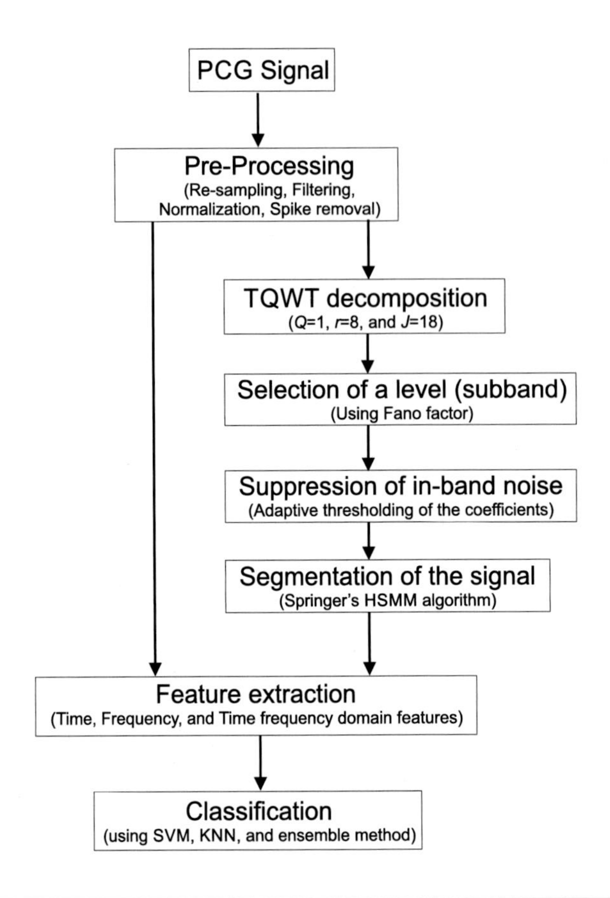

FIGURE 7.4

Proposed methodology.

Spike removal: A spike removal algorithm proposed by Schmidt et al. [58] is applied to each signal to remove the spike from the signal. These spikes are caused probably due to the movement of the stethoscope while recording the signal.

7.4.2 TQWT based denoising algorithm

As mentioned in the introduction section, the presence of noise and murmur makes the segmentation task challenging. Therefore, to suppress the in-band noise, a TQWT based adaptive thresholding method is proposed and described as follows.

7.4.2.1 Decomposition of the signal using TQWT

According to Eqs. (7.4) and (7.5), the center frequency (F_c) and bandwidth (BW) of a particular level depends on the value of the scaling parameters α and β and consequentially on the value of the quality factor Q and the redundancy factor r.

Therefore, the values of Q and r should be selected in such a way that the FHS and murmur sounds lie in different decomposed levels and hence a particular level with distinct FHS can be obtained. As observed in [33], the value of Q close to unity helps in separation of FHS and murmur, the proposed method set $Q = 1$. While the value of r is set equal to 8 to control the ringing effect and also to control the number of decomposition levels. Consequentially, according to Eq. (7.1), the value of α will be 0.875 and the value of β will be 1.

In the proposed method, the heart sound signal is decomposed using the TQWT decomposition up to eighteen ($J = 18$) levels. Thus, 18 detailed level and one approximation level coefficient vectors are obtained. Following the Eqs. (7.4) and (7.5), the bandwidth of the eighteenth detailed level will be from 12.91 Hz to 38.74 Hz and the bandwidth of the approximation level will be lower than 25.82 Hz. In the lower frequency range, the real-life noises dominate relatively more compared to the FHS and therefore approximation level is discarded from further consideration in the level selection process.

The next task after the decomposition of the signal is to select a level (subband) which have distinct FHS.

7.4.2.2 Selection of a level with emphasized FHS

Among all the decomposed levels obtained using TQWT decomposition, a particular level having the emphasized FHS has to be selected. For this purpose, the Fano factor is used as proposed in [18] and briefly described as follows.

Fano factor [92], a variance-to-mean ratio, is generally used for quantitative comparison of the observed occurrences to a standard statistical model in terms of clustering or dispersion. The Fano factor (F) is obtained as follows [92].

$$F = \frac{\sigma_w^2}{\mu_w} \tag{7.15}$$

where σ_w^2 and μ_w are the variance and the mean of a signal with finite window length w. The level where the noise or murmur is dominant, the absolute mean of the level will be larger than the clean signal. While in the case of a sharp peak murmur where the variance of the level will be high, the mean value will increase relatively more compared to the rate of increase of the variance. Thus, the Fano factor will be able to select a level with clean and emphasized FHS. Fig. 7.5 describing the behavior of the denominator and nominator term of the Fano factor with respect to the signal-to-noise ratio (SNR) of the signal. From the figure, it can be observed that while the numerator term (σ^2) decreases gradually, the denominator term (μ) increases exponentially as the SNR of the signal decreases. Thus, the Fano factor can also act as a measure of the SNR and hence helpful to select a level with a clean signal. Therefore, the Fano factor of the absolute values for each of the detailed level is obtained and the level which results in the maximum value of the Fano factor is selected.

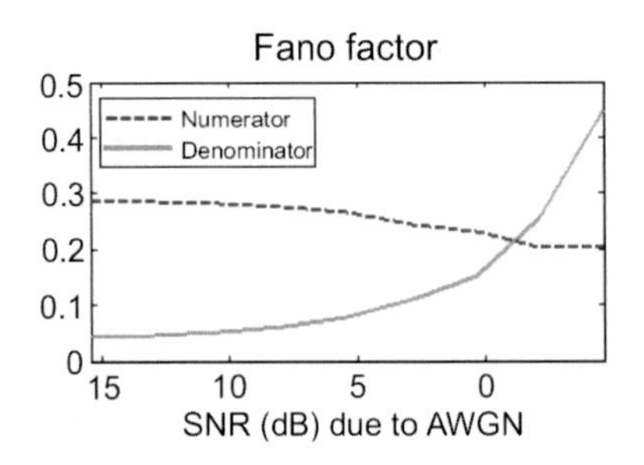

FIGURE 7.5

Characteristics of the numerator and denominator terms of Fano factor with the SNR of the signal.

7.4.2.3 Suppression of in-band noise components

The obtained signal from the previous step generally contains the in-band noise which may affect the performance of the segmentation algorithm. Therefore, a thresholding method proposed in the previous work [19] is used to suppress these components. The method adaptively estimates a threshold value based on three statistical parameters including mean, variance and a new parameter Med_{75}. The Med_{75} is calculated as 75th percentile value in the ascending sorted absolute values of the selected level. Instead of 50th percentile as measured for the median, 75th percentile is calculated to incorporate the domain knowledge that the ratio between the total length of FHS and length of a cardiac cycle is 1:3 [93]. Considering the behavior of all these three parameters with respect to the SNR of the signal, the threshold value is estimated as follows [19]:

$$T = \begin{cases} med_{75} \times [1 - (v - med_{75})] & \text{if } (med_{75} < v) \quad \textbf{Case:1} \\ med_{75} & \text{if } ((med_{75} > v) \\ & \&\& \ (med_{75} < m)) \quad \textbf{Case:2} \\ med_{75} + (med_{75} - m) & \text{if } (med_{75} > m) \quad \textbf{Case:3} \end{cases} \tag{7.16}$$

All three statistical parameters are calculated on the absolute values of the signal. In this equation, the first case is representing the low level of noise, the second case representing a moderate level of noise, and third case is representing a high level of noise. The estimated threshold value is applied to the signal using the soft threshold function [94] as follows.

$$x_s^T[n] = \begin{cases} sign(x[n])\,(|x[n]| - T) & \text{if } |x[n]| > T \\ 0 & \text{otherwise} \end{cases} \tag{7.17}$$

7.4.3 Segmentation using Springer's HSMM algorithm

Segmentation of the heart sound signal in the four periods namely, S1, S2, systole, and diastole helps to extract the diagnostically important features since the time and frequency characteristics of these components provide ample features to analyze the health of the heart [5]. Therefore, the obtained denoised signal from the previous step is used to segment the heart sound signal. Segmentation is performed using the HSMM based method proposed in [20] and briefly described in section 7.3. A Matlab (Mathworks, USA) software implementation of the algorithm is available at [88]. The result of this algorithm marks the onset of each component sequentially and stores in the array A, where

1 represents the onset of S1
2 represents the onset of systole period
3 represents the onset of S2
4 represents the onset of diastole.

7.4.4 Feature extraction

From the segmented components, features from time, frequency and time-frequency domain are extracted from the preprocessed signal. Here the preprocessed signal (signal before denoising step) is used instead of the denoised signal because, during the denoising, a lot of out-of-band as well as in-band components were suppressed to emphasize the FHS. However, these components have diagnostically relevant features in addition to noise. Various features extracted in these three domains are described as follows.

7.4.4.1 Time domain features

In the time domain, features related to the time and amplitude of the preprocessed signal are extracted. Table 7.1 describes the various time-domain features used in the study. The mean duration of all four components is considered while the standard deviation of duration is not considered due to the small length of the individual sample in the dataset ($\approx$3 seconds). Although these features are helpful to diagnose the arrhythmic abnormal cases when the samples with adequate length are available [95]. Further, the study of the relative energy of different components is also considered since the energy of S1, S2, systole, and diastole vary according to the pathological class [2]. Kurtosis reflects the distribution of signal whether Gaussian, SubGaussian or SuperGaussian. It measures whether the data are heavy-tailed or light-tailed relative to a normal distribution [96] and therefore, considered as a feature. Fano factor is a variance-to-mean ratio as discussed in section 7.4 and hence can be viewed as a SNR because it will be larger for the signal with noise and lower for a clean signal. SampEn of each component is also calculated to measure the complexity of the signal which is similar to the approximate entropy (ApEn) [97]. However, SampEn does not count self-matching and relatively invariant to the size of the dataset as compared to ApEn. Therefore, SampEn is more suitable for applications with relatively short data size.

Table 7.1 Time domain features: Onset of each component obtained by HSMM segmentation method and stored in a matrix A is used.

Features Sr No	Feature calculation	Feature description
1	$RR = S1^{i+1}_{onset} - S1^{i}_{onset}$	Length of one cardiac cycle, duration between consecutive S1.
2	$S1_{len} = \frac{1}{n}\sum_{i=1}^{n}(S1^{i}_{len})$	Mean length of S1 component of n cycles.
3	$S2_{len} = \frac{1}{n}\sum_{i=1}^{n}(S2^{i}_{len})$	Mean length of S2 component of n cycles.
4	$Sys_{len} = \frac{1}{n}\sum_{i=1}^{n}(Sys^{i}_{len})$	Mean length of systole component of n cycles.
5	$DiaS_{len} = \frac{1}{n}\sum_{i=1}^{n}(Dias^{i}_{len})$	Mean length of diastole component of n cycles.
6	$E_{S1} = \frac{1}{length(S1)}\sum_{i=1}^{length(S1)} \lvert x_{S1}(i)\rvert^2$	Normalized energy of S1
7	$E_{S2} = \frac{1}{length(S2)}\sum_{i=1}^{length(S2)} \lvert x_{S2}(i)\rvert^2$	Normalized energy of S2
8	$E_{sys} = \frac{1}{length(sys)}\sum_{i=1}^{length(sys)} \lvert x_{sys}(i)\rvert^2$	Normalized energy of systole
9	$E_{dias} = \frac{1}{length(dias)}\sum_{i=1}^{length(dias)} \lvert x_{dias}(i)\rvert^2$	Normalized energy of diastole
10	$E_{FHS:HM} = \frac{E_{S1}+E_{S2}}{E_{sys}+E_{dias}}$	Energy ratio of FHS to systole+diastole
11	$E_{S1:sys} = \frac{E_{S1}}{E_{sys}}$	Energy ratio of S1 duration to systole
12	$E_{S2:dias} = \frac{E_{S2}}{E_{dias}}$	Energy ratio of S2 duration to diastole
13	$E_{sys:dias} = \frac{E_{sys}}{E_{dias}}$	Ratio of energy of systole to diastole
14	$L_{sys:RR} = \frac{length(systole)}{length(cardiac\_cycle)}$	Ratio of duration of systole to heart cycle
15	$L_{dias:RR} = \frac{length(diastole)}{length(cardiac\_cycle)}$	Ratio of duration of diastole to heart cycle
16	$L_{sys:dias} = \frac{length(systole)}{length(diastole)}$	Ratio of duration of systole to diastole
17–20	$K_i = \frac{1/N\sum(x(n)-mean(x))^4}{\left(\sum(x(n)-mean(x))^2/N\right)^2}$	Kurtosis of ith component, i = S1, S2, systole and diastole.
21–24	$F_i = \frac{\sigma_w^2}{\mu_w}$	Fano factor of ith component, i = S1, S2, systole and diastole.
25–26	$ZCR_i = \frac{1}{w_l}\sum_{n=2}^{w_l}\lvert sign[x(n)] - sign[x(n-1)]\rvert$	Zero-crossing rate for i = systole and diastole period, higher ZCR is expected in case of pathological cases.
27	$SampEn_i = -\log\left(\left(\sum A_i\right)/\sum B_i\right) = -\log(A/B)$	Sample entropy of ith component, i = S1, S2, systole and diastole. Here, A_i is number of matches of length $m+1$ with $i'h$ template and B_i is number of matches of length m with $i'h$ template.

7.4.4.2 Frequency domain features

The frequency-domain analysis helps to identify the dominant frequency contents in the heart sound signal. For this, the signal is first converted from time-domain to frequency domain using the DFT technique as follows:

$$X(R) = \sum_{n=0}^{N-1} x(n) \cdot e^{-j2\pi Rn/N}, \quad R \in Z(integers) \tag{7.18}$$

where $X(R)$ represents the DFT coefficients at frequency R for the signal $x(n)$ with length. Then, the spectrum was segmented into various frequency bands as follows:

band-1: 10–110 Hz
band-2: 110–210 Hz
band-3: 210–270 HZ
band-4: 270–500 Hz

These frequency bands were selected empirically and considering the frequency band of FHS and murmur sounds [5]. From each band, the following features were extracted: area under curve, dominant frequency and kurtosis of the power spectrum. Thus, 12 features are extracted from the frequency domain.

7.4.4.3 Time-frequency domain features

As discussed in section 7.4, TQWT is helpful to separate the FHS with murmurs. Therefore, the preprocessed signal is decomposed signal using the TQWT method with the same values as used for the denoising purpose ($\alpha = 0.875$, $\beta = 1$, and the number of decomposition level ($J = 18$)). Thus, a time-frequency domain matrix X is obtained where rows are indicating the 18 detailed levels and one approximation level coefficient vectors. From each of the detailed level coefficient vector, excluding the approximation level, the following features are extracted:

For each detailed level coefficient vector, the energy of systole and diastole periods is calculated as:

$$E_{sys}(level) = \frac{1}{length(sys)} \sum_{i=1}^{length(sys)} \left|X_{sys}(level, i)\right|^2 \tag{7.19}$$

and

$$E_{dias}(level) = \frac{1}{length(dias)} \sum_{i=1}^{length(dias)} \left|X_{dias}(level, i)\right|^2 \tag{7.20}$$

To identify the systole and diastole period, the marking provided by the segmentation algorithm in matrix A is used.

The entropy of each detailed level is calculated as follows:

$$S(level) = -\int X(level, n) \log(X(level, n))\, dn \tag{7.21}$$

where the $X(level, n)$ is the probability distribution of the energy at a particular detailed level ($level$). In addition to these features, for each detailed level, the kurtosis and Fano factor of systole and diastole periods are also calculated. Thus, 36 energy features, 18 entropy features, 36 kurtosis values and 36 Fano factors are calculated and a total of 126 features from the time-frequency domain matrix are extracted.

Thus, the proposed method uses total of 165 features extracted from time-domain (27), frequency-domain (12), and time-frequency domain (126).

7.4.5 Classification of heart sound signal

Automatic classification of the heart sound signal is a challenging task due to the intersubject variation of the signature of the disease. In this work, the following three machine learning methods with the extracted features have been implemented.

SVM As discussed in the literature survey section, SVM is the most popular technique to classify the heart sound signal. It transforms the input feature vector into a high-dimensional space domain using kernel function where the best separating hyperplane is constructed. SVM with the linear kernel function, quadratic and cubic polynomial functions, and Gaussian function has been implemented. Since the SVM performs a binary classification, for the multiclass classification the one-vs-one technique is used according to which 10 ($n(n-1)/2$) SVM models are trained for the given five-class ($n = 5$) classification application. The value of slack variable ξ_i is set equal to 0.002.

KNN The KNN is implemented with various choices of value k and the distance metric. Following is the brief description of the four models implemented based on the KNN method:

1. *Fine KNN:* In this model, the value of k is set to 1 and Euclidean distance (Eq. (7.12)) is used to measure the distance between the training and testing feature vectors. Distance weight for each feature is set as equal.

2. *Medium KNN:* It is similar to the fine KNN, except the value of k is set equal to 10.

3. *Cosine KNN:* In the cosine KNN model, the value of k is set equal to 10 same as medium KNN, while the cosine function (Eq. (7.13)) as distance metric is used.

4. *Weighted KNN:* Weighted KNN is similar to medium KNN except for the weight assigned to the neighbors. In the case of equal weight, equal weight is assigned to each neighbor in majority voting irrespective of its distance to the testing sample. It may cause the wrong prediction if the nearest neighbors vary widely with respect to distance and the closest neighbors are more accurate indication of the class of the testing sample. On the other hand, in weighted KNN, more weight is given to the neighbors which are closer compared to the neighbors which are farther to the testing sample in terms of distance metric. Therefore, the squared inverse of the distance of each neighbor is considered to incorporate the distance information in the class prediction.

Ensemble methods As described in section 7.3, the ensemble method uses multiple machine learning methods and the final prediction of the class is performed based on the voting of the results obtained from each method. The following four ensemble methods have been implemented:

1. *Boosted Trees:* In this model, an aggregated bootstrapping using the Adaboost method is used on the decision tree as the basis model. The maximum number of splits is set equal to 20 and the number of learners (models) is set equal to 30. For each model, the learning rate γ is set equal to 0.1.

2 *Bagged Trees:* As the name is indicating, the bagging technique is used to ensemble the multiple decision trees. The number of maximum splits is set equal to 829 and the number of learner is set equal to 30.

3. *Subspace KNN:* In this method, instead of bagging or boosting, subsapce technique is used to implement the ensemble method as described in section 7.3. KNN is used as a base learning method with the euclidean distance metric. The number of subspace dimension is set equal to 78 and the number of learners is set equal to 30.

4. *Subspace Discriminant:* This is the same as subspace KNN except the base method is based on discriminant analysis [98] instead of KNN.

7.5 Results and discussion

For the experiment, a publicly available dataset of heart sound signal is used which contains 200 samples of five categories including the AS, MR, MS, MVP, and normal (N) [99]. Each sample is in the wav format and has 8 kHz sampling frequency. As mentioned in the data preprocessing section, each sample was downsampled to the 1 kHz sampling frequency. Moreover, signals with a very short duration with less than two cardiac cycles or less than 2 seconds were excluded in the study. The number of recordings in each category after the exclusion is as follows: AS (173), MR (156), MS (162), MVP (154), and N (185).

All the experiments are conducted using the Matlab (version R2020, MathWorks, USA) software on a desktop computer equipped with a Core-i5 64-bit processor and 16-GB RAM.

7.5.1 Performance evaluation metrics

For the quantitative analysis of the results, five performance evaluation metrics including sensitivity, specificity, precision, recall, F-score, and overall-accuracy (OAccuracy) are obtained as follows [100]:

$$\text{Sensitivity}_i(\%) = \frac{TP_i}{TP_i + FN_i} \times 100 \tag{7.22}$$

$$\text{Specificity}_i(\%) = \frac{TN_i}{FP_i + TN_i} \times 100 \tag{7.23}$$

$$\text{Precision}_i(\%) = \frac{TP_i}{TP_i + FP_i} \times 100 \tag{7.24}$$

$$\text{Recalll}_i(\%) = \frac{TP_i}{TP_i + FN_i} \times 100 \tag{7.25}$$

$$F - score_i(\%) = \frac{2 \times TP_i}{2 \times TP_i + FP_i + FN_i} \times 100 \tag{7.26}$$

$$OAccuracy(\%) = \frac{\sum_{i=1}^{5} C_{ii}}{\sum_{i=1}^{5} \sum_{i=1}^{5} C_{i,j}} \times 100 \tag{7.27}$$

where TP represent number of true positive, FN: false negative, FP: false positive, and TN: true negative. The C matrix represents the confusion matrix for the five class classification of heart signal as follows:

$$C = \begin{bmatrix} C_{11} & C_{12} & C_{13} & C_{14} & C_{15} \\ C_{21} & C_{22} & C_{23} & C_{24} & C_{25} \\ C_{31} & C_{32} & C_{33} & C_{34} & C_{35} \\ C_{41} & C_{42} & C_{43} & C_{44} & C_{45} \\ C_{51} & C_{52} & C_{53} & C_{54} & C_{55} \end{bmatrix}$$

where, C_{ij} represents the percentage of samples of original class i predicted as class j.

7.5.2 Results using the SVM method

SVM method is the most popular method for the classification of heart sound signals due to its ability to transform the features into a higher dimension and then classify the sample based on the support vectors. To analyze the impact of the kernel function, popular functions including linear, quadratic polynomial, cubic polynomial, and Gaussian function are tested. As mentioned in the proposed method, the one-vs-one method is used to perform the multiclass classification. The models are trained and tested on the dataset using 10-fold stratified cross-validation.

Fig. 7.6 shows the obtained results using the SVM technique with different kernel functions. It shows the result in the form of a confusion matrix for multiclass classification. From the figure, it can be observed that all four variants of SVM are able to classify the normal class with 100% accuracy, i.e. the classification of the signal as normal vs abnormal can be performed satisfactorily. The reason for this is the inclusion of time-frequency features based on TQWT which helps to separate FHS and murmurs and hence helps the SVM to discriminate the normal signal with abnormal one. For other classes, the true-positive rate (TPR) more than 97.4% and false-negative rate (FNR) lower than 2.6% is achieved using the quadratic kernel. While, SVM with other kernel functions is also producing satisfactory results. It in-

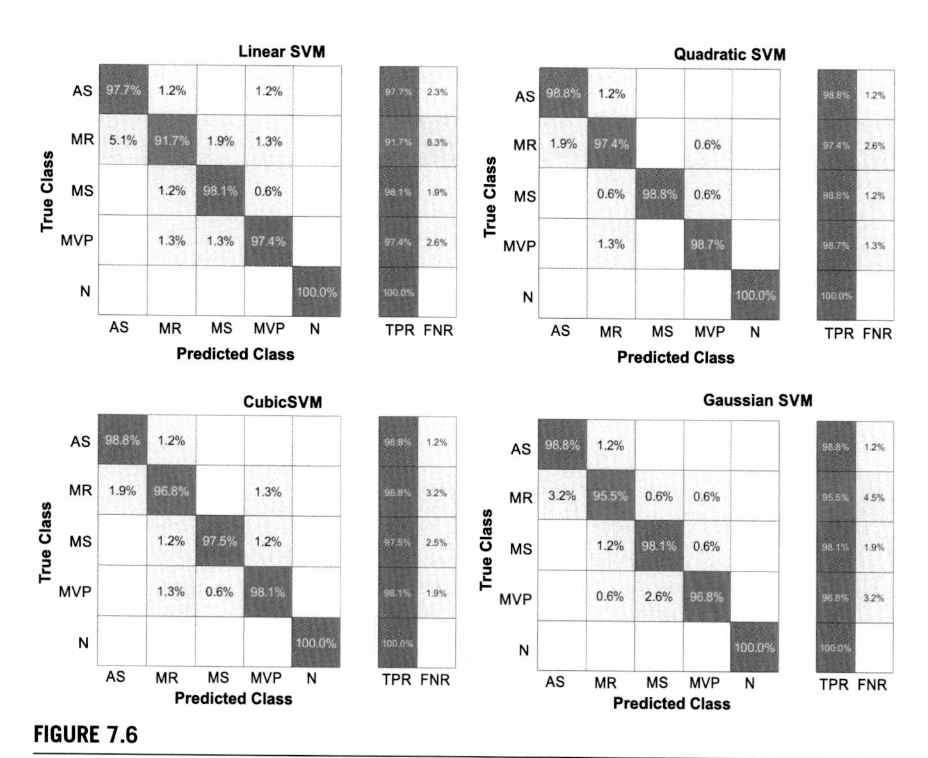

FIGURE 7.6

Obtained confusion matrices using linear SVM, quadratic SVM, cubic SVM, and Gaussian SVM.

dicates that the proposed feature set is robust to multiclass classification of the heart sound signal.

Table 7.2 is showing the various performance evaluation matrices for each class obtained using the SVM method with various kernel functions. Although best results are produced with the quadratic kernel, all the kernel functions are able to produce promising results. It shows the efficacy of the SVM method to classify the heart sound signal with the proposed feature set. The best accuracy of 98.88% is achieved using the quadratic kernel function. While the cubic kernel function produces 98.31%, Gaussian 97.95% and Linear produces the lowest among kernels 97.11%.

7.5.3 Results using KNN method

KNN method is a suitable method for an application where the size of the dataset is not large such as in the case of heart sound signal. To show the efficacy of the KNN method, four variants of it have been tested and the obtained results are shown in Fig. 7.7 in the form of a confusion matrix. From the figure, it can be observed that the fine KNN method is producing the best result among all other KNN methods. Fine KNN is producing TPR more than 96.2% and FNR lower than 3.8% for each of the

Table 7.2 Obtained performance evaluation metrics using the SVM based methods.

Kernel function	Diseases class	Sensitivity (%)	Specificity (%)	Precision (%)	Recall (%)	F-score (%)	Accuracy (%)
Linear	AS	0.9769	0.9878	0.9548	0.9769	0.9657	97.11
	MR	0.9167	0.9911	0.9597	0.9167	0.9377	
	MS	0.9815	0.9925	0.9695	0.9815	0.9755	
	MVP	0.9740	0.9926	0.9677	0.9740	0.9709	
	N	1.0000	1.0000	1.0000	1.0000	1.0000	
Quadratic	AS	0.9884	0.9954	0.9828	0.9884	0.9856	98.88
	MR	0.9744	0.9926	0.9682	0.9744	0.9712	
	MS	0.9877	1.0000	1.0000	0.9877	0.9938	
	MVP	0.9870	0.9970	0.9870	0.9870	0.9870	
	N	1.0000	1.0000	1.0000	1.0000	1.0000	
Cubic	AS	0.9884	0.9954	0.9828	0.9884	0.9856	98.31
	MR	0.9679	0.9911	0.9618	0.9679	0.9649	
	MS	0.9753	0.9985	0.9937	0.9753	0.9844	
	MVP	0.9805	0.9941	0.9742	0.9805	0.9773	
	N	1.0000	1.0000	1.0000	1.0000	1.0000	
Gaussian	AS	0.9884	0.9924	0.9716	0.9884	0.9799	97.95
	MR	0.9551	0.9926	0.9675	0.9551	0.9613	
	MS	0.9815	0.9925	0.9695	0.9815	0.9755	
	MVP	0.9675	0.9970	0.9868	0.9675	0.9770	
	N	1.0000	1.0000	1.0000	1.0000	1.0000	

class. While other variants are able to classify the signal as normal or abnormal with TPR higher than 97.8% and FNR lower than 2.2%, performance for other classes reduces significantly, specifically using the cosine KNN. It indicates that the euclidean function is suitable for the heart sound signal as a distance measurement metric.

Table 7.3 is showing the performance evaluation metrics for all the four variants of KNN used in the experiment. The best result of 97.71% accuracy is obtained using the Fine KNN, where the value of k is equal to 1. While other variants where k is equal to 10 are marginally lower compared to fine KNN. It shows that the inclusion of more number of data points in the majority voting reduces the accuracy of classification. The Fine KNN method is producing more than 96.15% sensitivity, 98.67% specificity, 94.41% precision, 96.15% recall, and 96.51% F-score for all classes and overall accuracy of 97.71%. The lowest accuracy of 92.89% is produced by the cosine KNN method.

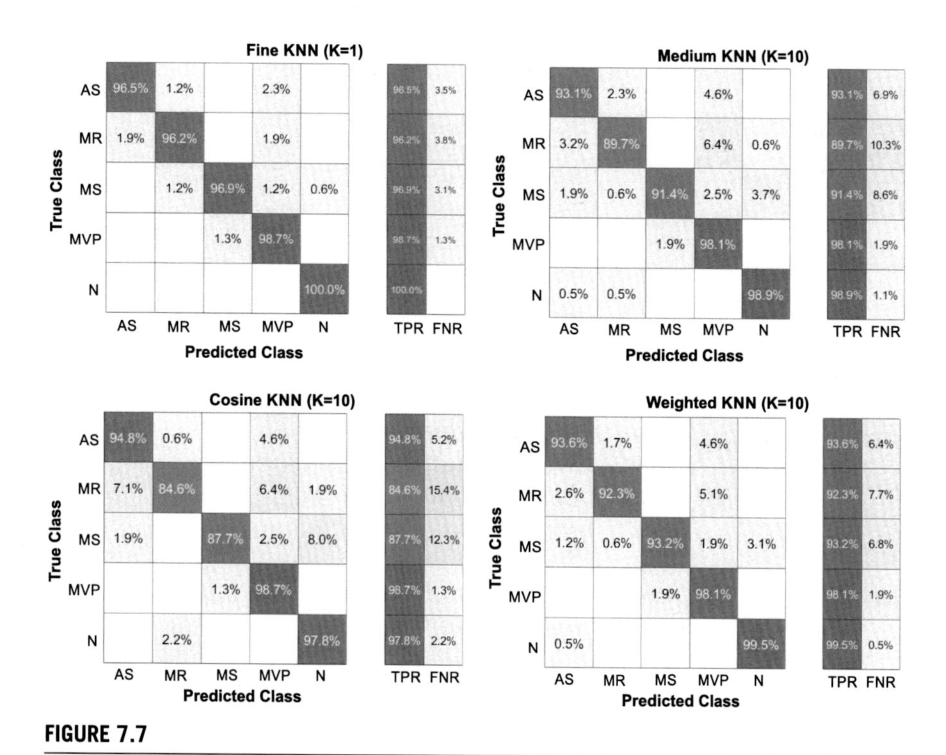

FIGURE 7.7

Obtained confusion matrices using fine KNN, medium KNN, cosine KNN, and weighted KNN.

7.5.4 Results using ensemble method

Fig. 7.8 is showing the results obtained using the four ensemble methods which are described in section 7.4. The confusion matrices in the figure are showing that the ensemble methods are producing a TPR higher than 92.3% and a FNR less than 7.7% for each class. The lowest TPR is observed for the MR class in most of the methods. The best results are obtained with the ensemble of subspace KNN method where TPR is higher than 98.7% and FNR is lower than 2.6% for each class. These results are showing that the ensemble of the multiple methods produces better results as compared to the individual method. Moreover, the subspace ensemble technique is producing better results compared to bagging and boosting approach.

Table 7.4 is showing the performance evaluation parameters obtained using all four ensemble methods. The overall accuracy obtained from the boosted tree, bagged tree and subspace discriminant is more than 96% which shows the adequacy of these methods to classify the heart sound signal. The best results are obtained using the ensemble of subspace KNN method with 99.04% accuracy. This method is producing more than 97.4% sensitivity, 99.54% specificity, 98.28% precision, 97.4% recall, and 98.04% F-score for all classes and overall accuracy of 99.04%. This result shows that

Table 7.3 Obtained performance evaluation metrics using the KNN based methods.

Kernel function	Diseases class	Sensitivity (%)	Specificity (%)	Precision (%)	Recall (%)	F-score (%)	Accuracy (%)
Fine KNN	AS	0.9653	0.9954	0.9824	0.9653	0.9738	97.71
	MR	0.9615	0.9941	0.9740	0.9615	0.9677	
	MS	0.9691	0.9970	0.9874	0.9691	0.9782	
	MVP	0.9870	0.9867	0.9441	0.9870	0.9651	
	N	1.0000	0.9984	0.9946	1.0000	0.9973	
Medium KNN	AS	0.9306	0.9863	0.9471	0.9306	0.9388	94.34
	MR	0.8974	0.9911	0.9589	0.8974	0.9272	
	MS	0.9136	0.9955	0.9801	0.9136	0.9457	
	MVP	0.9805	0.9675	0.8728	0.9805	0.9235	
	N	0.9892	0.9891	0.9632	0.9892	0.9760	
Cosine KNN	AS	0.9480	0.9787	0.9213	0.9480	0.9345	92.89
	MR	0.8462	0.9926	0.9635	0.8462	0.9010	
	MS	0.8765	0.9970	0.9861	0.8765	0.9281	
	MVP	0.9870	0.9675	0.8736	0.9870	0.9268	
	N	0.9784	0.9752	0.9188	0.9784	0.9476	
Weighted KNN	AS	0.9364	0.9893	0.9586	0.9364	0.9474	95.42
	MR	0.9231	0.9941	0.9730	0.9231	0.9474	
	MS	0.9321	0.9955	0.9805	0.9321	0.9557	
	MVP	0.9805	0.9719	0.8882	0.9805	0.9321	
	N	0.9946	0.9922	0.9735	0.9946	0.9840	

the ensemble of subspace KNN method can effectively classify the heart sound signal and can be used for the automatic analysis in telemonitoring systems.

7.5.5 Comparison of the proposed method with other methods

The proposed method is compared with the recently proposed methods for the classification of heart sound signal using the same dataset. Following are the methods.

Yaseen et al. 2018 [101]: the method extracts the MFCC and DWT based features and then analyzed the performance of the SVM, KNN, and deep neural network techniques.

S.K. Ghosh et al. 2019 [70]: the method calculates the magnitude and phase features from the time-frequency domain signal obtained using the WSST method and random forest for the classification.

S.L. Oh et al. 2020 [79]: the method uses a deep learning based deep wavenet method. Wavenet is a generative model which has been explored to create raw audio signals. It consists of a residual block with gated activation.

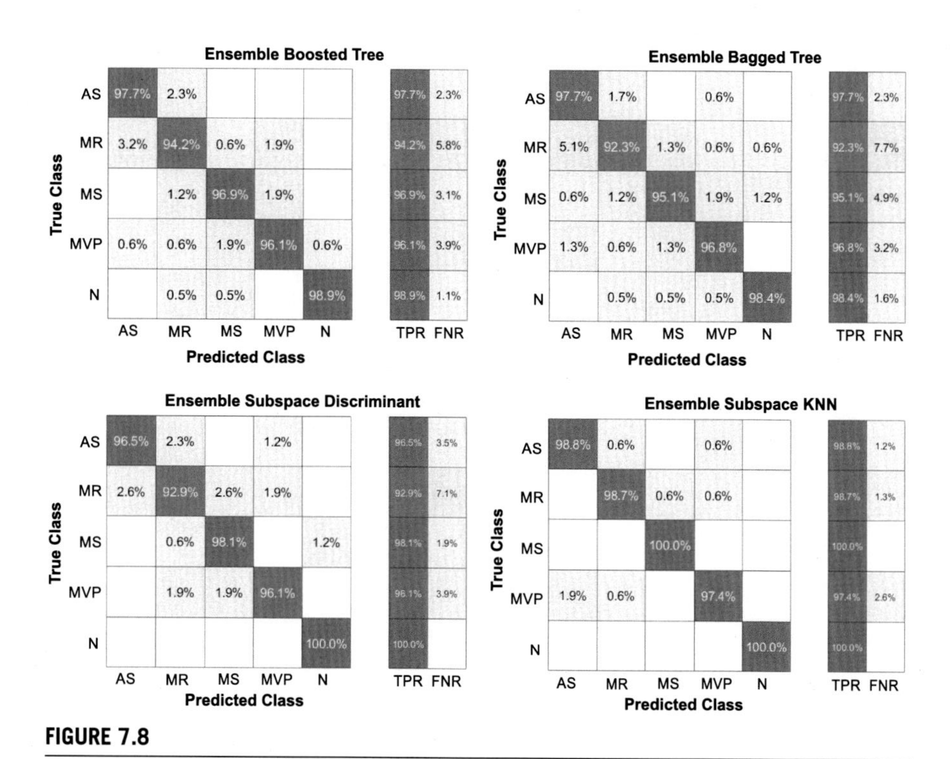

FIGURE 7.8

Obtained confusion matrices using ensemble boosted tree, bagged tree, subspace discriminant, and subspace KNN methods.

S.K. Ghosh et al. 2020 [66]: the method extracts the local energy and local entropy of a time-frequency domain matrix obtained using the chirplet transform. Then a multiclass composite classifier based on sparse representation of the features is used to classify the signal.

The obtained results using the proposed method and the state-of-the-art methods are provided in Table 7.5. The methods proposed in [70] and [66] are producing overall accuracy of 95.13% and 98.54%, respectively. However in the experiments, only four classes were considered. While the methods proposed in [101] and [79] considered all five classes and produced the overall accuracy of 97.6% and 97.0%, respectively. Compared to all these methods, the proposed method (ensemble subspace KNN) is producing an overall accuracy of 99.04%, which shows the superiority of the method compared to the state-of-art methods. The sensitivity produced by the proposed method is higher than 97.40% and specificity more than 99.54%. Moreover, it can be observed that the proposed method accurately classifying the normal category and abnormal category with 100% accuracy. Hence, the proposed method can be used for the telemonitoring systems where the first step is to detect any abnormality in the signal. In case of any abnormality detected, the signal can be transmitted to

Table 7.4 Obtained performance evaluation metrics using the ensemble based methods.

Kernel function	Diseases class	Sensitivity (%)	Specificity (%)	Precision (%)	Recall (%)	F-score (%)	Accuracy (%)
Boosted tree	AS	0.9769	0.9909	0.9657	0.9769	0.9713	96.87
	MR	0.9423	0.9881	0.9484	0.9423	0.9453	
	MS	0.9691	0.9925	0.9691	0.9691	0.9691	
	MVP	0.9610	0.9911	0.9610	0.9610	0.9610	
	N	0.9892	0.9984	0.9946	0.9892	0.9919	
Bagged tree	AS	0.9769	0.9833	0.9389	0.9769	0.9575	96.14
	MR	0.9231	0.9896	0.9536	0.9231	0.9381	
	MS	0.9506	0.9925	0.9686	0.9506	0.9595	
	MVP	0.9675	0.9911	0.9613	0.9675	0.9644	
	N	0.9838	0.9953	0.9838	0.9838	0.9838	
Subspace discriminant	AS	0.9653	0.9939	0.9766	0.9653	0.9709	96.87
	MR	0.9295	0.9881	0.9477	0.9295	0.9385	
	MS	0.9815	0.9895	0.9578	0.9815	0.9695	
	MVP	0.9610	0.9926	0.9673	0.9610	0.9642	
	N	1.0000	0.9969	0.9893	1.0000	0.9946	
Subspace KNN	AS	0.9884	0.9954	0.9828	0.9884	0.9856	99.04
	MR	0.9872	0.9970	0.9872	0.9872	0.9872	
	MS	1.0000	0.9985	0.9939	1.0000	0.9969	
	MVP	0.9740	0.9970	0.9868	0.9740	0.9804	
	N	1.0000	1.0000	1.0000	1.0000	1.0000	

medical fraternities and the patient can be suggested to visit the nearest health center for a detailed diagnosis by the medical experts in the field.

7.6 Conclusions

In this chapter, an algorithm to classify the heart sound signal based on the multidomain features and machine learning technique has been presented. Using the proposed multidomain features, accuracy of 99.0% is achieved using the ensemble of KNN method. Following reasons are observed for the satisfactory results; First, the TQWT based denoising method improves the segmentation performance of the HSMM based algorithm. Second, the inclusion of the time-frequency domain features extracted from the TQWT decomposed levels. The ability of TQWT to tune its quality and redundancy parameters helps to separate the FHS and murmurs. Thus, the extracted features were able to distinguish the normal from other pathological cases. In addition to it, an extensive list of machine learning techniques is experimented to

Table 7.5 Obtained performance evaluation metrics using the proposed method and the various methods proposed in the literature recently.

Authors, year	Feature extraction method	Classifier	Subject type	Sensitivity	Specificity	OA accuracy
Yaseen et al. 2018 [101]	DWT and MFCC	Deep Neural Network SVM KNN (Best result: SVM)	AS	99.00	98.25	97.6
			MR	94.00	99.88	
			MS	97.50	99.50	
			MVP	99.00	99.75	
			N	98.50	99.62	
S.K. Ghosh et al. 2019 [70]	Magnitude and Phase features from Time frequency matrix using WSST	Random Forest	AS	96.77 (Accuracy)		95.13
			MR	90.55 (accuracy)		
			MS	89.77 (accuracy)		
			N	98.55 (accuracy)		
Shu Lih Oh et al. 2020 [79]	Preprocessed signal	Deep WaveNet	AS	94.50	98.50	97.0
			MR	89.00	97.87	
			MS	96.50	98.12	
			MVP	88.50	96.87	
			N	94.00	99.25	
S.K. Ghosh et al. 2020 [66]	Local Energy and Local Entropy on Time frequency matrix using Chirplet transform	Multiclass Composite classifier	AS	99.66	99.04	98.54
			MR	96.33	99.49	
			MS	98.83	99.26	
			N	98.49	99.94	
Proposed method	Time domain frequency domain Time-frequency domain (TQWT)	SVM KNN Ensemble (Best result: Ensemble KNN)	AS	98.84	99.54	99.04
			MR	98.72	99.70	
			MS	100.0	99.85	
			MVP	97.40	99.70	
			N	100.0	100.0	

identify a suitable technique for the heart sound signal. Although obtained results are best for the ensemble of KNN method, the results using SVM are also close to it.

The efficacy of the proposed method shows that it can be used to detect heart valvular diseases automatically which will be very useful for a telemonitoring system. Moreover, it is observed that the classification of a heart sound signal as normal or abnormal can be performed more accurately compared to the classification of a specific disease. This is an important feature for a system that performs the screening at home without the intervention of a medical expert and once the abnormality get detected the patient can be recommended to visit the nearest medical facility where a detailed diagnosis can be performed. Thus, with the help of these systems at-home or out-of-clinic, long-term monitoring of the health of the heart can be performed and hence heart valve disease can be diagnosed at an early stage.

Further, the proposed system has to be tested on the signals contaminated with real-life noise and on a large dataset with a variety of pathological cases. Availability of a large dataset with a variety of pathological cases will open the exploration of neural network and deep neural network based machine learning techniques.

References

[1] D. Gradolewski, G. Redlarski, Wavelet-based denoising method for real phonocardiography signal recorded by mobile devices in noisy environment, Computers in Biology and Medicine 52 (2014) 119–129.
[2] P.K. Jain, A.K. Tiwari, Heart monitoring systems—a review, Computers in Biology and Medicine 54 (2014) 1–13, https://doi.org/10.1016/j.compbiomed.2014.08.014.
[3] S. Yuenyong, A. Nishihara, W. Kongprawechnon, K. Tungpimolrut, A framework for automatic heart sound analysis without segmentation, Biomedical Engineering Online 10 (2011) 13.
[4] A. Moukadem, A. Dieterlen, N. Hueber, C. Brandt, A robust heart sounds segmentation module based on s-transform, Biomedical Signal Processing and Control 8 (3) (2013) 273–281.
[5] A.K. Dwivedi, S.A. Imtiaz, E. Rodriguez-Villegas, Algorithms for automatic analysis and classification of heart sounds – a systematic review, IEEE Access 7 (2019) 8316–8345.
[6] H.J. Van Der Linde, B. Van Deuren, A. Teisman, R. Towart, D.J. Gallacher, The effect of changes in core body temperature on the qt interval in beagle dogs: a previously ignored phenomenon, with a method for correction, British Journal of Pharmacology 154 (7) (2008) 1474–1481, https://doi.org/10.1038/bjp.2008.265 [Online].
[7] D. Mozaffarian, et al., Heart disease and stroke statistics – 2015 update: a report from the American heart association, Circulation 131 (4) (2015) 29–322.
[8] C. for Disease Control and Prevention, State specific mortality from sudden cardiac death, www.cdc.gov/heartdisease/facts.htm, 2002. (Accessed 19 July 2016), Online.
[9] N. Vuyisile, G. Julius, S. Thomas, S.G. John, G.S. Christopher, E. Maurice, Burden of valvular heart diseases: a population-based study, The Lancet 368 (January 2006) 1005–1011.
[10] D.S. Bach, D. Siao, S.E. Girard, C. Duvernoy, B.D. McCallister, S.K. Gualano, Evaluation of patients with severe symptomatic aortic stenosis who do not undergo aortic valve replacement: the potential role of subjectively overestimated operative risk, Circulation: Cardiovascular Quality and Outcomes 2 (6) (2009) 533–539.
[11] D.S. Bach, Evaluation of patients with severe symptomatic aortic stenosis who do not undergo aortic valve replacement: the potential role of subjectively overestimated operative risk, Journal of Heart Valve Diseases 20 (3) (2011) 284–291.
[12] W.-C. Kao, C.-C. Wei, Automatic phonocardiograph signal analysis for detecting heart valve disorders, Expert Systems with Applications 38 (6) (2011) 6458–6468, https://doi.org/10.1016/j.eswa.2010.11.100 [Online].
[13] F. Safara, S. Doraisamy, A. Azman, A. Jantan, A.R.A. Ramaiah, Multi-level basis selection of wavelet packet decomposition tree for heart sound classification, Computers in Biology and Medicine 43 (10) (2013) 1407–1414.
[14] C. Ahlström, Nonlinear phonocardiographic signal processing, PhD thesis, Linköping University, 2008.
[15] S. Vaisman, S.Y. Salem, G. Holcberg, A.B. Geva, Passive fetal monitoring by adaptive wavelet denoising method, Computers in Biology and Medicine 42 (2) (2012) 171–179, https://doi.org/10.1016/j.compbiomed.2011.11.005 [Online].
[16] D. Song, L. Jia, Y. Lu, L. Tao, Heart sounds monitor and analysis in noisy environments, in: 2012 International Conference on Systems and Informatics (ICSAI), May 2012, pp. 1677–1681.

[17] I.W. Selesnick, Wavelet transform with tunable q-factor, IEEE Transactions on Signal Processing 59 (8) (Aug 2011) 3560–3575.
[18] P.K. Jain, A.K. Tiwari, A robust algorithm for segmentation of phonocardiography signal using tunable quality wavelet transform, Journal of Medical and Biological Engineering 38 (3) (2017) 396–410.
[19] P.K. Jain, A.K. Tiwari, An adaptive thresholding method for the wavelet based denoising of phonocardiogram signal, Biomedical Signal Processing and Control 38 (2017) 388–399, https://doi.org/10.1016/j.bspc.2017.07.002 [Online].
[20] D.B. Springer, L. Tarassenko, G.D. Clifford, Logistic regression-hsmm-based heart sound segmentation, IEEE Transactions on Biomedical Engineering 63 (4) (2016) 822–832.
[21] S. Leng, R. Tan, K. Chai, C. Wang, D. Ghista, L. Zhong, The electronic stethoscope, BioMedical Engineering OnLine 14 (1) (2015), https://doi.org/10.1186/s12938-015-0056-y [Online].
[22] Ying-Wen Bai, Chao-Lin Lu, The embedded digital stethoscope uses the adaptive noise cancellation filter and the type I Chebyshev iir bandpass filter to reduce the noise of the heart sound, in: Proceedings of 7th International Workshop on Enterprise Networking and Computing in Healthcare Industry, HEALTHCOM 2005, 2005, pp. 278–281.
[23] M. Sabarimalai Manikandan, K. Soman, Robust heart sound activity detection in noisy environments, Electronics Letters 46 (16) (August 2010) 1100–1102.
[24] S. Sanei, M. Ghodsi, H. Hassani, An adaptive singular spectrum analysis approach to murmur detection from heart sounds, Medical Engineering & Physics 33 (3) (2011) 362–367.
[25] V.S. Chourasia, A.K. Tiwari, R. Gangopadhyay, A novel approach for phonocardiographic signals processing to make possible fetal heart rate evaluations, Digital Signal Processing 30 (2014) 165–183, https://doi.org/10.1016/j.dsp.2014.03.009 [Online].
[26] F. Liu, Y. Wang, Y. Wang, Research and implementation of heart sound denoising, in: International Conference on Solid State Devices and Materials Science, April 1–2, 2012, Macao, Physics Procedia 25 (2012) 777–785, https://doi.org/10.1016/j.phpro.2012.03.157 [Online].
[27] L.H. Cherif, S. Debbal, F. Bereksi-Reguig, Choice of the wavelet analyzing in the phonocardiogram signal analysis using the discrete and the packet wavelet transform, Expert Systems with Applications 37 (2) (2010) 913–918, https://doi.org/10.1016/j.eswa.2009.09.036 [Online].
[28] S.M. Debbal, F. Bereksi-Reguig, Filtering and classification of phonocardiogram signals using wavelet transform, Journal of Medical Engineering and Technology 32 (1) (2008) 53–65.
[29] M.N. Ali, E.S.A. El-Dahshan, A.H. Yahia, Denoising of heart sound signals using discrete wavelet transform, Circuits, Systems, and Signal Processing 36 (2017) 4482–4497.
[30] D.L. Donoho, J.M. Johnstone, Ideal spatial adaptation by wavelet shrinkage, Biometrika 81 (3) (1994) 425–455 [Online], available: http://biomet.oxfordjournals.org/content/81/3/425.abstract.
[31] P.K. Jain, A.K. Tiwari, An adaptive method for shrinking of wavelet coefficients for phonocardiogram denoising, in: 2016 IEEE International Conference on Digital Signal Processing (DSP), 2016, pp. 1–5.
[32] K. Agrawal, A. Jha, S. Sharma, A. Kumar, V. Chourasia, Wavelet subband dependent thresholding for denoising of phonocardiographic signals, in: Signal Processing: Algorithms, Architectures, Arrangements, and Applications (SPA), 2013, Sept 2013, pp. 158–162.

[33] S. Patidar, R.B. Pachori, Segmentation of cardiac sound signals by removing murmurs using constrained tunable-q wavelet transform, Biomedical Signal Processing and Control 8 (6) (2013) 559–567.
[34] J. Oliveira, A. Castro, M. Coimbra, Exploring embedding matrices and the entropy gradient for the segmentation of heart sounds in real noisy environments, in: 2014 36th Annual International Conference of the IEEE Engineering in Medicine and Biology Society, Aug 2014, pp. 3244–3247.
[35] Z. Jiang, S. Choi, A cardiac sound characteristic waveform method for in-home heart disorder monitoring with electric stethoscope, Expert Systems with Applications 31 (2) (2006) 286–298.
[36] S.I. Malik, M.U. Akram, I. Siddiqi, Localization and classification of heartbeats using robust adaptive algorithm, Biomedical Signal Processing and Control 49 (2019) 57–77, https://doi.org/10.1016/j.bspc.2018.11.003 [Online].
[37] L.C. Springer, et al., An open access database for the evaluation of heart sound algorithms, Physiological Measurement 37 (12) (2016) 2181–2213.
[38] S. Ari, P. Kumar, G. Saha, A robust heart sound segmentation algorithm for commonly occurring heart valve diseases, Journal of Medical Engineering & Technology 32 (6) (2008) 456–465.
[39] A. Gavrovska, V. Bogdanović, I. Reljin, B. Reljin, Automatic heart sound detection in pediatric patients without electrocardiogram reference via pseudo-affine Wigner–Ville distribution and Haar wavelet lifting, Computer Methods and Programs in Biomedicine 113 (2) (2014) 515–528, https://doi.org/10.1016/j.cmpb.2013.11.018 [Online].
[40] H. Liang, S. Lukkarinen, I. Hartimo, Heart sound segmentation algorithm based on heart sound envelogram, in: Computers in Cardiology 1997, Sep 1997, pp. 105–108.
[41] S. Sun, Z. Jiang, H. Wang, Y. Fang, Automatic moment segmentation and peak detection analysis of heart sound pattern via short-time modified Hilbert transform, Computer Methods and Programs in Biomedicine 114 (3) (2014) 219–230, https://doi.org/10.1016/j.cmpb.2014.02.004 [Online].
[42] V. Kudriavtsev, V. Polyshchuk, D.L. Roy, Heart energy signature spectrogram for cardiovascular diagnosis, BioMedical Engineering OnLine 6 (1) (2007) 1–22, https://doi.org/10.1186/1475-925X-6-16 [Online].
[43] K. Bajelani, M. Navidbakhsh, H. Behnam, J.D. Doyle, K. Hassani, Detection and identification of first and second heart sounds using empirical mode decomposition, Proceedings of the Institution of Mechanical Engineers, Part H: Journal of Engineering in Medicine 227 (9) (2013) 976–987 [Online], http://pih.sagepub.com/content/227/9/976.abstract.
[44] Liang Huiying, L. Sakari, H. Iiro, A heart sound segmentation algorithm using wavelet decomposition and reconstruction, in: Proceedings of the 19th Annual International Conference of the IEEE Engineering in Medicine and Biology Society. 'Magnificent Milestones and Emerging Opportunities in Medical Engineering' (Cat. No. 97CH36136), vol. 4, 1997, pp. 1630–1633.
[45] H. Naseri, M. Homaeinezhad, Detection and boundary identification of phonocardiogram sounds using an expert frequency-energy based metric, Annals of Biomedical Engineering 41 (2) (2013) 279–292.
[46] D. Kumar, P. Carvalho, M. Antunes, J. Henriques, L. Eugenio, R. Schmidt, J. Habetha, Detection of s1 and s2 heart sounds by high frequency signatures, in: Engineering in Medicine and Biology Society, 2006, EMBS '06, 28th Annual International Conference of the IEEE, Aug 2006, pp. 1410–1416.

[47] J. Vepa, Classification of heart murmurs using cepstral features and support vector machines, in: 2009 Annual International Conference of the IEEE Engineering in Medicine and Biology Society, 2009, pp. 2539–2542.
[48] M.Z. Belmecheri, M. Ahfir, I. Kale, Automatic heart sounds segmentation based on the correlation coefficients matrix for similar cardiac cycles identification, Biomedical Signal Processing and Control 43 (2018) 300–310, https://doi.org/10.1016/j.bspc.2018.03.009 [Online].
[49] G. Zajic, V. Bogdanovic, I. Reljin, B. Reljin, Identification of s1 and s2 heart sound patterns based on fractal theory and shape context, in: Complexity – New Methods for Analyzing Complex Biomedical Systems and Signals, vol. 2017, 2017.
[50] M.V. Shervegar, G.V. Bhat, Automatic segmentation of phonocardiogram using the occurrence of the cardiac events, Informatics in Medicine Unlocked 9 (2017) 6–10, https://doi.org/10.1016/j.imu.2017.05.002 [Online].
[51] M. Mishra, S. Pratiher, H. Menon, A. Mukherjee, Identification of S_1 and S_2 heart sounds using spectral and convex hull features, IEEE Sensors Journal 20 (8) (2020) 4311–4320.
[52] M.A. Alonso-Arévalo, A. Cruz-Gutiérrez, R.F. Ibarra-Hernández, E. García-Canseco, R. Conte-Galván, Robust heart sound segmentation based on spectral change detection and genetic algorithms, Biomedical Signal Processing and Control 63 (2021) 102208, https://doi.org/10.1016/j.bspc.2020.102208 [Online].
[53] M. Mishra, H. Menon, A. Mukherjee, Characterization of s_1 and s_2 heart sounds using stacked autoencoder and convolutional neural network, IEEE Transactions on Instrumentation and Measurement 68 (9) (2019) 3211–3220.
[54] C.N. Gupta, R. Palaniappan, S. Swaminathan, S.M. Krishnan, Neural network classification of homomorphic segmented heart sounds, Applied Soft Computing 7 (1) (2007) 286–297, https://doi.org/10.1016/j.asoc.2005.06.006 [Online].
[55] L.G. Gamero, R. Watrous, Detection of the first and second heart sound using probabilistic models, in: Proceedings of the 25th Annual International Conference of the IEEE Engineering in Medicine and Biology Society (IEEE Cat. No. 03CH37439), vol. 3, 2003, pp. 2877–2880.
[56] F. Renna, J. Oliveira, M.T. Coimbra, Deep convolutional neural networks for heart sound segmentation, IEEE Journal of Biomedical and Health Informatics 23 (6) (2019) 2435–2445.
[57] S.E. Schmidt, E. Toft, C. Holst-Hansen, C. Graff, J.J. Struijk, Segmentation of heart sound recordings from an electronic stethoscope by a duration dependent hidden-Markov model, in: 2008 Computers in Cardiology, 2008, pp. 345–348.
[58] S.E. Schmidt, C. Holst-Hansen, C. Graff, E. Toft, J.J. Struijk, Segmentation of heart sound recordings by a duration-dependent hidden Markov model, Physiological Measurement 31 (4) (mar 2010) 513–529, https://doi.org/10.1088/0967-3334/31/4/004 [Online].
[59] J. Oliveira, T. Mantadelis, F. Renna, P. Gomes, M. Coimbra, On modifying the temporal modeling of hsmms for pediatric heart sound segmentation, in: 2017 IEEE International Workshop on Signal Processing Systems (SiPS), 2017, pp. 1–6.
[60] J. Oliveira, F. Renna, T. Mantadelis, M. Coimbra, Adaptive sojourn time hsmm for heart sound segmentation, IEEE Journal of Biomedical and Health Informatics 23 (2) (2019) 642–649.
[61] D.B. Springer, L. Tarassenko, G.D. Clifford, Support vector machine hidden semi-Markov model-based heart sound segmentation, in: Computing in Cardiology, 2014, pp. 625–628.

[62] S.-W. Deng, J.-Q. Han, Towards heart sound classification without segmentation via autocorrelation feature and diffusion maps, Future Generation Computer Systems 60 (2016) 13–21, https://doi.org/10.1016/j.future.2016.01.010 [Online].
[63] M. Zabihi, A.B. Rad, S. Kiranyaz, M. Gabbouj, A.K. Katsaggelos, Heart sound anomaly and quality detection using ensemble of neural networks without segmentation, in: 2016 Computing in Cardiology Conference (CinC), 2016, pp. 613–616.
[64] D. Boutana, M. Benidir, B. Barkat, Segmentation and identification of some pathological phonocardiogram signals using time-frequency analysis, IET Signal Processing 5 (6) (September 2011) 527–537.
[65] M. Altuve, L. Suárez, J. Ardila, Fundamental heart sounds analysis using improved complete ensemble emd with adaptive noise, Biocybernetics and Biomedical Engineering 40 (1) (2020) 426–439, https://doi.org/10.1016/j.bbe.2019.12.007 [Online].
[66] S.K. Ghosh, R. Ponnalagu, R. Tripathy, U.R. Acharya, Automated detection of heart valve diseases using chirplet transform and multiclass composite classifier with pcg signals, Computers in Biology and Medicine 118 (2020) 103632, https://doi.org/10.1016/j.compbiomed.2020.103632 [Online].
[67] S. Das, S. Pal, M. Mitra, Supervised model for cochleagram feature based fundamental heart sound identification, Biomedical Signal Processing and Control 52 (2019) 32–40, https://doi.org/10.1016/j.bspc.2019.01.028 [Online].
[68] O. El Badlaoui, A. Benba, A. Hammouch, Novel PCG analysis method for discriminating between abnormal and normal heart sounds, IRBM 41 (4) (2020) 223–228, https://doi.org/10.1016/j.irbm.2019.12.003 [Online].
[69] B. Bozkurt, I. Germanakis, Y. Stylianou, A study of time-frequency features for CNN-based automatic heart sound classification for pathology detection, Computers in Biology and Medicine 100 (2018) 132–143, https://doi.org/10.1016/j.compbiomed.2018.06.026 [Online].
[70] S.K. Ghosh, R.K. Tripathy, R.N. Ponnalagu, R.B. Pachori, Automated detection of heart valve disorders from the pcg signal using time-frequency magnitude and phase features, IEEE Sensors Letters 3 (12) (2019) 1–4.
[71] S. Patidar, R.B. Pachori, Classification of cardiac sound signals using constrained tunable-q wavelet transform, Expert Systems with Applications 41 (16) (2014) 7161–7170, https://doi.org/10.1016/j.eswa.2014.05.052 [Online].
[72] S.M. Debbal, F. Bereksi-Reguig, Detection of differences of the phonocardiogram signals by using the continuous wavelet transform method, International Journal of Biomedical Soft Computing and Human Sciences: the official journal of the Biomedical Fuzzy Systems Association 18 (2) (2013) 73–81.
[73] B. Ergen, Y. Tatar, H.O. Gulcur, Time–frequency analysis of phonocardiogram signals using wavelet transform: a comparative study, Computer Methods in Biomechanics and Biomedical Engineering 15 (4) (2012) 371–381.
[74] S.M. Debbal, A.M. Tani, Heart sounds analysis and murmurs, International Journal of Medical Engineering and Informatics 8 (1) (2016) 49–62.
[75] W. Zhang, J. Han, S. Deng, Heart sound classification based on scaled spectrogram and tensor decomposition, Expert Systems with Applications 84 (2017) 220–231, https://doi.org/10.1016/j.eswa.2017.05.014 [Online].
[76] M. Mohanty, S. Sahoo, P. Biswal, S. Sabut, Efficient classification of ventricular arrhythmias using feature selection and C4.5 classifier, Biomedical Signal Processing and Control 44 (2018) 200–208, https://doi.org/10.1016/j.bspc.2018.04.005 [Online].
[77] S. Latif, M. Usman, R. Rana, J. Qadir, Phonocardiographic sensing using deep learning for abnormal heartbeat detection, IEEE Sensors Journal 18 (22) (2018) 9393–9400.

[78] O. Faust, M. Kareem, A. Shenfield, A. Ali, U.R. Acharya, Validating the robustness of an internet of things based atrial fibrillation detection system, Pattern Recognition Letters 133 (2020) 55–61, https://doi.org/10.1016/j.patrec.2020.02.005 [Online].
[79] S.L. Oh, V. Jahmunah, C.P. Ooi, R.-S. Tan, E.J. Ciaccio, T. Yamakawa, M. Tanabe, M. Kobayashi, U. Rajendra Acharya, Classification of heart sound signals using a novel deep wavenet model, Computer Methods and Programs in Biomedicine 196 (2020) 105604, https://doi.org/10.1016/j.cmpb.2020.105604 [Online].
[80] T. Chen, S. Yang, L. Ho, K. Tsai, Y. Chen, Y. Chang, Y. Lai, S. Wang, Y. Tsao, C. Wu, S1 and s2 heart sound recognition using deep neural networks, IEEE Transactions on Biomedical Engineering 64 (2) (2017) 372–380.
[81] S.A. Pavlopoulos, A.C. Stasis, E.N. Loukis, A decision tree-based method for the differential diagnosis of aortic stenosis from mitral regurgitation using heart sounds, BioMedical Engineering OnLine 21 (3) (2004).
[82] G. Amit, N. Gavriely, N. Intrator, Cluster analysis and classification of heart sounds, Biomedical Signal Processing and Control 4 (1) (2009) 26–36, https://doi.org/10.1016/j.bspc.2008.07.003 [Online].
[83] A.F. Quiceno-Manrique, J.I. Godino-Llorente, M. Blanco-Velasco, G. Castellanos-Dominguez, Selection of dynamic features based on time-frequency representations for heart murmur detection from phonocardiographic signals, Annals of Biomedical Engineering 38 (1) (2010) 118–137.
[84] R.C. King, E. Villeneuve, R.J. White, R.S. Sherratt, W. Holderbaum, W.S. Harwin, Application of data fusion techniques and technologies for wearable health monitoring, Medical Engineering & Physics 42 (2017) 1–12, https://doi.org/10.1016/j.medengphy.2016.12.011 [Online].
[85] C. Potes, S. Parvaneh, A. Rahman, B. Conroy, Ensemble of feature-based and deep learning-based classifiers for detection of abnormal heart sounds, in: 2016 Computing in Cardiology Conference (CinC), 2016, pp. 621–624.
[86] I.W. Selesnick, Tunable q-factor wavelet transform (tqwt), https://eeweb.engineering.nyu.edu/iselesni/TQWT/, 2011. (Accessed 11 March 2021) [Online].
[87] L. Rabiner, A tutorial on hidden Markov models and selected applications in speech recognition, Proceedings of IEEE 77 (2) (1989) 257–286.
[88] D. Springer, Logistic regression-hsmm-based heart sound segmentation, https://physionet.org/content/hss/1.0/, 2019. (Accessed 11 March 2021) [Online].
[89] V. Vladimir, The Nature of Statistical Learning Theory, Springer-Verlag, New York, 2000.
[90] L. Breiman, Bagging predictors, Machine Learning 24 (1) (1996) 123–140, https://doi.org/10.1023/A:1018054314350 [Online].
[91] T.K. Ho, Nearest neighbors in random subspaces, in: Joint IAPR International Workshops on Statistical Techniques in Pattern Recognition (SPR) and Structural and Syntactic Pattern Recognition (SSPR), in: Lecture Notes in Computer Science, 1998, pp. 640–648.
[92] U. Fano, Ionization yield of radiations. II. The fluctuations of the number of ions, Physical Review 72 (Jul 1947) 26–29 [Online], http://link.aps.org/doi/10.1103/PhysRev.72.26.
[93] J. Singh, R.S. Anand, Computer aided analysis of phonocardiogram, Journal of Medical Engineering & Technology 31 (5) (2007) 319–323.
[94] G. Luo, D. Zhang, Wavelet denoising, in: Dumitru Baleanu (Ed.), Advances in Wavelet Theory and Their Applications in Engineering, Physics and Technology, intechopen, 2012.

[95] S. Vernekar, S. Nair, D. Vijaysenan, R. Ranjan, A novel approach for classification of normal/abnormal phonocardiogram recordings using temporal signal analysis and machine learning, in: 2016 Computing in Cardiology Conference (CinC), 2016, pp. 1141–1144.

[96] P. Westfall, Kurtosis as peakedness, 1905–2014. r.i.p, American Statistician 68 (3) (2014) 191–195.

[97] J.S. Richman, D.E. Lake, J. Randall Moorman, Sample entropy, Methods in Enzymology 384 (2004) 172–184.

[98] C.J. Huberty, Discriminant analysis, Review of Educational Research 45 (4) (1975) 543–598.

[99] Y. Khan, Classification-of-heart-sound-signal-using-multiple-features, https://github.com/yaseen21khan/Classification-of-Heart-Sound-Signal-Using-Multiple-Features-, 2018. (Accessed 11 March 2021) [Online].

[100] T. Kautz, B.M. Eskofier, C.F. Pasluosta, Generic performance measure for multiclass-classifiers, Pattern Recognition 68 (2017) 111–125, https://doi.org/10.1016/j.patcog.2017.03.008 [Online].

[101] Yaseen, G.-Y. Son, S. Kwon, Classification of heart sound signal using multiple features, Applied Sciences 8 (12) (2018) [Online], https://www.mdpi.com/2076-3417/8/12/2344.

CHAPTER

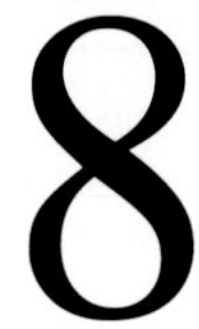

Efficient single image haze removal using CLAHE and Dark Channel Prior for Internet of Multimedia Things

Prateek Ishwar Khade and Amitesh Singh Rajput
Department of Computer Science & Information Systems, Birla Institute of Technology and Science, Pilani, Jhunjhunu, Rajasthan, India

8.1 Introduction

Nowadays, advanced Internet-of-Things (IoT) techniques with multimedia processing, referred to as the Internet-of-Multimedia-Things (IoMT) are emerging as a new paradigm. The technological advancements with the IoT have made the existing paradigm to expand with the multimedia by integrating multimedia sensors. However, efficient processing is required to handle the massive data generated from multimedia sensors so that their efficient handling can be performed with the IoT techniques. Consequently, timely delivery of multimedia data is required so that it can benefit the underlying scenarios, such as traffic monitoring systems, smart hospitals, driving assistance, etc. Considering this importance, we emphasize on one of the important problems of multimedia processing in smart city known as haze removal.

Images that are captured outdoors are mostly deteriorated because of light getting absorbed or scattered by fine suspended particles in the atmosphere. The degradation of images is visible during bad weather conditions such as mist, fog, etc. In such bad weather conditions, the captured images lose contrast and desirable color intensity. This happens due to the attenuation of irradiance received by the camera from the scene point along the line of sight [1]. Also, the mixing of incoming light with air light (a natural light reflected by atmospheric particles) is a major reason for getting degraded images [2,3]. In computer vision applications, there exist various techniques that analyze the contrast and intensity of the image for extracting scene features. The results provided by such techniques will be incorrect or will be deviated if extraction is performed on a hazy image. Thus, haze removal which is referred to as dehazing is highly desired in image processing and computer vision applications. Dehazing can significantly increase the visibility of the scene, making it visually pleasing. In addition to that, deteriorated and low-contrast scene radiance impacts the performance

Internet of Multimedia Things (IoMT). https://doi.org/10.1016/B978-0-32-385845-8.00013-7

of many computer vision tasks, where dehazing can be used to significantly improve the performance.

Broadly, image dehazing is categorized into daytime and night-time haze removal. The daytime haze images [4,5] consist of a single light source, i.e., sun and various schemes are available to process such images [6–11]. On the other hand, in night-time haze images [12], various light sources are available and are differently addressed by research communities [13–18]. Although, the way to encounter haze effects is different for both the categories, the image quality is still not considered to be of great importance. In this chapter, we address the daytime haze images with specific emphasis on the well-known Dark Channel Prior (DCP) method proposed by He et al. [19]. The DCP method considers the nonsky patches in outdoor images as special pixels, called dark pixels, with very low intensity values. The color images with Red-Green-Blue (RGB) color components are considered for this purpose with the fact that at least one color channel possesses dark pixels. Usually, the air light in haze images is very bright than dehazed images. Due to his, the haze images are affected by the higher pixel intensities of their dark channel in the patches with dense haze. This information is used during image dehazing, where pixels with the dark channel are well focused. However, if the color intensity distribution in the underlying image is considerably unequal, the dehazed image gets affected by artifacts. We address this problem and propose to preprocess the underlying haze image with adaptive histogram equalization. As a result, the overall quality of the processed image is improved.

8.1.1 Efficient multimedia processing for IoMT

Considering light-weight components that constitute IoT sensors, optimized multimedia processing leverages a mandatory need in IoMT. For example, a sensor-based video surveillance system may need to have a fast processing method so that the big pool of images can be quickly processed through IoT tunnel. To adapt with this requirement, the multimedia processing techniques should be expanded with their subsequent domains for timing constraints. On the other hand, the quality of the processed image should not be significantly affected while addressing the processing requirements. In this chapter, we analyze the trade-off relation between image quality and its computational complexity with respect to image dehazing. The well-known image dehazing schemes have been reconsidered for quality enhancement and optimized processing time. Since multimedia computing requires higher bandwidth, large memory, and faster computational resources, the analytical results presented in this chapter would help to improve the existing IoMT methods for better efficiency in terms of image quality and processing time. The major contributions of this chapter are summarized as follows:

- We analyze the well-known dark channel prior method and propose to further enhancement it using the contrast adaptive limited histogram equalization.

- The proposed enhancement is tested with the recent DehazeNet model [20] over five hundred images. The in-depth dehazing effects are analyzed with respect to the DCP.
- For computational efficiency, the resized dark channel based optimization is presented that improves the processing time and is able to achieve superior speed up as compared to the existing methods.

The rest of the paper is organized as follows: Section 8.2 discusses the background requirements for DCP and contrast enhancement methods, followed by section 8.3 for detailed description of the preprocessing method and its association with the DCP. Section 8.4 presents the detailed results and discussion. Finally, section 8.5 concludes the paper.

8.2 Background and related work

In the past, several schemes have been proposed for image dehazing by utilizing features of multiple input images. For example, the contrast restoration based methods [1,2], where multiple images of the same scene are captured under different weather conditions and then used as reference images with clear weather condition. Similarly, in the polarization-based methods [3,21,22], dehazing is performed through two or more images captured with varying degrees of polarization. Lately, the dehazing methods which require a single image as a reference were proposed. These methods rely on the use of effective priors. For our research, we have followed the novel DCP method proposed by He et al. [19]. The DCP is based on the fact that in the outdoor images most of the nonsky patches have some pixels, called dark pixels, with very low intensity values (close to zero) in at least one color channel among RGB. We have described the DCP based dehazing method in more detail in Section 8.2.1.

For single-image dehazing, the existing schemes such as Tan et al. [22], Fattal [23], and Tarel et al. [24] produced relevant results. However, the major drawback with these methods is their high processing time. On the other hand, the DCP method proposed by He et al. [19] is simple and effective, but produces artifacts around the image regions with an abrupt change in intensity. To solve the problem of artifacts, a refinement stage is required that we analyzed and reported in this chapter. We present a modification to the DCP to improve the quality of haze removal that can be adapted for further enhancement.

8.2.1 Dark Channel Prior (DCP)

Image dehazing is the process of removing haze effects from the captured image. Here, the main objective is to recover original details, including color intensity, illumination, and radiance in the underlying image. The haze image model [19] is described as

$$I(x) = J(x)t(x) + A(1 - t(x)) \tag{8.1}$$

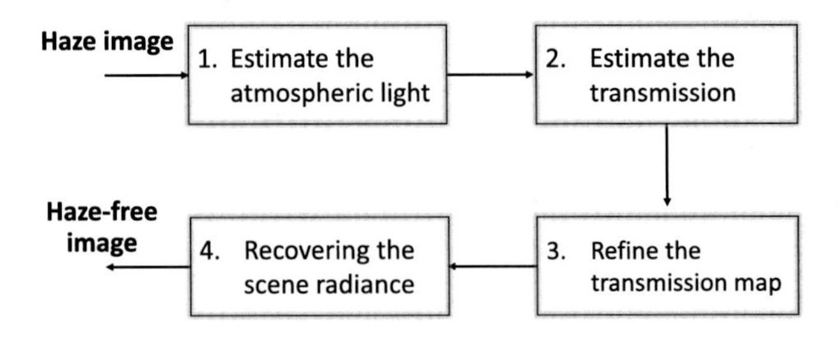

FIGURE 8.1

Steps involved in the Dark Channel Prior method [19].

where, I is the captured haze image, J is the scene light, A is the atmospheric light, and t is the medium transmission specifying the part of the light that is not scattered and reaches the camera. The purpose of haze removal methods is to regain J, A, and t from I. Among the existing schemes for haze removal, we found the DCP approach proposed by He et al. [19] as the most attractive and effective. The DCP method is based on statistics that in nonsky regions of the image, some pixels often have very low intensity values in at least one color channel. It translates that the minimum pixel intensity in such patches/regions is close to zero leading that the intensity of J's dark channel is low and tends to zero. This observation is called dark channel prior. The haze image because of the addition of air light is much brighter than its haze-free version and hence the intensity of the dark channel of a haze image gets higher in the patches having dense haze. The step-wise description of the DCP method proposed by He et al. [19] is shown in Fig. 8.1.

8.2.2 Contrast-Limited Adaptive Histogram Equalization (CLAHE)

Histogram equalization is an image processing technique, used to adjust the contrast of the input image. It uses the histogram information and disperses out the most common intensity values to produce a contrast improved image. Zuiderveld [25] proposed a novel adaptive histogram equalization method, Contrast-Limited Adaptive Histogram Equalization (CLAHE). Unlike histogram equalization, the CLAHE method computes multiple histograms, each belonging to a different segment of the image and utilizes them for redistribution of lightness value of the image. Due to this, the CLAHE finds a good scope for improving the local contrast and sharpness of edges in each region of an image. CLAHE limits the intensification by cutting the histogram at a predefined value before computing the cumulative distribution function, thus limiting the appearance of artifacts in those regions. We use CLAHE to improve the contrast of the underlying haze images in this paper, where each color channel is processed individually and then concatenated to form a single enhanced color image. Fig. 8.2 illustrates the steps involved in CLAHE [25].

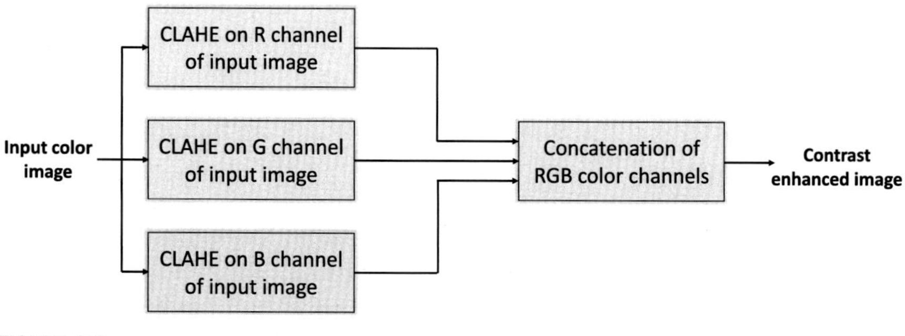

FIGURE 8.2

Contrast enhancement of input color image using the CLAHE technique [25].

8.3 Analysis of adaptive contrast enhancement with DCP and its optimization

8.3.1 Adaptive contrast enhancement with DCP

Our primary objective is to get the dehazing done while keeping minimum the unwanted artifacts using the DCP. We achieve this by first processing the underlying haze image for contrast enhancement, followed by haze removal suggested in DCP. The image contrast is enhanced with the help of adaptive histogram equalization, CLAHE. The graphical flow with detailed description of each step is illustrated in Fig. 8.3.

As mentioned in Fig. 8.3, the contrast of the original hazy image is first modified using an adaptive fashion using CLAHE. The processed haze image with adaptive contrast modification is shown in Fig. 8.4. Once the underlying haze image is processed for contrast modification, the next task is to recover its radiance using the DCP. The resulting effects are shown in Fig. 8.5, where it can be clearly observed that the dehazed image obtained using the contrast modification is visually appealing. For comparison, the dehazed image obtained by directly using the DCP method is shown in Fig. 8.5(a), where a significant difference between the preprocessed image and DCP image can be observed.

The complete stepwise results are shown in Fig. 8.6, where the underlying haze image is first processed for contrast enhancement, followed by DCP to verify its effectiveness.

8.3.2 Optimization for computational advantage

In DCP method, He et al. [19] refined the computed transmission map using a soft-matting technique. However, the soft-matting is computationally very expensive and consumes exponential time to process the underlying image. To overcome this problem, we use the Guided Image Filtering (GIF) [26,27] to refine the transmission map. The GIF is extremely fast and efficient for transmission refinement, and produces re-

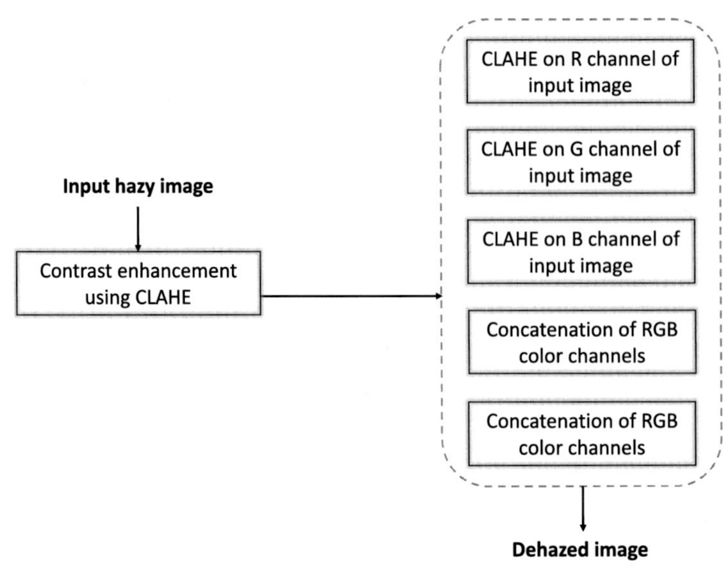

FIGURE 8.3

Overview of DCP and CLAHE based processing.

FIGURE 8.4

(a) Original haze image, (b) Contrast-enhanced haze image using CLAHE.

sults visually comparable to the soft-matting technique used by He et al. [19]. Though GIF significantly reduces the processing time, there is a need to further optimize it to serve real-time latency-sensitive applications.

To further reduce the computational time, we compute the dark channel and atmospheric light estimation on the resized (reduced size) input image and revert it back to the size of the original image before recovering the scene radiance. The methodology that we worked on for optimizing the computational time is illustrated in Fig. 8.7.

The resized dark channel based optimization significantly improves the processing time and is able to achieve superior speed up as compared to the existing methods

FIGURE 8.5

(a) Dehazed image using DCP, (b) Dehaze image using DCP + CLAHE.

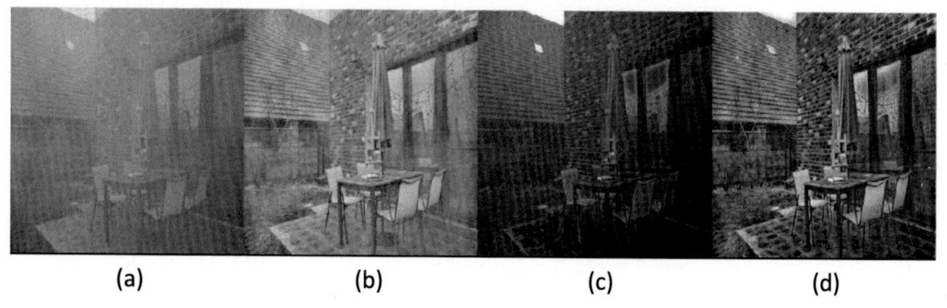

FIGURE 8.6

(a) Original haze image, (b) Contrast enhancement of original image using CLAHE, (c) Dehazed image using DCP method, (d) Dehazed image using DCP + CLAHE.

[19,26,27]. A comparative examination of the processing time with existing dehazing techniques has been done and results are discussed in the next section. Since we are able to get the reduced computational time, the color effects of the processed image become slightly affected resulting the processed image being a bit degraded as compared to the image obtained without resizing. However, the quality of the image is still better as compared to existing methods [19,26,27] The amount of speedup which we achieve through optimization is significant to ignore the minor image quality differences.

8.4 Results and discussion

For analysis purpose, the proposed modifications (CLAHE and optimization) are evaluated over the O-haze [28] image data set. The O-HAZE image dataset consists of 45 various outdoor pictures, captured both in hazy and normal conditions. The pictures in the dataset were recorded during cloudy days mostly at early morning or sunset. The images in the dataset are of size 5456 × 3632 in JPG and ARW (RAW)

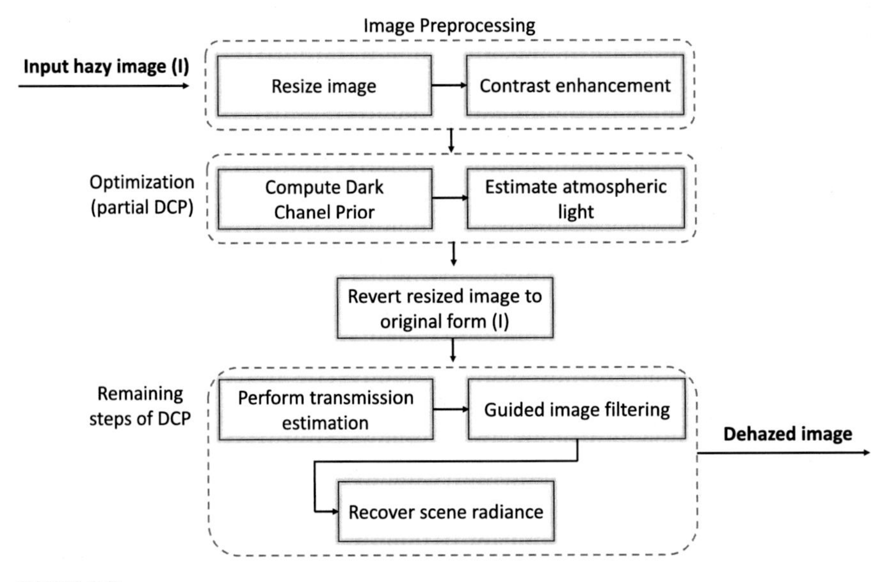

FIGURE 8.7

Optimization using the resized image sample.

formats with 24-bit depth. For analyzing CLAHE-DCP, a comparison is made against eight state-of-the-art methods. Two image quality metrics, SSIM and FADE, are used to accomplish this task.

The range of SSIM is $[-1, 1]$, with a maximum value of 1 for two identical structural images. On the other hand, the Fog Aware Density Evaluator (FADE), predicts the visibility of a foggy scene from a single image without reference to a corresponding fog-free image. It avoids the dependence on salient objects in a scene, geographical camera information, estimating a depth-dependent transmission map, and without training on human-rated judgments. A lower value of FADE implies better dehazing performance. The proposed optimization is assessed using computational time, and comparing it with two state-of-the-art methods.

8.4.1 Quality of haze removal

Initially, the quality of image dehazing is assessed by selecting a variety of eleven images from the O-Haze dataset. These images are assessed by computing the SSIM score between their ground truth and the dehazed results obtained using CLAHE-DCP combination. While experimentation, it has been found that the resulting structural similarity of the processed images using the CLAHE-DCP exhibits superior SSIM scores as compared to existing haze removal techniques proposed by He et al. [19], Meng et al. [29], Fattal [23], Cai et al. [20], Ancuti et al. [30], Berman et al. [31], and Ren et al. [32]. The individual SSIM per image along with the overall average SSIM scores is shown in Table 8.1, where higher SSIM scores achieved by the

Table 8.1 Comparison of SSIM scores between existing schemes and the CLAHE-DCP based processing over selected images from the O-Haze dataset [28].

Image_Name	[19]	[29]	[23]	[20]	[30]	[31]	[32]	CLAHE-DCP
01_Outdoor_hazy	0.9933	0.77	0.73	0.58	0.75	0.76	0.81	0.9972
06_Outdoor_hazy	0.9925	0.78	0.73	0.59	0.68	0.77	0.72	0.9975
10_Outdoor_hazy	0.9941	0.76	0.75	0.71	0.73	0.72	0.8	0.9972
19_Outdoor_hazy	0.9929	0.84	0.79	0.72	0.78	0.82	0.83	0.9988
20_Outdoor_hazy	0.9846	0.72	0.62	0.5	0.78	0.72	0.63	0.9990
21_Outdoor_hazy	0.9848	0.78	0.63	0.71	0.78	0.72	0.73	0.9992
27_Outdoor_hazy	0.9881	0.68	0.67	0.64	0.77	0.7	0.71	0.9972
30_Outdoor_hazy	0.9917	0.74	0.72	0.77	0.83	0.81	0.82	0.9981
33_Outdoor_hazy	0.9948	0.74	0.76	0.81	0.61	0.66	0.88	0.9972
41_Outdoor_hazy	0.9899	0.72	0.66	0.84	0.84	0.82	0.88	0.9958
42_Outdoor_hazy	0.9914	0.82	0.73	0.58	0.74	0.82	0.72	0.9981
Average	0.9907	0.75	0.70	0.67	0.75	0.75	0.77	0.9977

images processed by CLAHE-DCP are observed. The same results are found to be very close to the ground truth which indicates the effectiveness of the CLAHE-DCP based processing.

Furthermore, we compare quality of haze removal obtained by the CLAHE-DCP based dehazed images with the well-known original DCP method [19] using FADE score [33]. The resulting FADE scores are shown in Table 8.2, where it has been observed that the FADE score of the original image is improved when haze removal using DCP is performed. However, the FADE score is further improved when contrast enhancement using CLAHE is performed. The CLAHE-DCP based dehazing outperformed the benchmark method proposed by He et al. [19]. Here, the major reason for quality gain is that the original DCP do not take into consideration a very important observation that the haze images have very low contrast as compared to the nonhaze images of the same scene. This results the images with a low contrast. On the other hand, we preprocess the underlying haze images using an adaptive contrast enhancement, CLAHE. Due to this, a better reconstruction with fine details is clearly observed in the original colors.

Considering one of the state-of-the-art schemes, DehazeNet [20], another observation is found when we tested the CLAHE as a preprocessing method over the wide SOTS (Synthetic Object Testing Set) dataset [34]. Proposed by Cai et al. [20], the DehazeNet generates the medium transmission estimation to get haze-free image via atmospheric scattering model. A Convolutional Neural Network based architecture is used to specifically deal with the image dehazing. On the other hand, the SOTS dataset is a part of the standard RESIDE dataset, comprising of 500 outdoor hazy images. The hazy images in the dataset have been synthesized artificially from natural images. When these images are tested with the CLAHE prior to dehazing with

Table 8.2 Analysis of FADE scores between original and processed images obtained using different methods.

Image Name	FADE (Original Image)	FADE (CLAHE only)	FADE (DCP only)	FADE (CLAHE+DCP)
01_outdoor_hazy	2.391	1.4786	1.6269	1.1312
02_outdoor_hazy	2.3114	1.7253	1.5407	1.2773
03_outdoor_hazy	2.2171	2.2705	1.5551	1.491
04_outdoor_hazy	2.7888	2.3868	1.8374	1.4727
05_outdoor_hazy	3.0697	2.4232	2.0656	1.4404
08_outdoor_hazy	2.9687	2.114	1.9769	1.3064
10_outdoor_hazy	2.5178	1.7104	1.6622	1.1746
11_outdoor_hazy	2.1238	1.3607	1.4886	1.0319
12_outdoor_hazy	2.5568	1.4726	1.5037	1.0984
15_outdoor_hazy	2.9995	2.5138	1.6398	1.6003
18_outdoor_hazy	2.2567	1.5881	1.5191	1.1224
19_outdoor_hazy	2.8408	1.5004	1.5438	1.0425
23_outdoor_hazy	1.9271	1.1469	1.2932	0.8963
24_outdoor_hazy	2.1471	1.4687	1.4306	1.1366
28_outdoor_hazy	3.3212	2.249	1.764	1.4408
30_outdoor_hazy	2.9045	1.5104	1.6806	1.0377
31_outdoor_hazy	2.7284	1.4046	1.5404	1.0018
32_outdoor_hazy	2.674	1.6629	1.6102	1.1701
33_outdoor_hazy	3.2583	1.9596	1.8033	1.2454
34_outdoor_hazy	2.6873	1.6228	1.6074	1.163
35_outdoor_hazy	2.4791	0.936	1.0898	0.7177
36_outdoor_hazy	2.4098	1.67	1.5659	1.2721
37_outdoor_hazy	2.5491	1.5909	1.6569	1.1149
38_outdoor_hazy	2.1283	1.6338	1.4772	1.239
39_outdoor_hazy	2.1717	2.0699	1.5775	1.5282
40_outdoor_hazy	3.0482	2.2119	1.7042	1.4641
41_outdoor_hazy	3.078	1.885	1.8174	1.202
42_outdoor_hazy	3.1244	1.6138	1.7833	1.1441
43_outdoor_hazy	3.0318	1.8036	1.7441	1.3111
44_outdoor_hazy	2.9087	1.7491	1.6422	1.179
45_outdoor_hazy	2.4967	2.1832	1.6154	1.3905
Average score	2.6458	1.7835	1.6237	1.2257

Table 8.3 Analysis of FADE scores between original and processed images using the DehazeNet [20].

Original images (without processing)	Dehazed images (DehazeNet Only)	Dehazed images (CLAHE + DehazeNet)
1.5687	0.6011	0.2030

Table 8.4 Comparative analysis of computational time (in seconds).

Method	Image name (O-Haze dataset [28])				
	01	20	27	33	41
He et al. [19]	33.67	70.13	22.07	98.17	25.39
He et al. (DCP+GIF) [26]	7.53	10.89	4.45	12.99	5.76
DCP + CLAHE	7.74	11.35	4.42	12.61	5.71
DCP + CLAHE (Resized)	4.87	7.06	2.78	8.21	3.71

the DehazeNet, the FADE scores are found to be significantly improved. The average FADE scores computed over the images of SOTS dataset with and without inclusion of CLAHE with the DehazeNet are shown in Table 8.3, where superior results with the inclusion of CLAHE are clearly observed. However, the structural similarity is slightly affected from 0.73 to 0.47 without and with CLAHE. This indicates that the best combination one can achieve is with DCP and CLAHE.

8.4.2 Performance assessment

As mentioned in Section 8.3.2, the performance gain is achieved by first resizing the underlying haze image, processing it for dark channel, and then finally recovering the radiance in the original dimension. In this section, the optimized computational time has been compared with various dehazing techniques [19,26,27], with or without optimization. Table 8.4 shows the performance results, where it can be observed that the dehazing using dark channel computation over resized image leads 10 times computational efficiency gain as compared to the original DCP method proposed by He et al. [19]. However, the quality of dehazed image is slightly degraded as compared with the images directly processed using the DCP-CLAHE in original dimension.

8.4.3 Discussion

Considering the IoMT scenario, the proposed method is well analyzed over image quality and computational efficiency. For better image quality, it has been found that the DCP-CLAHE combination gives better visual results. Also, the SSIM score of the dehazed images obtained using DCP-CLAHE combination is found superior than seven existing image dehazing schemes as depicted in Table 8.1. The resulting images are visually appealing, have better contrast and colors as compared to the original method proposed by He et al. [19]. Moreover, the resulting FADE scores, depicted in Table 8.2 show superiority of the DCP-CLAHE combination for better dehazing.

On the other hand, for IoMT setting requiring optimized computational time, the low resolution based dark channel processing is found to reduce the processing time by 10 times as compared to the existing methods. This optimization is significant over the existing methods and best suits the IoMT scenario. However, it has been observed that the quality of images is slightly degraded due to low resolution features. Such degradation is negligible to fulfill the immediate IoMT requirements and can be further improved for visual quality in future. Here, the underlying client can choose among the two experiments, DCP-CLAHE and low resolution based optimization, its preference for efficient image quality or processing time.

8.5 Conclusion

To address optimized computational aspects of IoMT, we have analyzed well-known dark channel prior (DCP) based dehazing technique and proposed its more efficient refinements both quality-wise and computationally in this chapter. The dehazed image is obtained by enhancing the contrast of the haze image by using the CLAHE method as a preprocessing step before DCP. As a result, better quality of the dehazed image is obtained as compared to the existing dark channel prior methods. Furthermore, we are able to reduce the computational time, and hence the proposed optimization is found to be faster than existing dehazing schemes. The computational gain achieved by the optimization makes it possible to be used in applications with latency-sensitive real-time requirements of IoMT. In future, the proposed work can be further improved for visual quality of the processed images to address high-level computer vision tasks, where image dehazing plays a significant role. Also, the robustness of the DCP and CLAHE can be considered to address some complicated cases related to noisy patches.

References

[1] S.G. Narasimhan, S.K. Nayar, Contrast restoration of weather degraded images, IEEE Transactions on Pattern Analysis and Machine Intelligence 25 (6) (2003) 713–724.

[2] S.K. Nayar, S.G. Narasimhan, Vision in bad weather, in: Proceedings of the Seventh IEEE International Conference on Computer Vision, vol. 2, IEEE, 1999, pp. 820–827.

[3] Y.Y. Schechner, S.G. Narasimhan, S.K. Nayar, Polarization-based vision through haze, Applied Optics 42 (3) (2003) 511–525.

[4] A. Dudhane, H. Singh Aulakh, S. Murala, Ri-gan: an end-to-end network for single image haze removal, in: Proceedings of the IEEE/CVF Conference on Computer Vision and Pattern Recognition Workshops, 2019, pp. 159–170.

[5] A. Dudhane, S. Murala, Ryf-net: deep fusion network for single image haze removal, IEEE Transactions on Image Processing 29 (2019) 628–640.

[6] C.-H. Yeh, C.-H. Huang, L.-W. Kang, Multi-scale deep residual learning-based single image haze removal via image decomposition, IEEE Transactions on Image Processing 29 (2019) 3153–3167.

[7] X. Yang, H. Li, Y.-L. Fan, R. Chen, Single image haze removal via region detection network, IEEE Transactions on Multimedia 21 (10) (2019) 2545–2560.

[8] H. Dong, J. Pan, L. Xiang, Z. Hu, X. Zhang, F. Wang, M.-H. Yang, Multi-scale boosted dehazing network with dense feature fusion, in: Proceedings of the IEEE/CVF Conference on Computer Vision and Pattern Recognition, 2020, pp. 2157–2167.

[9] G. Saxena, S.S. Bhadauria, An efficient single image haze removal algorithm for computer vision applications, Multimedia Tools and Applications 79 (37) (2020) 28239–28263.

[10] T. Zhang, X. Yang, X. Wang, R. Wang, Deep joint neural model for single image haze removal and color correction, Information Sciences 541 (2020) 16–35.

[11] B. Li, J. Zhao, H. Fu, Dlt-net: deep learning transmittance network for single image haze removal, Signal, Image and Video Processing 14 (6) (2020) 1245–1253.

[12] M. Yang, J. Liu, Z. Li, Superpixel-based single nighttime image haze removal, IEEE Transactions on Multimedia 20 (11) (2018) 3008–3018.

[13] T. Yu, K. Song, P. Miao, G. Yang, H. Yang, C. Chen, Nighttime single image dehazing via pixel-wise alpha blending, IEEE Access 7 (2019) 114619–114630.

[14] J. Zhang, Y. Cao, S. Fang, Y. Kang, C. Wen Chen, Fast haze removal for nighttime image using maximum reflectance prior, in: Proceedings of the IEEE Conference on Computer Vision and Pattern Recognition, 2017, pp. 7418–7426.

[15] S. Kuanar, D. Mahapatra, M. Bilas, K. Rao, Multi-path dilated convolution network for haze and glow removal in nighttime images, The Visual Computer (2021) 1–14.

[16] J. Zhang, Y. Cao, Z.-J. Zha, D. Tao, Nighttime dehazing with a synthetic benchmark, in: Proceedings of the 28th ACM International Conference on Multimedia, 2020, pp. 2355–2363.

[17] Q. Tang, J. Yang, X. He, W. Jia, Q. Zhang, H. Liu, Nighttime image dehazing based on retinex and dark channel prior using Taylor series expansion, Computer Vision and Image Understanding 202 (2021) 103086.

[18] Y. Liu, A. Wang, H. Zhou, P. Jia, Single nighttime image dehazing based on image decomposition, Signal Processing 183 (2021) 107986.

[19] K. He, J. Sun, X. Tang, Single image haze removal using dark channel prior, IEEE Transactions on Pattern Analysis and Machine Intelligence 33 (12) (2010) 2341–2353.

[20] B. Cai, X. Xu, K. Jia, C. Qing, D. Tao, Dehazenet: an end-to-end system for single image haze removal, IEEE Transactions on Image Processing 25 (11) (2016) 5187–5198.

[21] Y.Y. Schechner, S.G. Narasimhan, S.K. Nayar, Instant dehazing of images using polarization, in: Proceedings of the 2001 IEEE Computer Society Conference on Computer Vision and Pattern Recognition, CVPR 2001, vol. 1, IEEE, 2001, pp. I–I.

[22] S. Shwartz, E. Namer, Y.Y. Schechner, Blind haze separation, in: 2006 IEEE Computer Society Conference on Computer Vision and Pattern Recognition (CVPR'06), vol. 2, IEEE, 2006, pp. 1984–1991.

[23] R. Fattal, Dehazing using color-lines, ACM Transactions on Graphics (TOG) 34 (1) (2014) 1–14.

[24] J.-P. Tarel, N. Hautiere, Fast visibility restoration from a single color or gray level image, in: 2009 IEEE 12th International Conference on Computer Vision, IEEE, 2009, pp. 2201–2208.

[25] K. Zuiderveld, Contrast limited adaptive histogram equalization, in: Graphics Gems, 1994, pp. 474–485.

[26] K. He, J. Sun, X. Tang, Guided image filtering, in: European Conference on Computer Vision, Springer, 2010, pp. 1–14.

[27] K. He, J. Sun, Fast guided filter, arXiv preprint, arXiv:1505.00996, 2015.

[28] C.O. Ancuti, C. Ancuti, R. Timofte, C. De Vleeschouwer, O-haze: a dehazing benchmark with real hazy and haze-free outdoor images, in: Proceedings of the IEEE Conference on Computer Vision and Pattern Recognition Workshops, 2018, pp. 754–762.

[29] G. Meng, Y. Wang, J. Duan, S. Xiang, C. Pan, Efficient image dehazing with boundary constraint and contextual regularization, in: Proceedings of the IEEE International Conference on Computer Vision, 2013, pp. 617–624.
[30] C. Ancuti, C.O. Ancuti, C. De Vleeschouwer, A.C. Bovik, Night-time dehazing by fusion, in: 2016 IEEE International Conference on Image Processing (ICIP), IEEE, 2016, pp. 2256–2260.
[31] D. Berman, S. Avidan, et al., Non-local image dehazing, in: Proceedings of the IEEE Conference on Computer Vision and Pattern Recognition, 2016, pp. 1674–1682.
[32] W. Ren, S. Liu, H. Zhang, J. Pan, X. Cao, M.-H. Yang, Single image dehazing via multi-scale convolutional neural networks, in: European Conference on Computer Vision, Springer, 2016, pp. 154–169.
[33] L.K. Choi, J. You, A.C. Bovik, Referenceless prediction of perceptual fog density and perceptual image defogging, IEEE Transactions on Image Processing 24 (11) (2015) 3888–3901.
[34] B. Li, W. Ren, D. Fu, D. Tao, D. Feng, W. Zeng, Z. Wang, Benchmarking single-image dehazing and beyond, IEEE Transactions on Image Processing 28 (1) (2019) 492–505.

CHAPTER

9

A supervised and unsupervised image quality assessment framework in real-time☆

Zahi Al Chami[a,b], Chady Abou Jaoude[a], and Richard Chbeir[b]

[a]*Antonine University, Faculty of Engineering – TICKET Lab, Beirut, Lebanon*

[b]*Universite de Pau et des Pays de l'Adour, E2S UPPA, LIUPPA, Anglet, France*

9.1 Introduction

With the ongoing advances in technology, there have been enormous growths in multimedia traffic on the global internetwork due to the huge interest in the development and usage of multimedia-based applications and services. Newsfeeds, podcasts, live interviews, and real-time content delivery are examples of real-time multimedia applications, systems, and solutions that have led to Internet multimedia traffic's rapid growth. As compared to scalar data collected by standard IoT devices, multimedia content, e.g., audio, video, and images, obtained from the physical environment has distinct characteristics. The introduction of these multimedia objects opens the door to a wide range of commercial and military uses. Some examples are real-time multimedia based security/monitory systems in smart homes, intelligent multimedia surveillance systems deployed in smart cities, transportation management, optimized using smart video cameras, etc. However, supplementing IoT applications with multimedia devices and content requires additional resources and functionalities of existing ones, resulting in a specialized subset of IoT known as 'Internet of Multimedia Things' (IoMT). In addition, and due to the massive amount of multimedia data received and produced at a rapid pace, traditional approaches to processing data content do not scale well for data streaming scenarios. These approaches require data to be stored first before being processed, which takes a significant amount of time.

More particularly, and because of the social media photo-sharing applications' growth, most of the shared data between users are images. According to [1], over 300 million images are posted daily to Facebook, while over 95 million photos are uploaded daily to Instagram. These streams could be distorted in various ways, such as using adaptation/protection functions to meet business needs. For example, blur-

☆ This study is supported by OpenCEMS Industrial Chair.

Internet of Multimedia Things (IoMT). https://doi.org/10.1016/B978-0-32-385845-8.00014-9

ring an individual's face in an interview to conceal his/her identity, deleting sensitive information from a Twitter stream, or highlighting only items of interest in a video because of certain hardware or network limitations. Hence, the outcome of content protection or adaptation functions must be assessed to ensure that the trade-off between the quality of the produced content and the intended result is satisfactory. More particularly, and in the past few decades, processing images that contain faces have received great attention due to its numerous applications, including video surveillance, entertainment, etc. Since low-quality images limit these applications, it is important to assess the images' visual features before publishing. This can be achieved by evaluating the contents' structure and semantic. For example, ensuring that the faces kept intact after applying content adaptation functions and that some useful information can still be extracted after using content protection functions.

Normally, the visual image quality is assessed by comparing the distorted image with an ideal imaging model or perfect reference image. But, in most scenarios, we do not have the source of the distorted image. Consequently, evaluating an image's quality blindly has been becoming increasingly important (a.k.a. No Reference Image Quality Assessment). In such cases, objective image quality measurement approaches (for example, SSIM [2], PSNR [3], Content-Based Image Retrieval [4], etc.) and face quality metrics (for example, face alignment [5]) can't be used to measure image quality deterioration since distortion-free images are needed. For this purpose, we need to directly quantify image degradations by exploiting features that are discriminant for image degradation. Several approaches have been suggested and received attention in this field. They extracted Natural Scene Statistics (NSS) using the wavelet transform [6–8] or the DCT transform [9]. These methods are very slow since costly image transformations were used. For this purpose, we will employ the convolution neural network (CNN) to learn discriminant features and to reduce the computation time by minimizing the image dimensionality while preserving its useful information. CNN has also shown superior performance due to its deeper structure [10]. However, most of the existing IQA techniques [11,12] that are based on CNN are bound to one or more specific types of distortion. More particularly, CNN can not accurately predict or classify the images' quality score having different levels in their distortions due to the limited number of neurons (at the last layer) responsible for representing the score. To clarify the previous points, we provide the following scenario.

9.1.1 Motivating scenario

Let us consider, for example, a photo-sharing company (shown in Fig. 9.1), which provides its customers with the ability to share and publish images online while offering additional services such as:

- Protecting his/her identity to avoid disclosing sensitive information using several techniques, such as masking functions.
- Adapting their images to meet the limitations imposed by the available resources—for example, image compression.

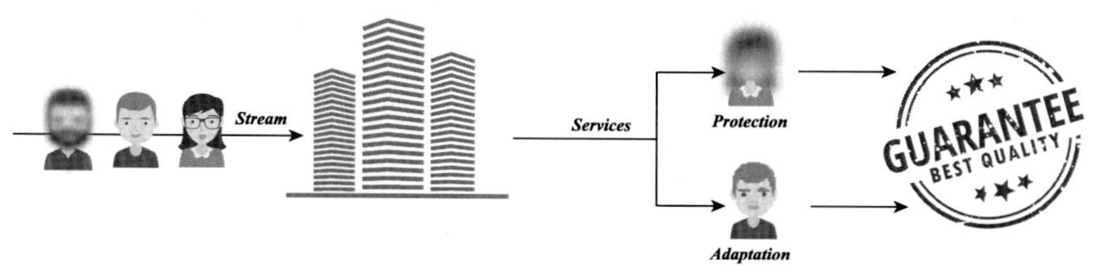

FIGURE 9.1

Scenario.

However, and despite the fact that the users are getting benefits from these services, they could remove or damage the images' information. Therefore, the company starts receiving complaints from its users about image quality degradation. Moreover, and as shown in Fig. 9.1, the company may also accept distorted images from the stream that were already affected during the transmission or processing phase. In addition, and since the company receives an unbounded stream of images from its users, it is demanded to instantly process these images. So, the company wants to start looking for a new solution to satisfy the needs of the users. In summary, the framework that should be implemented in the company must be able to:

- Preserve the images' quality: ensure that the modified images' features, such as color, texture, semantic, etc., are remained intact and can still be extracted from the output, especially the most challenging images that do not have access to their reference image.
- Handle the unbounded stream of images: processing the multimedia data content requires higher bandwidth, bigger memory, and faster computational resources. Therefore, it forces strict quality of service requirements and demands efficient network architecture. For this reason, we need to find a way to treat the significant number of images by ensuring a successful image processing and reliable delivery of the data instantly and in real-time, even though the previously mentioned services, along with the quality assessment process, could take much time.

Several existing works focused on solving these challenges using many methods/techniques cited, along with their limitations, in the next section before presenting our proposed approach.

9.2 Related work

In this section, we present some existing works that are compared to our approach based on many criteria: 1) The useful images' information that remained intact when addressing business or users' needs, such as adaptation or protection, 2) The metrics used to evaluate the image quality; this will indicate the number of features used to measure the quality as well as the metrics' accuracy, and 3) The amount of time it takes to process the images, especially the proposed approaches that work in real-time.

Based on the availability of nondistorted reference images, image quality measures are typically classified into three categories: Full-Reference Image Quality Assessment (FR-IQA), reduced-reference (RR) IQA, and No-Reference (NR) IQA. To assess the quality of the distorted images, FR-IQA requires reference images. To compute the quality measure in RR-IQA, partial information from a reference image is needed. However, in many practical applications, reference image information is not available; therefore, it is desirable to develop NR-IQA methods that perform image quality estimation without relying on a reference image. In the following sections, we will present the techniques used to assess the useful image quality information and their limitations.

9.2.1 Full-reference image quality assessment

The mean square error (MSE) between the reference and distorted images is the most basic and straightforward image quality metric. Despite its widespread use, it does not correlate with perceived visual consistency [13]. This resulted in developing a whole zoo of image quality metrics to improve the agreement with human perceptions of image quality. Machine learning was primarily used in no-reference image quality assessment (NR-IQA), in which researchers attempted to develop features that could distinguish distorted images from their original images. On the other hand, machine learning has been partially adopted in full-reference image quality assessment (FR-IQA). In [14], singular value decomposition features were extracted and regressed onto the quality score using support vector regression (SVR). In [15], multiple features were extracted from the difference of Gaussian frequency bands and regressed onto the quality score.

In [16], the authors introduce a compression method to maintain image quality, while in [17], the distorted images' quality is assessed in real-time using quality methods. They proposed a machine learning-based method to lossy image compression that generates files 2.5 times smaller than JPEG and JPEG 2000 while maintaining images' quality. While these techniques provide valuable results, they assessed the distorted images by analyzing only the color and structure features without considering the other features that may be damaged and led to content degradation.

The suggested solution in [18] defines a utility function to evaluate adaptation operators' usefulness by imposing semantic restrictions for video adaptation. This *utility function* is based on: 1) affected area, 2) affected priority area, and 3) the visual coherence of processed videos. In [19], the authors estimate the quality of the multimedia data in real-time by proposing an end-to-end framework over heterogeneous networks. They provided four services: 1) video assessment, 2) Quality of Service (QoS), 3) QoE-based mapping, and 4) adaptation procedures. In [20], The authors present the architecture of an adaptive multimedia learning service, where their engine enables users to identify the best combination of adaptive features of visual and audio content.

9.2.2 No-reference image quality assessment

In general, there are two types of no-reference IQMs: distortion-specific IQMs [21–25] and generalized IQMs [7,26–28]. Different IQMs have their pros and cons. The former category estimates the quality of an image by quantifying the particular artifacts induced by the distortion process and usually works well for one specific distortion type. The latter category often refers to the general-purpose IQA (we usually use an unsupervised network), which is a much more challenging task than the former category due to the lack of distortion information. However, they may not perform as good as distortion-specific IQMs for certain distortion since the neural network is trained using unlabeled data. In this paper, we focus on combining the two types of no-reference IQMs to overcome each type's drawbacks.

Most of the state-of-the-art IQA methods [29–31] train their network to predict the distorted images' quality using human subjective quality scores that are available in several datasets, e.g., the images in the TID2013 [32] and LIVE [33] databases. All the previously mentioned methods follow a two-step framework: feature extraction and model regression by human scores. The IQA metric in [31] extracts three sets of features based on natural images' statistics, distortion textures, and blur/noise. Then, three models are trained to evaluate the quality based on each feature, and finally, a weighted combination of them is used to estimate the image quality. There are several existing studies using no-reference IQMs to assess face sample quality. Abaza et al. [34] evaluated no-reference IQMs that can measure image quality factors in the context of face recognition. Then they proposed a face image quality index that combines multiple quality measures. Dutta et al. [35] proposed a data-driven model to predict the performance of a face recognition system based on image quality features. They modeled the relationship between image-based quality features and recognition performance measures using a probability density function. However, the studies mentioned above have a common shortage: the image-based quality attributes in these studies do not cover the most important face attributes such as semantic, structure, etc.

Motivated by the recent success of CNNs for image classification tasks, [36] presents a Deep Neural Network-based approach to image quality assessment. It can be used in a no-reference as well as in a full-reference IQA. They rely on extracting feature vectors from the distorted and reference image to be then concatenated together while assigning weights for each region. They showed superior performance compared to the state-of-the-art. However, the authors did not show the execution latency and the complexity as this method may take more time to assess an image since the authors consider each region and not only the distorted parts, which limited their use in real-time applications. In [37,38], the authors make use of full-reference IQA (FR-IQA) models for quality annotation. Their performance is directly affected by that of FR-IQA models, which may be inaccurate across distortion levels and types. The authors in [39] proposed an IQA metric based on deep metalearning. They first trained the model based on a number of distortion-specific NR-IQA tasks to learn a metamodel, then the latter can capture the humans' shared metaknowledge when evaluating images with various distortions, enabling fast adaptation to the NR-IQA

Table 9.1 Showing the content and the limitations of each approach.

<table>
<tr><th>Cited approaches</th><th>Content</th><th>Limitations</th></tr>
<tr><td colspan="3">Full-Reference IQA</td></tr>
<tr><td>[14,15]</td><td>They train a regression model to predict the image quality score using multiple features</td><td rowspan="2">They showed good results, but they only extract color and luminance features to learn the IQA regression model while ignoring several important features such as structure, texture, shape, etc.</td></tr>
<tr><td>[16,17]</td><td>They propose methods to maintain the image quality, while processing the images in real-time in the second approach</td></tr>
<tr><td>[18–20]</td><td>They provide an end-to-end quality assessment framework by enabling the users to choose the best combination of semantic, visual, and audio features.</td><td>Adaptation operators are different from protection operators in terms of content processing, and the use of gaps to calculate the visual coherence is not appropriate in content protection.</td></tr>
<tr><td colspan="3">No-Reference IQA</td></tr>
<tr><td>[29–31]</td><td>They used subjective image scores to train their models to predict the distorted images' quality</td><td>The trained model is heavily dependent on the human scores images, which are not as accurate as the objective scores.</td></tr>
<tr><td>[37,38]</td><td>The authors used the FR-IQA methods to annotate and train their models.</td><td>The use of the FR-IQA may be inaccurate and insufficient to predict images having different distortion levels and types.</td></tr>
</table>

task of unknown distortions. Even though the method is characterized by good generalization ability, it did not achieve good results than state-of-art methods.

The authors in [40] proposed a quality assessment method in live video streaming by first training a deep unsupervised network at the server-side, then making no-reference measurements at the client-side. Even though they achieved good results, they did not consider the processing time and the complexity, especially that they are working in real-time. We present, in Table 9.1, the previously mentioned approaches along with their limitations.

Our work aims to find a fair trade-off between the quality of the altered content and the users' expected outcome in real-time, as detailed in the next section.

9.3 Contributions

In this paper, we propose an adaptive faces quality estimation in the images by combining an unsupervised with a supervised neural network alongside the face alignment anomaly detection metric while processing these images in real-time. We choose the convolutional neural network as a supervised one since it is the most commonly used when it comes to analyzing and treating images. We assume that the faces are affected by adaptation or protection functions to be evaluated on the fly.

Our contributions can be summarized as follows:

- We propose an adaptive No-Reference faces quality estimation in the images by using a convolutional neural network (CNN) followed by an unsupervised neural network. The CNN is tuned according to the Structural Similarity Index (SSIM) [2], Content-Based Image Retrieval (CBIR) [4] and Perceptual Coherence Measure (PCM) [41]. To do so, we proceed as follows:
  - We first partition the image into patches before classifying each patch to a quality score group using the CNN. The partitioning process's main idea is that an image may not have a uniform distribution in its distortion and to have the input images' size the same as the first network layer, since we may lose important information in resizing the images.
  - We then use the unsupervised neural network by clustering those patches within each group to learn the sublevel quality score.

  This combination will provide us the ability to assess accurately not only specific types of distortion, but also the images having variations in their distortion. In addition, we come up with a second metric that will allow us to assess the facial features in a no-reference image using an unsupervised face alignment anomaly detection. It is worth mentioning that the main difference between our method and the state-of-arts methods is that the first time a combination of these metrics has been used to predict the images' quality while considering such a number of features to train the neural network.
- We develop a framework with the capability of evaluating a stream of images efficiently while estimating its quality.

The remainder of this article is organized as follows. Section 9.4 presents some definitions and terminologies used in our work. The neural network for image quality assessment and the face alignment metric are described in section 9.5, while the proposed framework is then detailed in section 9.6. We evaluate our proposed approach in section 9.7 through a set of experiments. Conclusions and future work are summarized in section 9.8.

9.4 Definitions

In this section, we present the data model and data manipulation functions needed to fully understand the proposed framework.

9.4.1 Data model

Definition 1 (Image). An image, denoted by **im**, is a basic data structure consisting of attributes that provide information about its content. It is written as follows:

$$im \prec \boldsymbol{DESC}, \boldsymbol{F}, \boldsymbol{SO} \succ$$

where,

- ***DESC*** is a user-provided set of textual descriptions, keywords, or annotations.
- ***F*** is the set of features that depicts an image. It can describe an entire image or a feature at a specific location.
- ***SO*** is a set of salient objects representing objects of interest in an image detailed in the following definition.

Definition 2 (Salient object). It is an object of interest, denoted by **so**, in an image. For example: a person's face. It is defined as:

$$so \prec \boldsymbol{w}, \boldsymbol{h}, \boldsymbol{coord}, \boldsymbol{DESC}, \boldsymbol{F} \succ$$

where,

- ***w*** and ***h*** are the width and height of the salient object.
- ***coord*** indicates the coordinates of so.
- ***DESC*** is a set of textual descriptions related to so.
- ***F*** is the set of features revealing a salient object's visual content such as color, texture, and shape.

Definition 3 (Entity). It is a semantic object that exists by itself (e.g., person, vehicle) and is expressed as **e**. Each entity is represented by a set of salient objects, which can be either distorted or not. A relationship, $e \rightarrow \{so_1, so'_2, ..., so_n\}$, done via manual or automatic annotation, shows the salient objects $\{so_1, so'_2, ..., so_n\}$ that are associated with entity e, where so' represents a modified salient object.

Definition 4 (Multimedia data stream). It represents an infinite sequence of images, designated as **mds**, that may contain a mix of distorted and distortion-free images. It is formally defined as follows:

$$mds = im_1, im'_2, ..., im_k \quad \text{where } k \in \mathbb{N}^*$$

9.4.2 Data manipulation functions

This section defines the functions used to modify the images in the multimedia data stream; either protect or adapt the salient objects in these images. Our presumptions focus mainly on the identification of the salient objects that might be protected or adapted based on predefined rules in authorization or adaptation schemes, which are out of the scope of this paper. We assume that the protection and adaptation functions are known and that they can be called implicitly on a subset of specified entities.

Definition 5 (Image manipulation function). It is a low-level function, designated by **imf**, that modifies, suppresses, or removes a set of features attributed to so in an image im. $imf(so, im)$ takes a salient object so, the image im that contains so, and returns a modified salient object denoted by **so'**.

As previously mentioned, we focus on two types of functions: a protection function and an adaptation function. The first type is used to hide the images' content by deleting some of its features to conceal some sensitive information related to an entity (for example, his/her identity). The second type is used to meet resource constraints such as hardware limitations.

A group of manipulation functions could be applied on the entity that exists in the images. In our work, this group is known as entity manipulation function, and it is formally defined as follows;

Definition 6 (Entity manipulation function). It is denoted by **emf** and defined as:

$$emf(e, mds) = (imf_1(so_1, im_1) \circ ... \circ imf_i(so_n, im_n))$$

where, i and $n \in \mathbb{N}^*$. emf combines several image manipulation functions $(imf_1(so_1, im_1), ..., imf_i(so_n, im_n))$, which modifies the salient objects representing entity e in the multimedia data stream mds, by altering their features. As a result, this function returns a set of modified salient objects ***SO'*** that represents entity ***e***.

9.5 Data quality

9.5.1 Neural network architecture for image quality assessment

An overview of our Neural Network image quality assessment architecture is shown in Fig. 9.2. It is composed of two main submodules:

- The supervised Neural Network.
- The unsupervised Neural Network.

But before going into the details of each module, and since an image may be partially or fully distorted, we first partition the image into K patches, denoted by im'_{p_i}, with $i \in 1, 2, ..., K$. Moreover, instead of resizing the images conforming to the first network layer and losing important information, we use these small patches as input to our neural network. In the following, we will present in detail the architecture's modules.

9.5.1.1 Supervised neural network module

The number of parameters in a neural network grows rapidly with the increase in the number of layers. Hence, tuning so many parameters can be a huge task, and training such models would become computationally heavy. But, the Convolutional Neural Networks (CNNs) reduce the time required to tune these parameters without losing

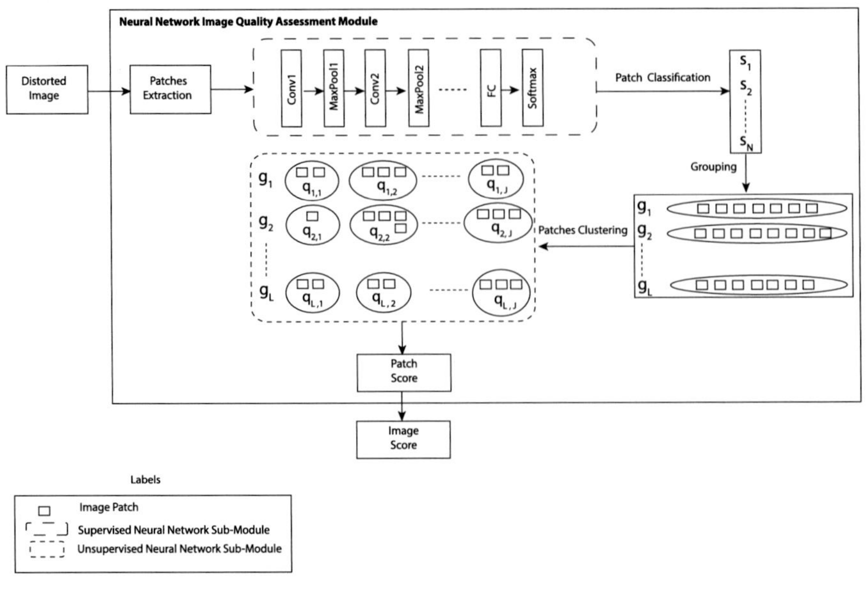

FIGURE 9.2

The neural network architecture for image quality assessment.

the quality of models. Therefore, this type of network is the most commonly used when processing and analyzing images since they have high dimensionality. For this reason, we employ the Convolutional Neural Networks as a supervised network. This module's input is the extracted patches of the distorted image, while the output is the predicted quality score of each patch. This module is trained using three objective quality methods:

- Structural Similarity Index (SSIM) [2].
- Perceptual Coherence Measure (PCM) [41].
- Content-Based Image Retrieval (CBIR) [4].

The Convolutional Neural Networks will first extract the images' features and then predict the quality score through its layers: the convolutional, the max-pool, the fully connected, and the softmax. The CNN uses the softmax layer to predict an input image's probabilities belonging to each given class. In our case, we define eleven classes that represent the predicted quality scores and expressed as $S = [s_1, ..., s_N]$ with $0 \leq s \leq 1$, the margin between the scores is equal to 0.1 and N is the total number of classes; hence the predicted output vector of probabilities, denoted as $Pr = [pr_{s_1}, ..., pr_{s_N}]$, will be assigned to the image quality scores with $\sum_{i=1}^{N} pr_{s_i} = 1$ where pr_{s_i} represents the probability of a quality score falling in the ith bucket.

9.5.1.2 Unsupervised neural network module

After predicting each patch's classes, we grouped the patches of similar quality into the same group, denoted by g_l, where each group has the same value as the previous classes. We then cluster those patches in the same quality group into different clusters based on their:

- Color feature: Color is the most critical and straight-forward feature that humans perceive when viewing an image. The human vision system is more sensitive to color information than gray levels, so color is the first candidate used for feature extraction. There are many color spaces used globally. Each of those color spaces was designed for certain applications and based on certain considerations. The most common one used for images is the RGB color space. So, we first extract these colors from each patch and then normalize each color's value between 0 and 1 by performing two consecutive divisions: 1) first by 255 on each pixel color, 2) then by M × N (the size of the patch) on the sum of the pixels value. As a result, we will obtain a 3D color vector denoted by $\mathbf{C}_{\mathbf{im'}_{p_i}} = [\mathbf{R}_{\mathbf{im'}_{p_i}}, \mathbf{G}_{\mathbf{im'}_{p_i}}, \mathbf{B}_{\mathbf{im'}_{p_i}}]$, where **R**, **G**, and **B** represent the red, green, and blue colors. However, and in order to reduce the dimension of the vector, we find the gray-scale of the patch by calculating the mean of $\mathbf{C}_{\mathbf{im'}_{p_i}}$.
- Contrast feature: Contrast is the difference in intensity that makes the object distinguishable from other objects within the same field of view. Several methods are used to quantify the contrast. In our work, we use the RMS contrast because, as mentioned in [42], it is the best index of detectability. This feature can be computed as follows:

$$RMS_{contrast}(im'_{p_i}) = \sqrt{\frac{1}{MN}\sum_{k=0}^{N-1}\sum_{j=0}^{M-1}\left(\frac{I_{kj}-\bar{I}}{\bar{I}}\right)^2} \tag{9.1}$$

  where I_{kj} is the intensity of the kth and jth element in a patch of size M by N, $\bar{I}$ is the average intensity of all pixel values in the patch as we assume that the values are normalized in the range [0, 1]. As a result, $RMS_{contrast}(im'_{p_i})$ will return a value in the same range.
- Texture feature: Texture is a very useful characterization for a wide range of images. It is generally believed that human visual systems use texture for recognition and interpretation. Texture is the key function used in image processing and computer vision to describe a given object's structure, which is a vital step in ensuring that an object (such as a face in our work) can still be detected. According to [43], the majority of methods are categorized under statistical and transform-based methods or a combination of these methods. The authors show that the Local Binary Patterns (LBP) method achieves high classification accuracy at low computational costs. Given a centered pixel, the operator makes use of its 3 × 3 square neighborhood. Then, it gives each of the neighboring pixels a value equals to 0 or 1 (as shown in Eq. (9.3)), where the resulted values are read as clockwise

or anticlockwise. In our work, the result will be obtained by reading the values in the clockwise direction, starting from the upper left neighbor. The calculation of LBP is shown in the equation below:

$$LBP(x_c, y_c) = \sum_{p=0}^{7} S(l_p - l_c)^{2P} \tag{9.2}$$

Where:

$$S(l_p - l_c) = \begin{cases} 1, & \text{if } l_p - l_c > 0. \\ 0, & \text{otherwise.} \end{cases} \tag{9.3}$$

With x_c and y_c are the central pixel coordinates, l_p denotes the gray-level value of the neighboring pixel p, and l_c is the gray-level value of the central pixel. $LBP(x_c, y_c)$ will return a decimal value between 0 and 255 as we divided by 255 to normalize the value between 0 and 1. This process will be applied to the entire patch, and as a result, we will obtain a vector of size $M \times N$ where we find the mean of this vector by dividing with the patch size; hence, we will obtain a value between 0 and 1.

The filtering outputs of im'_{pi} are then concatenated to form a 3D feature vector, denoted by F'_i. The quality clustering of $im'_{p_i} \in g_l$ is then performed by applying the K-mean clustering to F'_i using the Euclidean Distance:

$$\min_{\mathbf{q_{l,j}}} \sum_{im'_{p_i} \in g_{l,j}} \sum_{j=1}^{J} \sqrt{(F'_i - q_{1,j})^2} \tag{9.4}$$

where $g_{l,j}$ is the jth cluster in group g_l. We note that we use the Euclidean distance due to its low complexity cost. As a result, we get a set of centroids $\{q_{l,j}\}$ representing the quality levels vectors within each group g_l, where $j = 1, 2, ..., J$. Finally, we have L set of centroids on l different groups.

After learning the set of centroids $\{q_{l,j}\}$, we can easily find, for each given distorted image im', its quality by following the procedures: 1) patch partition and feature extraction, 2) cluster assignment to different quality groups, 3) patch quality score estimation, 4) and final image score.

- *Patch partition and feature extraction:* For the distorted image im', we partition it into K patches im'_{pi} (with $i \in 1, 2, ..., K$) and use the filters to extract the feature vector, denoted by F'_i, of each im'_{p_i}.
- *Cluster assignment:* By assuming that patches with similar structural, color and contrast features would have a similar quality, we find the closest centroid on each quality group l to the feature vector F'_i of patch im'_{p_i}. Then, we allocate im'_{pi} to L clusters by locating the nearest centroid, defined by q_{l,j_i}, inside each cluster. The quality of patch im'_{p_i} can be computed as the weighted average of these centroids' quality groups.

- *Patch quality estimation:* The distance between F_i' and its nearest centroid q_{l,j_i} within each group g_l is: $\min_{j \in 1...J} \sqrt{(F_i' - q_{l,j_i})^2}$. The shorter the distance is, the more likely patch im'_{p_i} should have the same quality as that of centroid q_{l,j_i}. Therefore, we use the following weighted-average rule to determine the final quality score of patch im'_{p_i}:

$$\mathcal{Q}_i = \frac{\sum_{l=1}^{L}(pr_{g_l} \times \overline{q_{l,j_i}})}{\sum_{l=1}^{L} pr_{g_l}} \tag{9.5}$$

where pr_{g_l} is the probability of a patch im'_{p_i} belonging to group g_l and $\overline{q_{l,j_i}}$ is the mean vector of the nearest quality level to F_i' within each group g_l.
- *Final image quality score:* After estimating the quality $\mathcal{Q}_i$ of each patch im'_{p_i}, we can find the final quality score, denoted by $\mathcal{S}$, of image im' by simply calculating the average of the patches' quality scores as follows:

$$\mathcal{S} = \frac{1}{K}\sum_{i=1}^{K}\mathcal{Q}_i \tag{9.6}$$

As a result, this function will return a score between 0 and 1. Higher scores mean quality preservation.

9.5.2 Face alignment anomaly detection

The goal of face alignment is to localize a set of predefined facial landmarks (eye corners, mouth corners, etc.) in an image containing faces. Face alignment is an important component of many computer vision applications, such as face verification, facial emotion recognition, human-computer interaction, and facial motion capture. In our work, we used the face alignment metric to ensure that the semantic information can still be preserved from a modified face affected by a manipulation function.

For this purpose, we train an unsupervised neural network to be able to detect anomalies in the face alignment. Anomaly detection is the identification of data, events, or observations that may significantly differ from the majority of the data. Therefore, we first extract the 68 facial landmark points from the face as we denote a facial landmark point by $l(x, y)$, with x and y the landmark point's coordinates. Some facial landmarks points are considered more representative than others. As stated in [44], the eyes, the nose, and the mouth are very representative parts of a person's face. For this reason, we refer to the points that are relative to these parts as *key points*, denoted by $k(x, y)$.

We then build a feature vector of a face, which can be compared with other faces' features, by calculating the distances between key points and facial landmarks. More specifically, we create a 5-points feature using five main key points, as shown in Fig. 9.3a: the centroids of the two eyes (k_1 and k_2), the center of the nose (k_3), and the sides of the mouth (k_4 and k_5). The centroids of the two eyes are calculated from

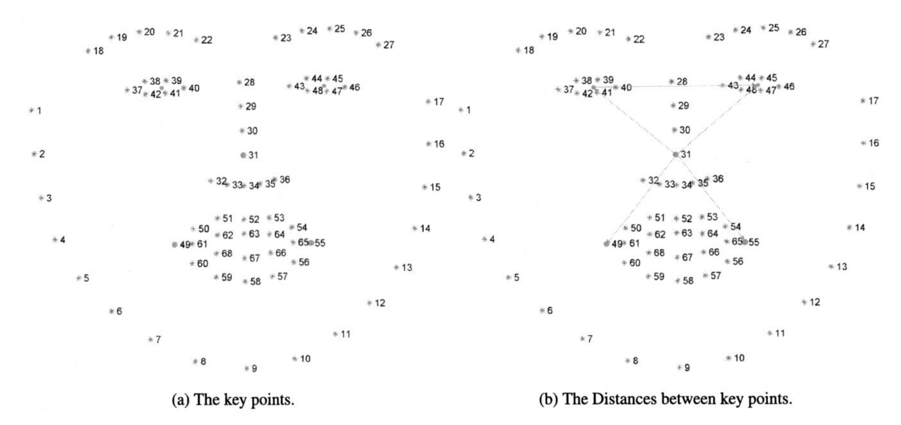

(a) The key points. (b) The Distances between key points.

FIGURE 9.3

The key points and the distances between them.

the six facial landmarks of each eye. For the nose's key point, we directly use the landmark #31 while selecting the landmarks #49 and #55 for the mouth's key points. We used these key points to compute the distance between the left eye centroid and the right eye centroid (Fig. 9.3b) as follows:

$$\mathcal{D}_{1,2}(so) = \frac{\sqrt{(k_1.x - k_2.x)^2 + (k_1.y - k_2.y)^2}}{h} \tag{9.7}$$

where h is the diagonal bounding box used to normalize the value and computed as: $h = \sqrt{so.w^2 + so.h^2}$. Similarly, we calculate the following four distances:

- $\mathcal{D}_{1,3}$: left eye centroid – nose.
- $\mathcal{D}_{2,3}$: right eye centroid – nose.
- $\mathcal{D}_{3,4}$: nose – left mouth.
- $\mathcal{D}_{3,5}$ nose – right mouth.

As a result, this produces a 5-dimensional float vector, denoted by FA, that we used as a 5-point feature of the face so. Then, we apply k-means on these vectors to create K clusters, denoted by $C = \{_1, c_2, ..., c_K\}$, where $K \in \mathbb{N}$. Therefore, we get a feature vector $\mathcal{F}_i$ that represents the centroid of cluster c_i. After completing the training phase, we assessed the facial landmarks distribution using the Euclidean Distance as follows:

$$d(so') = \min_{i \in 1...K, j \in 0...4} \sqrt{(FA'(j) - \mathcal{F}_i(j))^2} \tag{9.8}$$

The closer the distance is to 0, the more likely the facial landmarks will be aligned and will have the same distribution as cluster c_i. Moreover, if Eq. (9.8) returns a value greater than the standard deviation s_i of the closest cluster c_i, then it means that the positioning of the facial landmarks is not preserved. Thus, and after calculating the distance for a face, we can then check the face alignment for all of the faces contained in a distorted image im' as follows:

$$\mathcal{A}_{im'}(SO') = \frac{1}{count(SO')} \sum_{so' \in SO'} d(so'_m) \tag{9.9}$$

Finally, we subtracted the resulting value from 1 (as shown in the equation below) to adjust the measure with the previous subsection's outputted score. Therefore, higher scores will lead to face alignment preservation.

$$\mathcal{FA} = 1 - \mathcal{A}_{im}(SO') \tag{9.10}$$

9.5.3 Image score

Finally, we find the final image quality score by combining the previous methods. We compute the Image Score, denoted by IS, as follows:

$$IS = \frac{w_1 \times \mathcal{FA} + w_2 \times \mathcal{S}}{\sum_{i=1}^{2} w_i} \tag{9.11}$$

where w_1 and w_2 are weights between 0 and 1 with their sum equal to 1. The administrator chooses these weights to indicate the importance of each method based on his preferred features. The value of IS has a range from 0 to 1, and higher scores indicate quality preservation.

9.6 Framework

An overview of our framework is shown in Fig. 9.4. It consists of two main modules:

- Stream Processing.
- Back-end.

In the following, we present in detail the framework's modules.

9.6.1 Stream processing module

In this module, administrators query continuous data streams and detect conditions within a short amount of time from the date of receipt of these data. In our work, we took Twitter as a source for multimedia data streaming while processing only images.

As shown in Fig. 9.4, our framework has the ability to process two kinds of images:

- A distorted image.
- A distortion-free image.

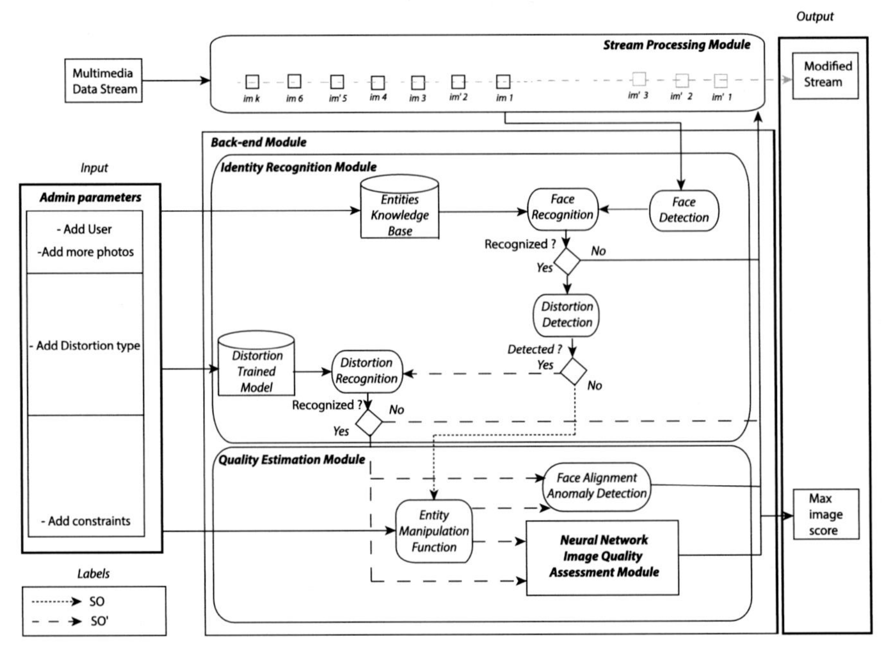

FIGURE 9.4

Framework.

Moreover, the images in the stream are marked from one to k, where k is equal to an infinite number, indicating that we are processing images without any bounds. In the end, im' is the resulted images that are returned from the back-end module.

9.6.2 Back-end module

It consists of two main submodules: a) Identity Recognition and b) Quality Estimation.

The first one is responsible of detecting and recognizing:

- The entities.
- The distortion if available.

These previous works are achieved using six different components: 1) *Face Detection*, 2) *Face Recognition*, 3) *Entities Knowledge Base*, 4) *Distortion Detection*, 5) *Distortion Recognition*, and 6) *Distortion Trained Model.*

The second submodule's task is to estimate the image quality with the aid of the *Neural Network Image Quality Assessment Module* and the remaining components: 7) *Entity Manipulation Function*, and 8) *Face Alignment Anomaly Detection*. We will detail each component in the upcoming sections.

9.6.2.1 Identity Recognition Module

9.6.2.1.1 Face Detection

Face Detection is an essential component for face recognition. While receiving the multimedia data stream, this component is responsible for detecting the entities' faces who appeared in the images. However, before detecting these images, we applied a reduction in dimensionality, which reduces the image to more manageable groups that will decrease the time required for processing without missing essential details. The detected faces are then sent to the *Face Recognition* component.

9.6.2.1.2 Face Recognition

This component is used to recognize the faces by comparing the extracted face features with those stored in the *Entities Knowledge Base*, which will be detailed in the next section. If a match is found, the image is assigned to the *Distortion Detection*. Otherwise, the image will go back to the *Stream Processing* module.

9.6.2.1.3 Entities Knowledge Base

This component reflects the database in which the trained entities reside. An administrator can add new entities to the database to build his own schema and train more images to current entities. This will provide us with the opportunity to increase the recognition accuracy.

9.6.2.1.4 Distortion Detection

This component is responsible for checking if the image is distorted or not. If distortion is detected, the image is sent to the *Distortion Recognition* component. Otherwise, it will be headed to the *Entity Manipulation Function.*

9.6.2.1.5 Distortion Recognition

After detecting the distortion, this component has the ability to identify the type of this distortion by comparing the latter with those stored in the *Distortion Trained Model.* This component has a lot of importance in order to accurately predict the distorted image in the *Quality Estimation Module.* However, if the distortion is not recognized, it will be redirected to the *Stream Processing Module.*

9.6.2.1.6 Distortion Trained Model

This component represents the repository where different types of distortion are trained and stored. Our model can recognize five kinds of distortions: Median blurring, Gaussian blurring, Pixelate, additive Gaussian noise, and compression. Besides, an administrator can add more distortion types.

9.6.2.2 Quality Estimation Module

9.6.2.2.1 Entity Manipulation Function

This component's role is to alter the salient objects of the entities, which are considered the faces in our work. This component takes the entities as inputs and applies a series of image manipulation functions on the salient objects. As a result, it will return a set of modified salient objects. As shown in Fig. 9.4, the constraints, such as protecting/adapting the entities, could be added/selected by the administrator and applied using the image manipulation functions. We used four main protection functions: Pixelate, Gaussian blurring, Median blurring, and additive Gaussian noise. As for the adaptation functions, we used two compression techniques: lossy and lossless. In fact, the functions differ by means of the features that they preserve. For example, a median blur returns a modified image so that certain visual, semantic, and multimedia features are damaged while metadata and audio features remain intact. Considering the fact that each function maintains certain features, we use the previously mentioned list of functions and apply them to find the most suitable one by ensuring an acceptable image quality using the quality assessment functions.

9.6.2.2.2 Face Alignment Anomaly Detection

After modifying the faces, the first metric used to assess the image is the extraction and estimation of the facial features through the use of the *Face Alignment Anomaly Detection component*. In this component, we measure the facial landmarks' divergence between the distorted face and the normal facial landmarks points distribution to ensure that the semantic information is preserved and can still be extracted. As a result, the function will return a value between 0 and 1 that will be aggregated with the *Neural Network Image Quality Assessment Module*.

9.6.2.2.3 Neural Network Image Quality Assessment Module

This module represents a model that will help us predict the images' quality, especially those containing faces. We recall that we trained this model using three quality metrics: SSIM, CBIR, and PCM, before clustering the images into subquality groups. When the training phase is finished, the module will have the ability to determine the image quality by assessing several features from an image. For each manipulation function, it will return a score between 0 and 1. This module will allow us to select the manipulation function with the highest score and apply it to the image.

9.6.2.2.4 Max Image Score

The final component is the *Max Image Score* that is responsible of:

- Aggregating the scores, which are returned from the previous *Neural Network Image Quality Assessment Module* and the *Face Alignment Anomaly Detection* component while assigning them weights based on the administrator preferred features selection.
- Displaying the highest image score, which is calculated using Eq. (9.11).

Simultaneously, the No-Reference distorted images (im') coming from the stream and the recently modified images will return to the *Stream Processing Module* to be then published.

9.7 Experiments

In this section, we first present the experimental setup and protocol before testing our framework's efficiency.

9.7.1 Experimental setup

First of all, we start by training the CNN using two different scenarios:

- In the first one, we used the CSIQ [44] dataset. It comprises 30 reference images and 866 distorted images, with five distortions' types having each one five different levels: JPEG compression, JPEG-2000 compression, global contrast decrements, additive pink Gaussian noise, and Gaussian blurring. We focused in our work on Gaussian blurring, JPEG compression, and additive pink Gaussian noise. Since our goal here is to verify the validity of our approach by comparing the Neural Network prediction accuracy with the state-of-art methods, we train the CNN using SE-ResNeXt-101(32x4d) classifier due to its highest accuracy as highlighted in the dark gray color in Table 9.2.
- In the second one, we used CelebA [45] dataset. It contains 202,599 face images with 10,177 identities. In order to prepare our training set, we select 52,800 images. We then applied three manipulation functions on these images while considering many distortion levels: Pixelation (a.k.a. mosaicking), Gaussian blurring, and Median blurring. Those three functions are also the most commonly encountered distortions in practical applications. Our goal here is to use this network to predict faces' quality in real-time. For this reason, we train our CNN using ResNeXt-101(32x4d), which has a remarkable trade-off between accuracy, inference time, and memory consumption as shown in Table 9.2 in the light gray color.

In both scenarios, we partition the images into 64 patches and distribute the patches to the eleven classes according to their quality score calculated using the objective quality methods in section 9.5. We ensure that each manipulation function will have 1000 patches per class; hence, each class will contain 3000 distorted patches for training and 1/5 for validating. We recall that we choose eleven classes ranging from 0 to 1 while keeping a margin of 0.1 between each class. We tried several scenarios by minimizing and maximizing the margin between the classes, and we noticed that the margin of 0.1 gives us the best result in terms of accuracy.

Secondly, and to train the unsupervised neural network, we create in each quality group 15 clusters (we select this number of clusters after performing several scenarios and finding a good trade-off between accuracy and inference time). As a result, and in each quality group, we will get 15 centroids representing the subquality levels. Finally, we select 100 celebrities, each having 50 images, from the CelebA dataset and extract their facial landmarks to train our Face Alignment Anomaly Detection Network with a normal facial landmarks points distribution. We create in this training 2 clusters by referring to two main metrics: Inertia (Intra Distance) and Dunn-index (Inter Distance). Lower Inertia values mean that clusters are internally coherent by

Table 9.2 Benchmark analysis of convolutional neural network architectures [46].

Classifier	Accuracy in %	Inference time in ms	Memory consumption in GB
ResNet-18	71	0.41	0.75
ResNet-34	74.4	0.71	1.09
ResNet-50	77.4	1.37	1.28
ResNet-101	78.4	2.29	1.37
ResNet-152	80.6	3.30	1.43
ResNeXt-101(32x4d)	80.4	3.25	0.87
ResNeXt-101(64x4d)	80.7	5.62	1.19
SE-ResNeXt-101(32x4d)	80.9	4.53	1.16
Inception v4	80.0	3.77	0.9
Inception-ResNet v2	80.3	4.88	0.95

ensuring that the distance between the points and the centroids is small. The more the Dunn-index value is, the better will be the clusters by taking into account the distance between two clusters. We notice that the Dunn-index value drops by 0.57 between clusters 2 and 3, while the Inertia value decreases from 12.145 to 10.742. For this purpose, the optimal number of clusters is 2.

Finally, we had to collect images for each of the celebrities to train an identity recognition model. However, the CelebA dataset is not intended for face recognition tasks. Thus, we first grouped images of the same individuals based on the identity annotations provided by the CelebA dataset. We then considered the celebrities that have 30 images and more, which resulted in almost 2600 identities.

After finishing the training above, a Java software is developed using eclipse on a desktop computer having a 2.66 GHz core 2 duo and 4 GB RAM running Linux Ubuntu 14.04 64 bit. After running the program on one computer, the framework described in section 9.6 is deployed in a distributed environment called Apache Storm [47]. In order for the storm cluster to run successfully, we must implement all of its components. To do so, we show in Tables 9.3 and 9.4 the number of computers used as well as the necessary libraries.

9.7.2 Experimental protocol

We conducted three sets of experiments as shown below:

- Firstly, we assessed our Neural Network's accuracy by predicting the quality score of the no-reference images from two datasets: LIVE [50], and TID2013 [32]. A subjective quality score, i.e., the mean opinion score (MOS), is assigned for each image in the dataset. The LIVE database consists of 29 original images along with their 779 distorted versions having five types of distortions on various levels: JPEG2000 compression (JP2K), JPEG compression, additive white Gaussian noise (WN), Gaussian blurring (GB), and simulated fast fading Rayleigh channel

Table 9.3 Showing the Apache Storm Configuration.

Apache Storm Configuration	
Machine	**Service**
Client node [48]	It tests the framework locally before deploying it to the cluster
Nimbus node [49]	It is the master in a Storm cluster that is responsible for distributing the application code across various worker nodes.
Three Zookeeper nodes [49]	They handle the communication between the Nimbus and the supervisors.
Eleven Supervisor nodes [49]	Each worker node runs a daemon called the Supervisor, which listens for work assigned to its machine and starts and stops worker processes based on what Nimbus has assigned to it.

Table 9.4 Showing the needed libraries for Apache Storm.

Implemented Libraries	
Libraries' name	**Service**
Apache Storm 0.9.3 and zookeeper 3.4.6	They should be distributed and installed on all nodes to successfully run the storm cluster.
OpenCV 3.4.3 API, Python dlib library and MTCNN	We use the manipulation functions from OpenCV, facial landmarks extraction using dlib and performing face detection via MTCNN.
Trained ResNet Model	It is used to recognize the distortion and the faces.

(FF). The TID2013 database is composed of 25 original images and 3000 distorted images with 24 types of distortions. But, and as in many previous works [36,51,52], we only consider three types of distortions that are common to the two databases: JPEG compression, additive Gaussian noise, and Gaussian blurring. We then compare our results with the state-of-art methods.

- Secondly, we assessed the quality of the images from Twitter Stream that may be affected after applying a manipulation function. We limit the processed images' size to 9000 as our goal is to determine the image quality using the Neural Network and the Face Alignment defined in section 9.5. To do so, we started our scenario by filtering the number of faces contained in the images from 1 to 3 and applying several manipulation functions to find the suitable one that will return the best score in terms of quality. In these two scenarios, the prototype is tested only on a local cluster without being uploaded to the distributed system.
- Thirdly, we implemented our framework on apache storm distributed system as we evaluated in real-time its performance in terms of:

1. Execution latency: The average amount of time it takes for an image to be executed.
2. Number of nodes: Number of supervisors engaged in processing the images.

To do so, the following scenario was executed:

1. We distributed the libraries on all nodes.
2. We processed 50,000 images from Twitter Stream.
3. Finally, we uploaded the framework to the cluster.

We start by uploading the framework on two nodes as the number of nodes will be incremented by 2 to evaluate Apache Storm's performance, mainly focusing on the execution latency by comparing each run.

9.7.3 Results

9.7.3.1 Performances comparison

The objective of this test is to evaluate the performance of our neural network by comparing its prediction scores of the no-reference images to the subjective ratings. As mentioned before, the two largest publicly available subject-related databases used are: LIVE [50], and TID2013 [32]. Two correlation coefficients between the prediction results and the subjective scores have been adopted to evaluate the performance of our method: the Spearman Rank Order Correlation Coefficient (SROCC), which is a measurement of the strength and direction of the relationship between two variables using a monotonic function, and the Pearson Correlation Coefficient (PCC), which measures the linear correlation between two variables. A high correlation coefficient (close to 1) with the subjective score MOS indicates a good method. After processing the images from LIVE and TID2013 datasets, we obtain the results along with the state-of-arts methods shown in Tables 9.5 and 9.6.

We start first by comparing our method to the FR IQA metrics: PSNR and SSIM. In general, Table 9.5 shows that our neural network has better results than the FR metrics in the overall LIVE dataset, especially for the Gaussian blurring and Gaussian noise. Table 9.6 shows that our approach gives a good SROCC and PCC in the overall TID2013 dataset and outperforms the PSNR metric on JPEG compression. Moreover, our approach has almost the same overall SROCC and PCC as SSIM. Furthermore, and in both datasets, we can notice that our results are highly correlated with the SSIM metric since the latter was taken into consideration when we trained our neural network.

As for the NR IQA metrics, we can notice that our approach achieves state-of-the-art performance on the two databases. While our method has a slight advantage towards the JPEG compression and Gaussian blurring, our approach reaches better average values for all the distortions. Second, our method outperforms CORNIA (the best method in Table 9.5), which is considered as an unsupervised feature-learning IQA framework using codebook representations. Third, our approach has slightly better results than the deepIQA. We believe the performance improvement arises in both datasets because:

- Our proposed method classifies first the images using a supervised network (the CNN in our work) before clustering those images into groups. This process assisted in the improvement of the unsupervised neural network accuracy.

Table 9.5 Image Quality Assessment on LIVE. Italicized are Full-Reference methods.

Manipulation functions	Gaussian blurring		Gaussian noise		JPEG compression		Average values	
Correlation coefficients	**PCC**	**SROCC**	**PCC**	**SROCC**	**PCC**	**SROCC**	**PCC**	**SROCC**
PSNR	0.812	0.814	0.987	0.984	0.896	0.894	0.898	0.897
SSIM [2]	0.951	0.952	0.972	0.970	**0.980**	**0.974**	0.967	0.965
CORNIA [28]	**0.968**	**0.969**	0.987	0.976	0.965	0.955	0.973	0.967
DIIVINE [6]	0.923	0.921	0.988	0.984	0.921	0.910	0.944	0.938
BRISQUE [7]	0.943	0.947	**0.992**	**0.987**	0.966	0.961	0.967	0.965
ILNIQE [8]	0.936	0.927	0.976	0.979	0.966	0.944	0.959	0.950
dipIQ [38]	0.948	0.940	0.983	0.975	**0.980**	0.969	0.970	0.961
BLISS [37]	0.948	0.936	0.978	0.967	0.972	0.956	0.966	0.953
Our approach	0.965	0.961	0.989	0.985	0.978	0.973	**0.977**	**0.973**

Table 9.6 Image Quality Assessment on TID2013. Italicized are Full-Reference methods.

Manipulation functions	Gaussian blurring		Gaussian noise		JPEG compression		Average values	
Correlation coefficients	**PCC**	**SROCC**	**PCC**	**SROCC**	**PCC**	**SROCC**	**PCC**	**SROCC**
PSNR	**0.958**	0.965	**0.963**	**0.942**	0.925	0.929	**0.948**	**0.945**
SSIM [2]	**0.958**	**0.969**	0.902	0.896	0.968	0.935	0.942	0.933
CORNIA [28]	0.934	0.934	0.778	0.798	0.960	0.912	0.890	0.881
DIIVINE [6]	0.860	0.859	0.882	0.879	0.696	0.680	0.812	0.806
BRISQUE [7]	0.884	0.886	0.886	0.889	0.950	0.894	0.906	0.889
ILNIQE [8]	0.816	0.815	0.899	0.890	0.944	0.873	0.886	0.859
dipIQ [38]	0.928	0.922	0.906	0.905	**0.973**	0.932	0.935	0.919
BLISS [37]	0.897	0.910	0.943	0.938	0.960	0.921	0.933	0.923
Our approach	0.952	0.951	0.905	0.901	0.962	**0.958**	0.939	0.936

- Patches instead of images are used as inputs to train the neural network, which lowers the quality score classification error by learning the distorted patches' score.

9.7.3.2 Evaluating the images quality after applying a manipulation function

This test aims to evaluate the images' quality that may be affected after applying a manipulation function as our goal to find the appropriate function, which will return the highest quality score based on the constraints imposed by an administrator/user. We use in this study three manipulation functions, which are mainly considered as protection functions: Pixelation (a.k.a. mosaicking), Gaussian blurring, and Median

Table 9.7 Parameters list.

Manipulation functions parameters list			
Manipulation functions	**Gaussian Blur**	**Median Blur**	**Pixelate**
Kernel size	31×31	31×31	–
standard deviation	5	5	–
Pixel size	–	–	10
Quality methods weights			
Quality methods	$\mathcal{S}$	$\mathcal{FA}$	–
Weights	0.4	0.6	–

Table 9.8 Quality scores when applying a manipulation function, where * represents the number of persons.

Quality scores before and after adding weights for each manipulation function						
Manipulation functions	**Gaussian Blur**		**Median Blur**		**Pixelate**	
Applying weights	**Before**	**After**	**Before**	**After**	**Before**	**After**
$\mathcal{S}$ 1*	0.9851	0.3940	0.9838	0.3935	0.9841	0.3936
$\mathcal{S}$ 2*	0.9794	0.3917	0.9779	0.3911	0.9777	0.3911
$\mathcal{S}$ 3*	0.9779	0.3911	0.9771	0.3908	0.9770	0.3908
$\mathcal{FA}$ 1*	0.9742	0.5845	0.9264	0.5558	0.9515	0.5709
$\mathcal{FA}$ 2*	0.9646	0.5787	0.9124	0.5474	0.9374	0.5624
$\mathcal{FA}$ 3*	0.9351	0.5610	0.8904	0.5342	0.9284	0.5570

blurring. We choose fixed parameters for each manipulation function while providing the users the possibility to select a weight for each quality method based on their preferred image features. Table 9.7 shows the manipulation functions' parameters and the quality methods weights.

According to the literature, we selected the parameters of the manipulation functions since they are considered average intensity values for each manipulation function while giving a higher priority for the face alignment over the remaining features due to its high importance in face detection and facial expression analysis. In order to find the most appropriate manipulation function, we treat 9000 images from Twitter Stream as we partition them into three equal parts based on the number of persons (from one to three persons) contained in each image. After applying the manipulation functions on each image, we obtain Table 9.8 and the graph shown in Fig. 9.5. These results represent the predicted quality from the Neural Network ($\mathcal{S}$) and the face alignment ($\mathcal{FA}$) average values for each manipulation over 9000 images.

According to the below graph (Fig. 9.5), the manipulation function with the highest image score is the Gaussian blur. Moreover, and as stated in Table 9.7, this function will meet the users' needs since it preserves the structure, contrast, color features, and most importantly the location of the facial landmarks points that are assessed by the Neural Network ($\mathcal{S}$) and face alignment ($\mathcal{FA}$).

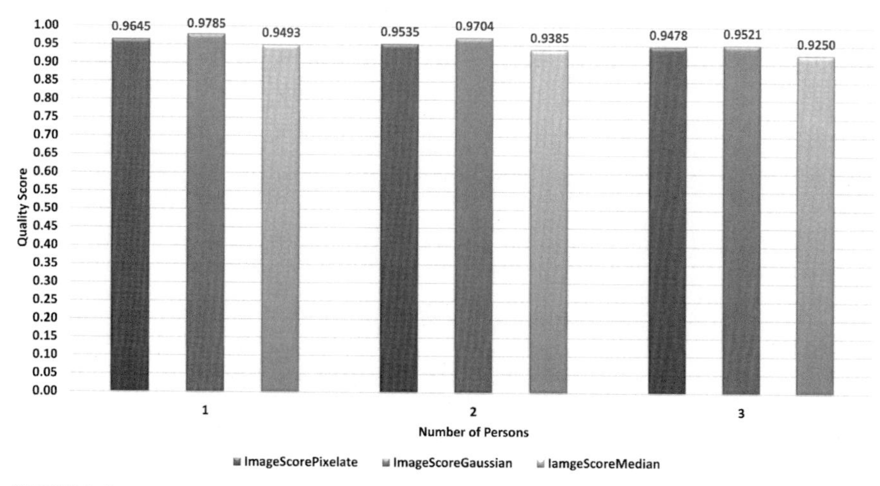

FIGURE 9.5

The dependence between a manipulation function and the image quality score.

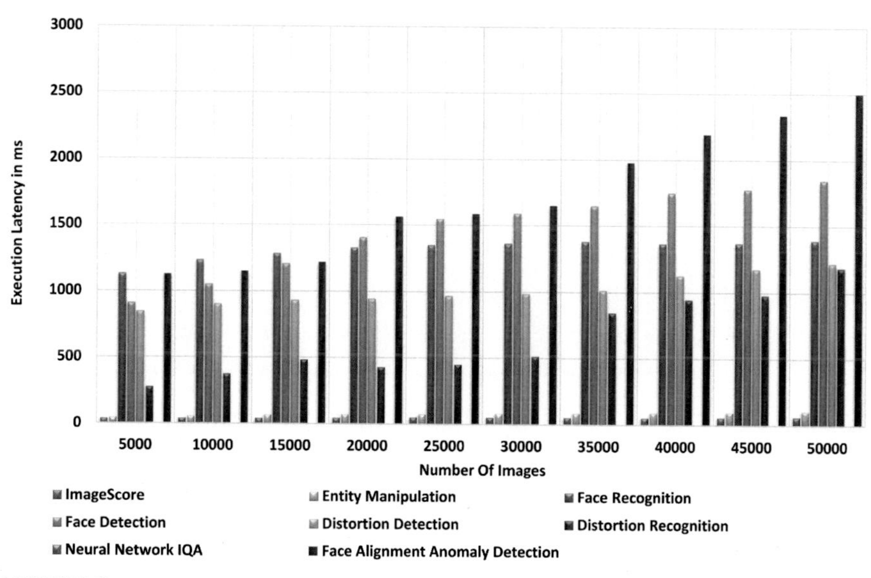

FIGURE 9.6

The dependence between execution latency and number of images at each component.

9.7.3.3 Evaluating the framework in real-time

Since IQA measures are often used in real-time applications, speed is an important issue in determining whether an IQA measure can be used in these applications. For this purpose, we treat 50 000 images from Twitter stream. We deployed our framework in an apache storm distributed system to measure the execution latency at each

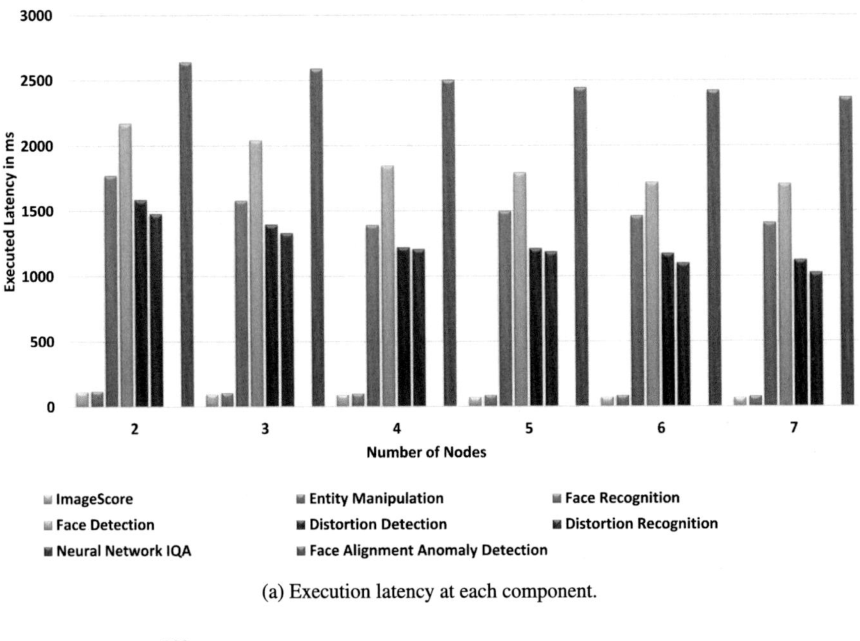

(a) Execution latency at each component.

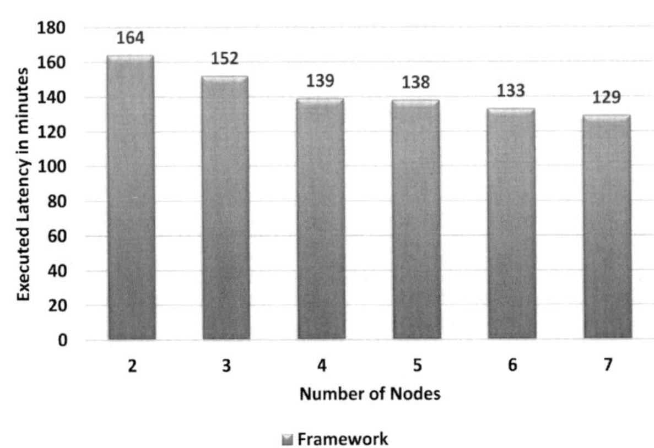

(b) Framework execution latency.

FIGURE 9.7

Execution latency related to the number of nodes.

component by conducting two sets of experiments. In the first one, we vary the number of images from 5000 to 50 000 while fixing the number of nodes to 4. As a result, we obtain the graph shown in Fig. 9.6.

In this figure, we can see that when the number of images is incremented by 5000, the executed latency gets higher values due to the fact that a node needs to execute more images in each run.

In the second set, we vary the number of nodes from 2 to 7 while fixing the number of images to 50 000. Therefore, the results are shown in Fig. 9.7a and 9.7b.

We notice that while incrementing the number of nodes, the time required to process the images has decreased due to the fact that the number of workers is increasing, which may lead to improve the performance of Apache Storm.

9.8 Conclusion

In this paper, we presented a framework that intends to achieve a No-Reference and Full-Reference images' quality estimation that may be distorted during the processing and transmission phase while treating these images in real-time. We focus on assessing the faces' quality in the images using a Convolutional Neural Network that represents the supervised network followed by an unsupervised network; we first trained the CNN by using three quality metrics before clustering the images into groups using the unsupervised network. We also used face alignment anomaly detection as a second metric to estimate the faces' quality. A set of experiments has been tested in order to evaluate our approach.

In future work, we intend to provide a framework that can process a wide range of multimedia data, such as videos and audio. Moreover, we aim to consider more features in assessing the images to improve quality prediction accuracy.

References

[1] Social media statistics in 2020, https://dustinstout.com/social-media-statistics/#instagram-stats. (Accessed 24 January 2020).

[2] Zhou Wang, A.C. Bovik, H.R. Sheikh, E.P. Simoncelli, Image quality assessment: from error visibility to structural similarity, IEEE Transactions on Image Processing 13 (4) (2004) 600–612.

[3] Q. Huynh-Thu, Scope of validity of psnr in image/video quality assessment, Electronics Letters 44 (June 2008) 800–801 [Online], available: https://digital-library.theiet.org/content/journals/10.1049/el_20080522.

[4] V.N. Gudivada, V.V. Raghavan, Content based image retrieval systems, Computer 28 (9) (1995) 18–22.

[5] S. Lai, Z. Chai, S. Li, H. Meng, M. Yang, X. Wei, Enhanced normalized mean error loss for robust facial landmark detection, in: BMVC, 2019.

[6] A.K. Moorthy, A.C. Bovik, Blind image quality assessment: from natural scene statistics to perceptual quality, IEEE Transactions on Image Processing 20 (12) (2011) 3350–3364.

[7] A. Mittal, A.K. Moorthy, A.C. Bovik, Blind/referenceless image spatial quality evaluator, in: 2011 Conference Record of the Forty Fifth Asilomar Conference on Signals, Systems and Computers (ASILOMAR), 2011, pp. 723–727.

[8] L. Zhang, L. Zhang, A.C. Bovik, A feature-enriched completely blind image quality evaluator, IEEE Transactions on Image Processing 24 (8) (2015) 2579–2591.

[9] M.A. Saad, A.C. Bovik, C. Charrier, Blind image quality assessment: a natural scene statistics approach in the dct domain, IEEE Transactions on Image Processing 21 (8) (2012) 3339–3352.

[10] D. Ciregan, U. Meier, J. Schmidhuber, Multi-column deep neural networks for image classification, in: 2012 IEEE Conference on Computer Vision and Pattern Recognition, 2012, pp. 3642–3649.

[11] S. Bianco, L. Celona, P. Napoletano, R. Schettini, On the use of deep learning for blind image quality assessment, Signal, Image and Video Processing 12 (2018) 355–362.

[12] W. Hou, X. Gao, D. Tao, X. Li, Blind image quality assessment via deep learning, IEEE Transactions on Neural Networks and Learning Systems 26 (6) (2015) 1275–1286.

[13] B. Girod, What's Wrong with Mean-Squared Error?, MIT Press, Cambridge, MA, USA, 1993, pp. 207–220.

[14] M. Narwaria, W. Lin, Svd-based quality metric for image and video using machine learning, IEEE Transactions on Systems, Man and Cybernetics Part B Cybernetics 42 (2) (2012) 347–364.

[15] S. Pei, L. Chen, Image quality assessment using human visual dog model fused with random forest, IEEE Transactions on Image Processing 24 (11) (2015) 3282–3292.

[16] O. Rippel, L. Bourdev, Real-time adaptive image compression, 2017.

[17] Q. Li, W. Lin, Y. Fang, No-reference quality assessment for multiply-distorted images in gradient domain, IEEE Signal Processing Letters 23 (4) (2016) 541–545.

[18] V. El-Khoury, N. Bennani, D. Coquil, Utility function for semantic video content adaptation, in: Proceedings of the 12th International Conference on Information Integration and Web-Based Applications and Services, ser. iiWAS '10, Association for Computing Machinery, New York, NY, USA, 2010, pp. 921–924 [Online], available: https://doi.org/10.1145/1967486.1967649.

[19] M. Mu, E. Cerqueira, F. Boavida, A. Mauthe, Quality of experience management framework for real-time multimedia applications, IJIPT 4 (2009) 54–64.

[20] A. Rueangprathum, S. Limsiroratana, S. Witosurapot, User-driven multimedia adaptation framework for context-aware learning content service, Journal of Advances in Information Technology 7 (2016) 182–185.

[21] J. Caviedes, S. Gurbuz, No-reference sharpness metric based on local edge kurtosis, in: Proceedings, International Conference on Image Processing, vol. 3, 2002, pp. III–III.

[22] P. Marziliano, F. Dufaux, S. Winkler, T. Ebrahimi, A no-reference perceptual blur metric, in: Proceedings, International Conference on Image Processing, vol. 3, 2002, pp. III–III.

[23] Z. Chami, B. Al Bouna, C. Abou Jaoude, R. Chbeir, A weighted feature-based image quality assessment framework in real-time, in: Transactions on Large-Scale Data- and Knowledge-Centered Systems XLV: Special Issue on Data Management and Knowledge Extraction in Digital Ecosystems, vol. 12390, 2020, p. 85.

[24] V. Hosu, H. Lin, T. Sziranyi, D. Saupe, Koniq-10k: an ecologically valid database for deep learning of blind image quality assessment, IEEE Transactions on Image Processing 29 (2020) 4041–4056.

[25] H. Lin, V. Hosu, D. Saupe, Deepfl-iqa: Weak supervision for deep iqa feature learning, 2020.

[26] L. He, D. Tao, X. Li, X. Gao, Sparse representation for blind image quality assessment, in: 2012 IEEE Conference on Computer Vision and Pattern Recognition, 2012, pp. 1146–1153.

[27] A.K. Moorthy, A.C. Bovik, A two-step framework for constructing blind image quality indices, IEEE Signal Processing Letters 17 (5) (2010) 513–516.

[28] P. Ye, J. Kumar, L. Kang, D. Doermann, Unsupervised feature learning framework for no-reference image quality assessment, in: 2012 IEEE Conference on Computer Vision and Pattern Recognition, 2012, pp. 1098–1105.
[29] M.A. Saad, A.C. Bovik, C. Charrier, A dct statistics-based blind image quality index, IEEE Signal Processing Letters 17 (6) (2010) 583–586.
[30] M.A. Saad, A.C. Bovik, C. Charrier, Dct statistics model-based blind image quality assessment, in: 2011 18th IEEE International Conference on Image Processing, 2011, pp. 3093–3096.
[31] H. Tang, N. Joshi, A. Kapoor, Learning a blind measure of perceptual image quality, in: CVPR 2011, 2011, pp. 305–312.
[32] N. Ponomarenko, L. Jin, O. Ieremeiev, V. Lukin, K. Egiazarian, J. Astola, B. Vozel, K. Chehdi, M. Carli, F. Battisti, C.-C. Jay Kuo, Image database tid2013: peculiarities, results and perspectives, Signal Processing Image Communication 30 (2015) 57–77, https://doi.org/10.1016/j.image.2014.10.009 [Online].
[33] Live image quality assessment database, [Online], available: http://live.ece.utexas.edu/research/quality/subjective.htm.
[34] A. Abaza, Design and evaluation of photometric image quality measures for effective face recognition, IET Biometrics 3 (December 2014) 314–324 [Online], available: https://digital-library.theiet.org/content/journals/10.1049/iet-bmt.2014.0022.
[35] A. Dutta, R.N.J. Veldhuis, L.J. Spreeuwers, Predicting face recognition performance using image quality, CoRR, arXiv:1510.07119, 2015 [Online].
[36] S. Bosse, D. Maniry, K. Müller, T. Wiegand, W. Samek, Deep neural networks for no-reference and full-reference image quality assessment, IEEE Transactions on Image Processing 27 (1) (2018) 206–219.
[37] P. Ye, J. Kumar, D. Doermann, Beyond human opinion scores: blind image quality assessment based on synthetic scores, in: Proceedings of the IEEE Conference on Computer Vision and Pattern Recognition (CVPR), June 2014.
[38] K. Ma, W. Liu, T. Liu, Z. Wang, D. Tao, dipiq: blind image quality assessment by learning-to-rank discriminable image pairs, IEEE Transactions on Image Processing 26 (8) (2017) 3951–3964.
[39] H. Zhu, L. Li, J. Wu, W. Dong, G. Shi, Metaiqa: deep meta-learning for no-reference image quality assessment, in: Proceedings of the IEEE/CVF Conference on Computer Vision and Pattern Recognition (CVPR), June.
[40] M.T. Vega, D.C. Mocanu, J. Famaey, S. Stavrou, A. Liotta, Deep learning for quality assessment in live video streaming, IEEE Signal Processing Letters 24 (6) (2017) 736–740.
[41] Z. Chami, B. Al Bouna, C. Abou Jaoude, R. Chbeir, A real-time multimedia data quality assessment framework, 11 2019, pp. 270–276.
[42] P.J. Bex, W. Makous, Spatial frequency, phase, and the contrast of natural images, Journal of the Optical Society of America A 19 (6) (Jun 2002) 1096–1106 [Online], available: http://josaa.osa.org/abstract.cfm?URI=josaa-19-6-1096.
[43] L. Armi, S. Fekri-Ershad, Texture image analysis and texture classification methods – a review, 2019.
[44] Z. Lian, Y. Li, J. Tao, J. Huang, M.-Y. Niu, Expression analysis based on face regions in read-world conditions, International Journal of Automation and Computing 17 (2019) 04.
[45] Z. Liu, P. Luo, X. Wang, X. Tang, Deep learning face attributes in the wild, in: Proceedings of International Conference on Computer Vision (ICCV), December 2015.
[46] S. Bianco, R. Cadène, L. Celona, P. Napoletano, Benchmark analysis of representative deep neural network architectures, IEEE Access 6 (10 2018) 64270–64277.

[47] Apache storm – concepts, [Online], available: http://storm.apache.org/releases/current/Concepts.html, 2015.

[48] Setting up a development environment, [Online], available: http://storm.apache.org/releases/1.0.6/Setting-up-development-environment.html, 2015.

[49] Apache storm cluster architecture, [Online], available: http://storm.apache.org/releases/1.0.6/Setting-up-development-environment.html, 2018.

[50] H.R. Sheikh, M.F. Sabir, A.C. Bovik, A statistical evaluation of recent full reference image quality assessment algorithms, IEEE Transactions on Image Processing 15 (11) (2006) 3440–3451.

[51] X. Liu, J. van de Weijer, A.D. Bagdanov, Rankiqa: learning from rankings for no-reference image quality assessment, in: Proceedings of the IEEE International Conference on Computer Vision (ICCV), Oct. 2017.

[52] K.-Y. Lin, G. Wang, Hallucinated-iqa: no-reference image quality assessment via adversarial learning, in: Proceedings of the IEEE Conference on Computer Vision and Pattern Recognition (CVPR), June 2018.

CHAPTER

10 A computational approach to understand building floor plan images using machine learning techniques

Shreya Goyal[a], Chiranjoy Chattopadhyay[a], and Gaurav Bhatnagar[b]
[a]*Department of Computer Science and Engineering, Indian Institute of Technology Jodhpur, India*
[b]*Department of Mathematics, Indian Institute of Technology Jodhpur, India*

10.1 Introduction

In the modern digital age, the use of paper based documents, still remains popular, despite the existence of electronic documents. Paper based documents are digitized in the form of images by scanning, for storage, future retrieval purpose, and reduce the amount of paper used. However, with growth in the digital data, the complexity in their storage and understanding has increased accordingly. These digitized documents or document images are the electronic form of paper based documents and easy to store, retrieve, and transmit. Due to growing popularity of digitized paper based documents, Document Image Analysis (DIA), became an important area of research. DIA systems are the applications of computer vision, machine learning, and image processing and carry their importance in understanding and recognition of digitized documents. With growth in the digital document images, a human like automatic document understanding is need of the hour.

Digital documents consist of textual and graphical components, which are required to be recognized. The textual part is analyzed with Optical Character Recognition (OCR) techniques and graphic recognition require extensive signature based or machine learnt methods, which makes it another important research area. The graphical components/graphical images are the binary-valued symbols having pixel values 0 and 1, which makes them different from natural images and require special feature descriptors and methods for their accurate recognition. A few examples of graphical documents are, engineering drawings, architectural floor plans, circuit diagrams, maps, flow diagrams, musical notations, optical characters, mathematical expressions, and other form of line drawings. These graphical documents can be recognized using computer vision techniques and converted into other modalities for interpreting them in common user readable form.

Internet of Multimedia Things (IoMT). https://doi.org/10.1016/B978-0-32-385845-8.00015-0

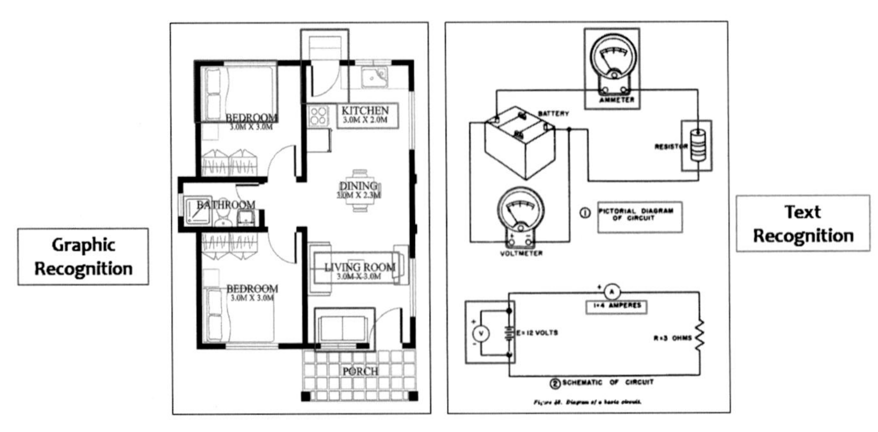

FIGURE 10.1

Illustration of various subareas of document image analysis (DIA) on a sample scanned document [1].

Fig. 10.1 depicts two examples of tasks which are performed in graphical document images. The image shows two different kinds of document images, architectural floor plan, and circuit diagram. Document recognition covers different research aspects such as graphical symbol recognition and classification. Text recognition is another research aspect covered under document image analysis. Graphical documents such as architectural floor plans and circuit diagrams contain a wide range of symbols and textual components including numerical figures and mathematical symbols.

Graphic recognition primary deals with recognizing the symbols, the line drawings, and the graphical language the document communicating. Graphical language is a collection of complex symbols, rules, and the relations between them in a document image. Hence a graphical document becomes a way of representing an information or a phenomenon using the graphical language. In order to recognize a graphical document, it is required to accurately identify all the graphical components, and the underlying syntax & semantics. One of such graphical document is architectural floor plan which is the heart of all the construction drawings and represents all the necessary information about the layout of a house. Floor plan is a blueprint of the house or any other building which contains complex graphical symbols and design in architectural language making it difficult for a common buyer to understand without having engineering knowledge. It is a mandatory document for a building which shows the design before the actual built and a buyer would definitely want to have a clarity about the property under construction by understanding it. However because of its complexity it is difficult for a common user to get a thorough understanding of it and hence an automation model for floor plan understanding and recognition is an immediate solution. A learning based model which could recognize every element of a floor plan and interpret it in a human readable form, provides a solution for architect and potential buyer for making the process of construction easier.

For understanding and describing a scene or any other image, a human brain functions in two steps, first understand its visual components by taking complete look and then explain it in the form of verbal speech or textual description. Hence, the most natural form of understanding and interpreting a natural or document image would be by generating a textual description for it in natural language. The systems which describe an image in textual form are called as Image Captioning systems. These captions could be single sentenced or multisentenced paragraph form. However, systems designed for generating captions for natural images, fail to perform on document images because of the different nature of document images. Graphical documents images lacks information at every pixel when compared to natural images and are very sparse in nature. Hence, special algorithms are required to be designed for document images to recognize them and generate a holistic description. Image captioning system are essentially divided into two categories.

- Visual feature based image captioning
- Retrieval based image captioning

The visual feature based image captioning system functions in a way human would describe a scene. It extracts visual features from the image by recognizing objects, faces, attributes, scenes using the computer vision techniques and use natural language generation (NLG) methods to narrate them. These NLG methods could be template based, n-gram or grammar rules for generating description from the identified visual features. The other category of solutions take image to text generation as a retrieval problem. They look for similar images in the database and the new description is generated by re-using the descriptions for most similar images present in the database. However, these techniques are not efficient for describing a document image or a graphical image such as a floor plan. These images have very different properties than a natural image, hence there is a requirement of system which could deal with graphical documents in an efficient manner. In the next section, motivation for the proposed problem is discussed in detail.

10.2 Motivation of the problem

The motivation behind the problem of understanding a floor plan and generating its textual narration originated from the present day requirement of real-estate industry and online-rental platforms. In modern times, finding the required property visiting door to door, has become a thing of the past. Hence, online rental portals are gaining popularity, since required properties can be searched with a click of the mouse. The web-based real estate business is not only useful for the potential buyers, but also for the sellers who are looking buyers for renting or selling their properties. These web-based portals not only have pictures of the house but there are categorical details for the same. However, given the bulk posting of ads for housing rentals, it is a difficult task for website administrators to observe all the details through images and manually write a comprehensive detailed description which cover all the image connecting

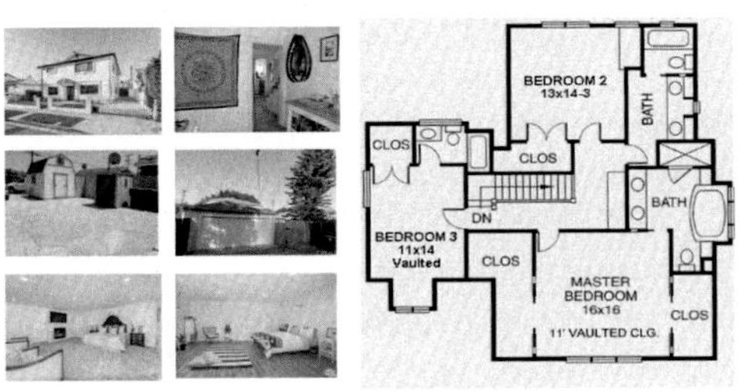

FIGURE 10.2

Motivation of the problem statement.

This house has a living room located n the north direction. The connected rooms are bathroom and hall. There is a kitchen in the south east direction. The connected rooms are hall and a sitting area.
(a) Indoor Navigation
(b) Textual Description

FIGURE 10.3

Indoor navigation in smart buildings using textual description.

them with each other. However, looking at the scene images for a house, it is difficult for a system to infer the global connections between each image and generate a holistic description. A floor plan of a building is the heart of all the construction drawings and gives the best representation for a house or a building. It captures all the intricate details which needs to be explained while giving a description about the property. Hence, the proposed problem targets this requirement to automatically generate a holistic description of a property, through its floor plan image.

Fig. 10.2 depicts the motivation behind the work presented in this chapter. Given the floor plan of a house or any other building, its narration in textual form can be generated, which may be useful in describing it in online rental platforms, or used as indoor navigation for robots or visually impaired.

In the context of smart building in modern era, textual description is a very useful tool to generate a navigation path of a building. Fig. 10.3 shows an illustration of indoor navigation in smart building which gives a navigation path using the textual description for the floor plan of the building.

Thus, upon analyzing the existing work in the area of indoor space understanding and architectural floor plans, it was observed that online rental platforms, buyers and sellers are ubiquitous and hence automating their requirements is the need of the hour. Hence, the work proposed in this chapter focuses towards building a bridge between the modern age requirements and part of their solutions.

10.3 Problem statement

The goal of the work proposed in this chapter is to design a floor plan understanding and interpretation system to perform the following task:

Given the floor plan of a house or any other building, a system should be able to understand the image and generate a textual narration in natural language (refer Fig. 10.2).

As proposed in this chapter, understanding of indoor spaces and their interpretations, the input to the system would be a floor plan of indoor spaces and the expected output would be its interpretation in natural language as written by a human. Appropriate feature is proposed which represent an entire floor plan to capture the intricate details in the description. Also, methods are proposed to generate a description which is close enough to human written sentences capturing all the details and grammatically well placed.

10.3.1 Brief description of the work done

The proposed work in this chapter, provides a solution to generate a textual narration of a floor plan image in natural language using machine learning techniques. This method uses a multistaged pipeline for understanding a floor plan image, where each stage identifies a different component. Signature based decor characterization is used for decor identification, and classification. This work proposes a machine learning based scheme for room label identification in contrast with the state of the art techniques, which make use of OCR based methods for identifying room labels. The work proposes a feature vector Bag of Decor (BoD) which gives a room wise floor plan representation. BoD is essentially a sparse vector which captures the frequency count of each decor object present in a room of a floor plan. Floor plan image is very different from a natural image where each pixel may not carry useful information. Hence, special features are required to generate a meaningful representation of a floor plan image. BoD is used with the classical machine learning classifiers and Artificial Neural Network (ANN) for classifying room images into 5 predefined classes which are BEDROOM, BATHROOM, HALL, LIVING ROOM, AND KITCHEN. This chapter discusses step by step solution for understanding each major component of the floor plan, by describing room semantic segmentation, decor identification and room label classification. In the next section, several state of the art methods for graphic recognition, floor plan analysis, symbol spotting and image to text generation schemes are discussed. The novel contributions discussed in this chapter are listed below:

- A framework for generating textual interpretations from floor plan images is proposed.
- Novel feature vector Bag of Decor (BoD) is proposed which compactly represents the floor plan image.
- Classical machine learning classifiers and Artificial Neural Network are used for generating region keywords for floor plan image.

- Holistic interpretation for an entire house is generated which captures all the required details.

10.4 Literature survey

In this section, the state of the art in area of document image processing in the context of graphic recognition is discussed. With the growing number of digital documents these days, graphic recognition has become an important task and gained popularity. This section discusses the work related to graphic recognition, which includes graphical documents such as circuit diagrams, engineering drawings, architectural diagrams, musical notations, mathematical operators etc. These documents are digitally created or scanned version of paper based documents. This section discusses, related work for floor plan recognition, symbol spotting in graphical documents, interpretation generation from images in textual form. Further, it will discuss the popular metrics used for evaluating these generated description.

10.4.1 State of the art in graphic recognition

In the DIA research area, graphics recognition is widely explored. In the literature, authors have explored various learning and inference based methods for localizing and classifying graphical symbols. In this context authors in the work [2], have provided a survey in graphical symbol spotting in a variety of documents images, and discussed general issues addressed in this area. In the work in [3], authors have given a review for spotting symbols and words in document images and their various geometric descriptors. In the document, [4], author have presents algorithm to separate text and graphical regions from engineering drawings by erasing graphical components and leaving only text by performing geometric operations on graphical components. In the document proposed in [5], authors have proposed an algorithm to detect dashed line segments by comparing the slopes of head and tail of each segment of the same. In another work, [6], authors have proposed a method to extract character from gray scale images instead of binary image to save information loss. The work in [7], has presented an algorithm for text string separation from graphics in document images by extracting connected components and using Hough transform for grouping them together. In the area of text localizing, authors in, [8] have spotted and recognized words in images by representing both word and text strings in a common vectorial space and casting recognition and retrieval as a nearest neighbor problem. In the work, [9], authors have extended previous work by discussing the right choice of threshold and postprocessing steps for extracting text components. The work presented in [10] describes various encoding schemes and geometric features for line drawings representations. In another work [11], text extraction in document images is done by grouping most dense group of points in the image and finding connectivity between them. In another work, [12] authors present a method to vectorize graphics in line drawings by separating each layer of the input binary image, followed by segmentation of each layer by random sampling based method.

10.4.2 State of the art in floor plan analysis

Floor plan recognition is one of the widely explored application due to its variety of applications in various real world areas. Hence, in the recent years, numerous systems have been proposed which explore floor plan recognition for tasks such as, retrieval, 2D to 3D reconstruction, graphical elements recognition, etc. With the development in recent artificial intelligent algorithm, the existing approaches can be enhanced by making them more accurate and efficient. However, working with the architectural documents has always been challenging due to inherent heterogeneity in these plans and lack of domain knowledge. In the following paragraph, the most recent techniques for floor plan recognition and generating narration from them using classical machine learning methods and recent AI based models are discussed.

The work in, [13] proposed a system to convert hand drawn floor plans drawings to CAD format by proposing solutions to positional ambiguity and structural distortions in hand drawn floor plans. In the same line [14] present a complete system for analyzing the architectural drawings, by low level preprocessing and recognizing symbols and later performing a 3D reconstruction of the plan. Also, in the work [15] authors proposed an end to end system to analyze architectural drawings by their segmentation, vectorization, detection of arcs for specific symbols and structural and symbol analysis.

In another work, [16] authors have performed text recognition in floor plan images after removing the walls and detecting connected components. In the same line, authors in [17] have introduced method for preprocessing the floor plan image by differentiating walls with their thickness and extracting other symbols. As continuation, in the work [18] authors have performed room detection and labeling in floor plan images by taking polygon approximation over detected walls and closing gaps and assigned labels using OCR. In the work proposed in, [19], authors performed image segmentation on ancient topographic maps and floor plans by removing non textual part and using OCR. In the similar line, [20] propose a statistical patch based segmentation method which is used for detecting the symbols and walls, and graph based method is explored for identifying rooms in the floor plans. In [21] this work, authors detected rooms by recursively decomposing image into nearly convex regions after detecting walls and doors in the floor plans. In the work [22], authors have introduced a new floor plan dataset and used fully convolutional network for wall segmentation. In work in, [23], present an unsupervised method to detect walls in floor plans with different textures without the use of any annotated data. In [24] authors have proposed a new floor plan dataset and CNN based algorithm to retrieve similar floor plan images from dataset. In the work [25] room type and shape identification is done using linear regression model. In another work [26] recognize and build vector representations of the scanned floor plans by using statistical methods and deep learning models for semantic segmentation and decor symbol detection.

Table 10.1 Details of publicly available existing floor plan datasets.

Floor plan Datasets		
Dataset	**Count**	**Remarks**
CVC-FP [28]	122	4 Subcategories, varying in wall textures, to study graphical notations in floor plans
FPLAN-POLY [1]	42	Used for floor plan analysis and room analysis
SESYD [33]	1000	100 layouts/ class, differ in arrangement of symbols, used for symbol spotting tasks
ROBIN [24]	510	Used for retrieval and symbol spotting tasks
CubiCasa5K [31]	5000	Used for floor plan object detection tasks
BRIDGE [30]	~ 13000	Used for information extraction and description generation

The work proposed in [27] proposed two training strategies, separate training and the use of a weighted loss function, to recognize and parse complex floor plan designs.

10.4.3 Publicly available floor plan datasets

In the literature, the publicly available datasets are: ROBIN [24], CVC-FP [28], SESYD [29], BRIDGE [30], CubiCasa5K [31], and FPLAN-POLY [32]. Table 10.1 describes the various datasets available in literature with the sample of floor plan images in them and purpose. The ROBIN dataset is a handcrafted dataset, primarily constructed for retrieval purpose, contains 510 samples of floor plan images with 3 broad categories based on several rooms. Another popular dataset in floor plan images is SESYD which is synthetically generated for symbol spotting task, which has 1000 samples of floor plans divided into 10 classes. Also, CVC-FP contains 122 samples of floor plans divided into 4 different layout classes, which has scanned documents. Another dataset, FPLAN-POLY, contains 42 floor plan vectorized images, and its primary purpose was to evaluate different symbol spotting method. Another dataset CubiCasa5K contains 5000 floor plan images annotated into 80 floor plan object categories and targeted to recognize different elements in the floor plan. The BRIDGE dataset contains 13000+ images and corresponding annotations, which include decor symbol annotations, region-wise captions, and paragraph based annotations. This dataset was primarily created for extracting information from floor plan images and generating textual descriptions from them. Table 10.1 contains the details of some publicly available floor plan datasets.

10.4.4 Symbol spotting in document images

Symbol spotting in floor plan images explores the information extraction from floor plans for understanding its components. Symbol spotting in document images has been explored using hand-crafted features and classical machine learning algorithms. The proposed work in, [34] presents a scheme for symbol spotting by subgraph matching and reduced computational complexity by using graph serialization. In another work, [35] authors have introduced a new image representation as "integral image" and a learning algorithm which selects small number of critical visual features for efficient classification. In the same context, [36] have posed symbol spotting as a substitution-tolerant subgraph isomorphism problem formulating it as integer linear programming. In another work, [32], authors have performed symbol spotting in floor plans by converting them in Region Adjacency Graphs where each symbol is represented by a node. In the work [37] authors have performed symbol spotting in architectural line drawings by partitioning them into shapes and finding salient convex groups of geometric primitives. In another work in, [38], dynamic time wrapping based, rotation invariant algorithm is presented for handwritten symbol recognition. In [39] the work authors proposed a exact and inexact graph matching network based on Messmer's network, where symbols are represented as constraint on segments and arcs of symbols. In the same like, [40], proposed a method to recognize hand written symbols in architectural drawings by finding a best fit between input image of symbol and deformable templates of symbols. In the context of symbol recognition, [41] authors have proposed a recurrent convolutional neural network based baseline for recognizing hand written music scores. In [42] authors proposed method for graph based representation of document images and spotting, recognizing symbols as subgraphs of that document.

10.4.5 Image description generation

Describing images in text is a challenging task in computer vision and natural language processing, which has been explored using traditional techniques of language generation such as template based, retrievals, n-grams, grammar rules, etc. and recent deep neural networks for example RNN, LSTM, GRU, etc. These methods use the features of image modality by extracting information related to image using conventional methods of object/scene recognition or CNN based architectures. Some of the initial work in this direction were proposed in the documents, [43], [44], [45], [46] which are using computer vision methods for extracting features or visual information from image and generated textual description using retrieval and n-gram techniques. While using deep neural networks, caption generation for natural images, Densecap [47] has proposed an algorithm which generates region wise captions in natural images. Prior to Densecap, image captioning was dealt over the entire image instead of regions.

Table 10.2 describes the salient properties of different state of the art methods in image description generation using classical machine learning methods and deep learning methods. The classical machine learning schemes uses conventional classifiers in computer vision to extract attributes from images. The textual description generation scheme varies from retrieval based method to similarity matching and graph label prediction. The deep learning approaches use end to end pipeline, where CNN acts as encoder and used for extracting feature while variants of RNN act as decoder, generating textual description for images.

Table 10.2 Different state of the art schemes in image to text generation.

Work	Category	Generation technique
[43]	Classical machine learning	Similarity Mapping between text and image attributes
[44]	Classical machine learning	Conditional Random Field and graph labeling prediction
[45]	Classical machine learning	Web-scale n-grams
[46]	Classical machine learning	Caption generation by retrieval
[47]	Deep learning	End to End encoder decoder based

10.4.6 Evaluation of text generation

Evaluation of the generated description from images is an important step to identify the accuracy of the generation. For that purpose several metrics for example BLEU (BiLingual Evaluation Understudy) as proposed in [48], ROUGE (Recall Oriented Understudy of Gisting Evaluation) as proposed in the document [49], METEOR (Metric for Evaluation of Translation with Explicit ORdering) presented in the document [50], CIDEr (Consensus-based Image Description Evaluation) as proposed by [51], SPICE (Semantic Propositional Image Caption Evaluation) as proposed by [52], BERTscore as presented by [53], and their several variations have been proposed in the existing literature. In another work presented by [54], authors provided a correlation between automatic metrics and human judgments, using previously mentioned metrics and their variants. These metrics compare the generated text with the human written text on parameters such as, uni-gram, bi-gram or n-gram and give the similarity score for each token between candidate sentence and reference sentence in terms of precision, recall, and F-score.

10.5 Descriptive narration generation from floor plan images

In this chapter we propose a scheme of generating descriptive narration of floor plan images in text using classical machine learning methods. Description generation from natural image is a widely explored research area which narrates the contents in a natural images and its subjectivity. However, narrating a document image or a graphical image in natural language, still remains an open ended area. A graphical document image is very different from natural images in terms of the subjectivity involved and information at each pixel level. Hence, conventional description generation schemes are not very effective on document images. Learning based systems require an internal representation for an image in the form of features extracted. There are various features in literature such as SIFT, SURF, for representation information in a natural images. However, document images lack information at every pixel, hence conventional features fails to provide a unique representation for floor plan images. There

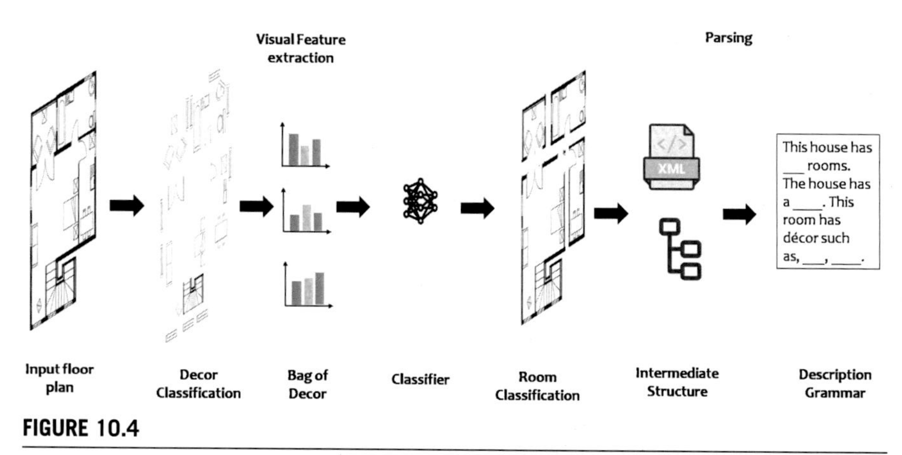

FIGURE 10.4

Framework for the proposed description generation scheme.

are several components in graphical images depicting different information and each component require separate attention and dealt in a different manner. Hence, in this work we propose a feature Bag of Decor (BoD) which provide a unique representation for each floor plan image. This feature captures the frequency count of each component present inside a room boundary and gives a unique representation for a floor plan image. Hence BoD is essentially a sparse matrix representation of a floor plan where each row vector is a representation of room having frequency count of the decor elements. This chapter proposes an end-to-end framework for extracting information from a floor plan image in a multistaged manner and generates semistructured textual narration for a floor plan image.

10.5.1 System overview

Fig. 10.4 shows the framework diagram for the proposed work. The proposed multistaged scheme works in several stages where first stage localizes and classifies the decor components. Further to which feature representations are generated for the floor plan image which is the proposed Bag of Decor. Bag of Decor is generated for each room in the floor plan after semantic segmentation. The feature representation for the floor plan images is used for appropriate classifiers and room labels are learnt. The visual information extracted at each step is stored in an XML format and XML tree is parsed to generate a semistructured textual narration for each floor plan image. This textual narration/description can be converted to text to speech synthesis system making it useful for navigation for visually impaired people. This is also useful for a customer who is willing to buy a house, to understand the given floor plan in a human readable form. For validation purpose, we have text annotated the dataset ROBIN by collecting 4 different descriptions from volunteers in order to compare the machine generated descriptions with human written descriptions.

10.5.2 Room annotation learning model

In day to day life, for understanding a visual scene and interpreting in another modality a human brain requires to look at every part carefully, understand it and then speak or write about it. Similarly, in order to understand a floor plan image, it is important to look at every component of it and understand it by learning methods or rule based methods before describing them. Rooms/regions are undoubtedly the most important constituent of a floor plan and to understand an entire plan, it is important to infer each room name correctly. In literature, there are schemes available which use OCR based solutions to detect room names/labels from annotated floor plan images. However, with the growing amount of data and variety in representation of floor plan images, it is very difficult to use a basic OCR system to predict room labels. Also, in case of un-annotated floor plans, OCR based solutions become insufficient to detect the room names. Hence, there is requirement of learning based methods, which could represent the entire floor plan with features and predict the room names accordingly. In this work, a machine learning based system is proposed which use BoD feature to represent a floor plan and use Artificial Neural Network (ANN) and other classification functions to predict the room names. The rooms are classified in 5 different classes, which are, BEDROOM, BATHROOM, HALL, KITCHEN, & LIVING ROOM. In the next sections, schemes described for segmenting the rooms from the floor plans, detection and classification of decor components, use of feature representation for floor plans and classification of room images, will be discussed in detail.

10.5.2.1 Room semantic segmentation

The technique proposed in [55] is adopted in this work, for the room segmentation task. Fig. 10.5 depicts the entire segmentation process on real world floor plan image. Walls in a floor plan are detected by performing morphological closing operation on the input floor plan image. To delineate room boundaries, doors are detected using scale invariant features proposed in [56] and gaps are closed in wall image corresponding to the door locations. In order to extract the rooms, the connected components were identified in the wall image by applying flood fill technique. The obtained connected components are the required rooms in the floor plans and their locations are obtained accordingly. Also, the areas are calculated for respective rooms (polygon area), converted them into square feet (taking 100 pixels = 1 feet) and all the information obtained, that is neighborhood, room area, room location coordinates, are stored in a separate data structure. Fig. 10.5(c), shows the segmented rooms in labeled colors, while Fig. 10.5(b) describes the extracted information from Fig. 10.5(a).

10.5.2.2 Decor characterization

In this section, decor characterization algorithm is discussed in detail. Fig. 10.6 shows the 12 decor symbols used in the dataset [24]. The technique proposed in [55] is improved by modifying the sequence of the morphological operations. The method proposed in [55] uses a signature based method for decor identification, where the

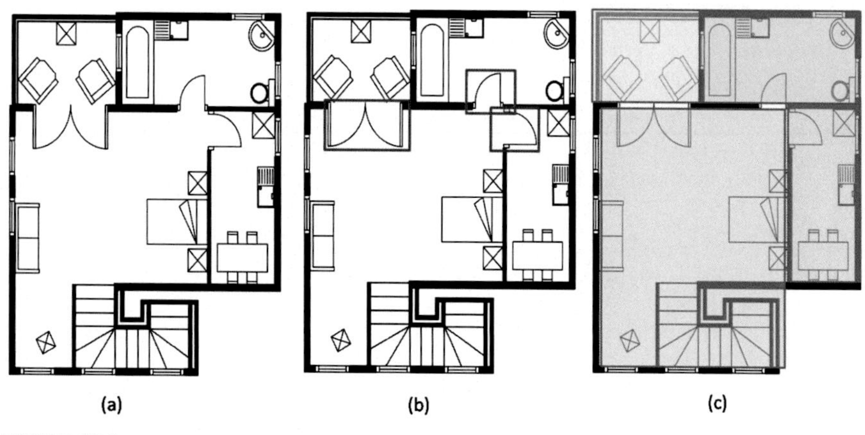

FIGURE 10.5

Schematic diagram of Bag of Decor feature vector for two room image samples.

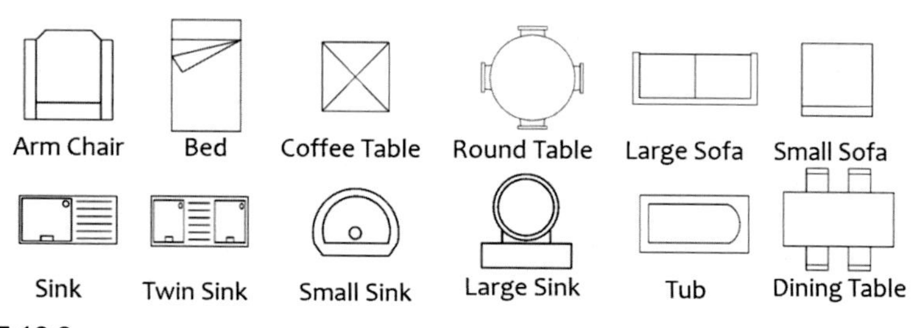

FIGURE 10.6

Twelve classes of Decor models used in the experiments.

signature- Unique Decor Identifier (UDI), is a feature for each decor item. The feature used is a normalized area ratio of largest three components of a decor symbol for classification and characterization of decors. The technique is improved by considering the variation in the symbols across different floor plan images and collecting 10 different signatures for each symbol. This step is followed by taking a mean over them (symbols with different orientations), as shown in Eq. (10.1), where, F_i is the mean UDI for decor i & f_j is UDI for one of the jth symbol taken across the dataset and stored them in a signature library ($\mathscr{F}$). Before classification, the symbol image is preprocessed by applying a sequence of morphological operations (erosion and dilation), so that the symbol do not have broken lines followed by application of blob detection over the image. Each blob is taken from the image and decor symbol is compared for its UDI with the available UDI signature in the library, where closest one is classified in its respective category. The proposed modification in the method, improved the classification accuracy for some symbols. The signature of the decor symbols if represented by a proposed feature Unique Decor Identifier (UDI), calculated by Algorithm 1. This UDI feature is a set of area ratios of three largest connected

Algorithm 1 UDI Computation.

1: **procedure** SIGNATURE($\mathscr{I}$)
2: $\mathscr{C} = CC(\mathscr{I})$ ▷ CC:Connected components
3: $Count = |\mathscr{C}|$ ▷ | |: Cardinality of connected components
4: **for** $k = 1$ to Count **do**
5: $A_k = Area(c_k)$, where $c_k \in \mathscr{C}$
6: **end for**
7: $\mathscr{A} = Sort_{desc} A_k$
8: $\mathscr{F}(\mathscr{I}) = \{(\mathscr{A}_1/\mathscr{A}_3), (\mathscr{A}_2/\mathscr{A}_3), 1\}$ ▷ $\mathscr{F} : Signature$
9: **end procedure**

components in a decor symbol. The decor library is created using the signature function by computing UDI of 10 different symbols and taking an average over it. The averaging of UDI over different floor plan images and decor symbols and storing into symbol library is governed by following equation. This modification in the algorithm captures the variability in the decor symbol representation over the entire floor plan image dataset.

$$F_i = \sum \frac{f_j}{10}, \quad \text{where } j = 1...10, \ i = 1...12 \tag{10.1}$$
$$\mathscr{F} = \{F_1, F_2, ..., F_{12}\}$$

The decor characterization method proposed, calculates UDI feature of the decor items in sample room images and compares it with the UDI feature present in decor symbol library. The decor having closest UDI with decor item present in the library, corresponding decor class is assigned to it. The cropped room image is given as input and walls are removed from it. After that morphological filling is applied to join the broken lines of the symbols. Each decor symbol is cropped from the image and given for UDI identification, which is further compared with UDI present in decor library and closest decor is assigned to it. Before evaluating UDI of decor items, blob detection is applied over image and number of decor items present is calculated.

Fig. 10.7 shows qualitative comparison between the proposed technique and that proposed in [55]. Fig. 10.7(b)–(d) show the intermediate processing stages of the proposed improved model. Erosion operation is applied with structuring element of

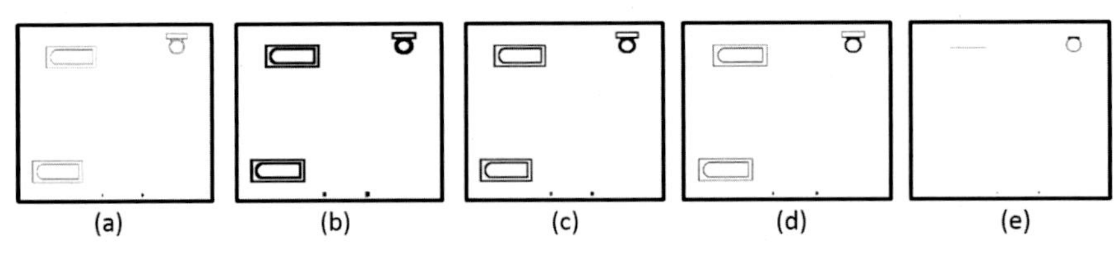

FIGURE 10.7

Qualitative comparison between ours and [55] for the task of Decor identification.

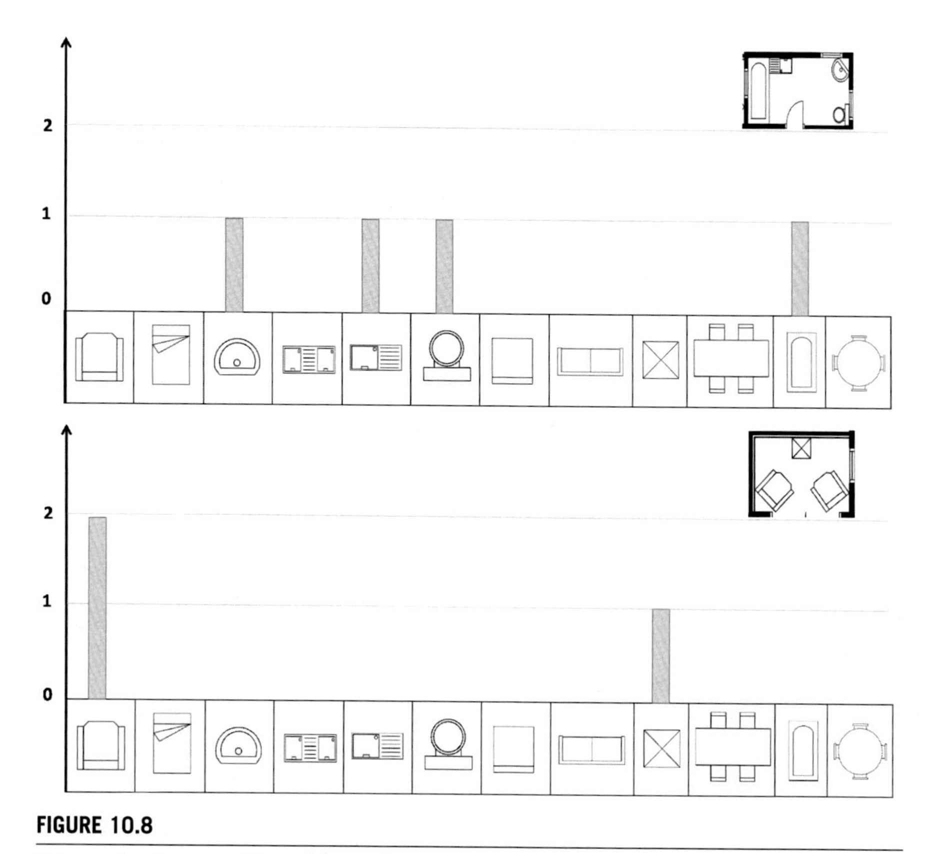

FIGURE 10.8

Schematic diagram of Bag of Decor feature vector for two room image samples.

size 8, which is followed by dilation operation. However, the sequence and number of times should these morphological operations be applied has an effect on the overall result. Fig. 10.7(e) shows the effects of over application of filling operations which results in fading of symbols.

10.5.3 Bag of decor (BoD)

The decor detection is performed for all the individual room images, followed by feature extraction for each room image with the proposed feature Bag of Decor (BoD). Fig. 10.8 shows a schematic representation of BoD vector for two example room images in a floor plan. The sparse vector contains the frequency count for 12 decor objects in ROBIN floor plan dataset. Each cell of the vector represents a decor object and stores its frequency count. BoD represent the entire floor plan images with a matrix of dimension $N_r \times 12$, where N_r is the number of rooms in a floor plan image. Put up an example and explain that.

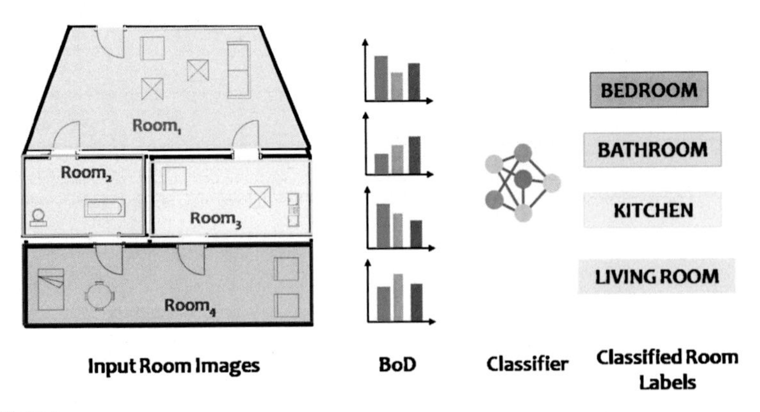

FIGURE 10.9

Schematic diagram of the room image classification process for a 4 room floor plan image.

10.5.3.1 Room annotation learning

Room labels for 1355 training samples are learned by BoD feature and classifier. Fig. 10.9 shows a schematic representation of the entire learning process. Room samples from the floor plans are labeled and used for training the classifier. Several classifiers including Artificial Neural Network and variants of SVM are experimented for training. For each room images taken from floor plan image, BoD classifier is generated and machine learning models are trained. The classifier with the best training accuracy is taken for testing purpose with new floor plan images.

10.5.4 Semistructured description generation

In the previous stages, all the visual information is extracted from the floor plan images and stored in XML format. This XML tree is parsed in order to generate the description of the floor plan. There is information such as room name, area, neighboring rooms, global position and the types of decor present inside the room along with their relative position in the room which is extracted by parsing the XML tree. For generating semistructured description from the visual element information we have defined a sentence model with 6 production rule (see Table 10.3).

Table 10.3 Sentence model based on proximity.

Sentence	Rule
S_1	This floor plan has N_r rooms
S_2	There is **DT** O_i
S_3	It has an area of **AREA**
S_4	Its neighboring room{s} **AUX** NR_j
S_5	It is located in the **LOC**
S_6	This room has {**C** $D\{s\}$ at the **DLOC**}$_k$

The first sentence (S_1) of description for every floor plan is a general sentence stating the number of rooms (N_r) in the floor plan. In S_2, DT is a determiner, which takes its value from the set $\{a, an\}$. Also, O_i is an object, which takes its value from level 1 nodes (Room names). Here value of the index i varies from 1 to N_r. In S_3, AREA takes its value from the $RoomArea$ tag in the XML. In S_4, the variable s takes its value from the set $\{s, \phi\}$, which is a proximity based value depends upon its previous word. Value s is chosen if the word in proximity (room) is a plural and ϕ otherwise. Also, AUX is an auxiliary verb, which takes its value from $\{is, are\}$, depending upon its proximity word and NR_j takes its value from Neighbors tag (neighboring rooms) of the XML file. Here, value of j varies from 1 to NN_r, which is the number of neighboring rooms. In the sentence S_5, LOC is the global position of room which takes its value from the predefined set of directions $\{North, North\ East, East, South\ East, South, South\ West, West, North\ West\}$ described by binning. In S_6, the value of k varies from 1 to DC i.e. decor count. Here, C is the count of individual decor item, D takes its value from the Decor tag in XML file, s takes its value from $\{s, \phi\}$, and $DLOC$ is the relative location of decor in the room which takes its value from the set of directions $\{North, North\ East, East, South\ East, South, South\ West, West, North\ West\}$ described by binning.

10.5.5 Experimental findings

We have performed our experiments on a platform with the following configurations: Intel core i5 (1st generation), with a 2.53 GHz processor (m460), 8 GB of RAM, Windows 7 operating system, and Matlab® $16a$.

10.5.5.1 Results of decor characterization

Table 10.4 shows the quantitative comparison between proposed, [55], [57] and our technique using LBP (local binary pattern) feature [58]. The maximum accuracy for a given symbol is highlighted using **bold** face numbers. It can be seen that performance is comparable between ours and [55]. However, for large sink and tub, a significant improvement in accuracy can be seen. Also for round table, there was no recognition using [55], while with our method accuracy of 82.35% could be achieved. Also in [57], recognition is 0 for many decor items with low accuracy for others. However, with our technique implemented using LBP feature, round table could be recognized with 100% accuracy, while comparatively much lower accuracy for others. Every decor item in each room image is classified in one of the 12 decor categories.

10.5.5.2 Results of room annotation learning

A forward pass neural network is trained with Sigmoid activation function, with 10 hidden nodes and 5 output nodes for our room annotation learning model. Different experimentation were performed with various classifiers. Table 10.6 shows the results of training and testing of room image samples taken from 2 datasets, ROBIN (R) [59] and SESYD (S) [33]. In the first column, training has been done using 1355 room

Table 10.4 Comparison of recognition accuracy (%) of proposed decor characterization vs others.

Symbol	[55]	[57]	[58]	Ours
Bed	**99.01**	80.39	44.60	98.5
Arm Chair	**100**	77.77	63.88	**100**
Coffee Table	99.15	0.004	11.44	**99.57**
Dining Table	**98.76**	77.77	66.66	**98.76**
Small Sofa	**100**	96.77	0.00	**100**
Large Sofa	98.06	0.009	65.44	**99.35**
Small Sink	83.33	0.00	72.22	**88.88**
Twin Sink	**95.23**	0.00	71.42	**95.23**
Sink	100	0.00	63.73	100
Large Sink	55.69	0.00	51.89	**67.08**
Tub	61.16	74.75	61.16	**97.08**
Round table	0.00	0.00	**100**	82.35

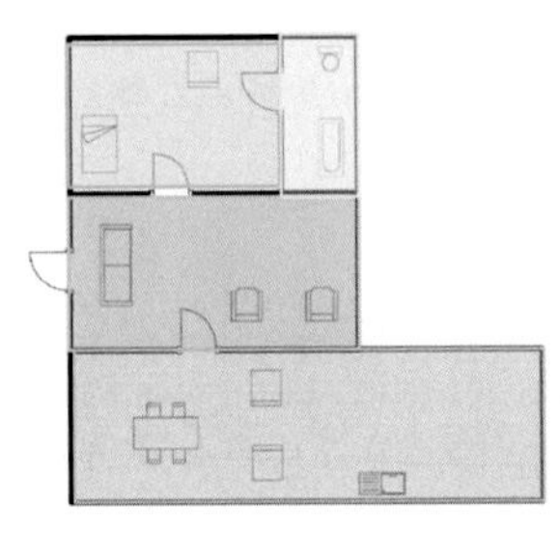

In this architectural floor plan there are 4 rooms. There is a BATHROOM. It has an area of 32.23 sq feet. Its neighbouring room is BEDROOM. It is located in the north east side of the house. This room has 1 tub in the south side of the room, 1 large sink in the north side of the room. There is a BEDROOM It has an area of 51.87 sq feet. Its neighbouring rooms are BATHROOM, LIVING ROOM. It is located in the north west side of the house. This room has 1 Bed in the south west side of the room, 1 small sofa in north side. There is a LIVING ROOM. It has an area of 83.26 sq feet. Its neighbouring rooms are BEDROOM, KITCHEN. It is located in the north side of the house. This room has 1 large sofa in west side of the room, 2 arm chair in the south west side of the room. There is a KITCHEN. It has an area of 140.67 sq feet. Its neighbouring room is LIVING ROOM. It is located in the south side of the house. This room has 1 dining table in east side of the room, 2 small sofa in north side of the room, 1 sink in south side of the room.

(a)

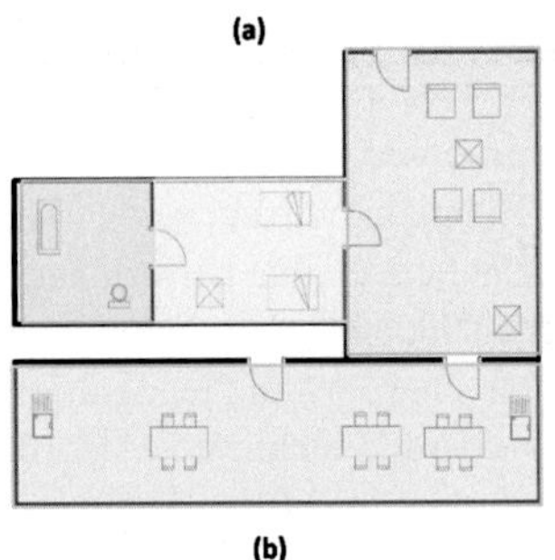

In this architectural floor plan there are 4 rooms. There is a LIVING ROOM. It has an area of 65.13 sq feet. Its neighbouring room is BEDROOM. It is located in the north east side of the house. This room has 2 small sofa in north side of the house, 1 coffee table in north side of the house, 2 small sofa in the north side of the house, 1 coffee table in south west side of the house.
There is a BEDROOM. It has area of 41.88 sq feet. Its neighbouring room is LIVING ROOM, BATHROOM. It is located in the north side of the house. This room has 2 bed in the west side of the house, 1 coffee table in the south east side of the house. There is a BATHROOM. It has area of 37.12 sq feet. Its neighbouring room is BEDROOM. It is located in the west side of the house. This room has a tub in the west side of the room, 1 large sink in the south east side of the room.
There is a KITCHEN. It has area of 138.76 sq feet. Its has neighbouring room is LIVING ROOM. It is located in the south side of the house. This room has 1 sink in east side of the room, 2 dining table in east side of the room, 1 dining table in west side of the room, 1 sink in the west side of the room.

(b)

FIGURE 10.10

Generated descriptions for four different floor plan images from ROBIN dataset.

images taken from ROBIN dataset, where linear SVM which gave highest training accuracy. In the second column, testing is done using samples taken from ROBIN. In the third column, test results are shown for SESYD samples, by trained model of ROBIN images which are comparatively low. Although, experiments are done by training the classifier using mixed samples from both datasets using 1940 images and

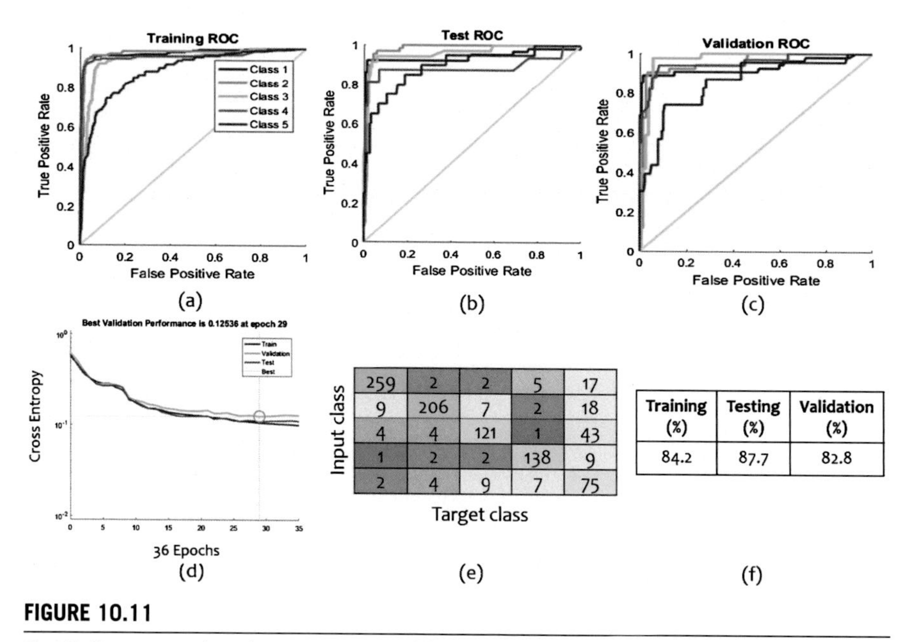

Training (%)	Testing (%)	Validation (%)
84.2	87.7	82.8

FIGURE 10.11

Qualitative comparison between ours and [55] for the task of Decor identification.

training and testing accuracies are shown in fourth and fifth columns respectively. This can be noted that testing accuracy is considerably enhanced with this model. Fig. 10.11 shows the experiment of room annotation classification with neural network showing the accuracy achieved in Fig. 10.11(f). Also Fig. 10.11(a)–(e) show the ROC curves for training, testing and validation, validation performance, and the confusion matrix respectively.

This can be seen that best validation is achieved at 30th epoch of the training cycle. It can be concluded from the test ROC that class 5 samples has minimum testing accuracy as the curve moves much away from the upper left corner and goes below the diagonal line (moving towards false positive rate axis). Also from the training ROC curve it can be seen that class 1 samples has maximum training accuracy as the curve remains concentrated in the upper left corner (moving towards true positive rate axis). Low training and testing accuracy for class 5 samples results from the low number of training samples it contained.

10.5.6 Results for description generation

10.5.6.1 Qualitative results

Rooms are classified into 5 classes using the trained model and give output as room labels. These labels are mapped with the information extracted from room segmentation process and stored in an XML file. This XML file is parsed and a description is generated. Fig. 10.10 depicts the resultant description of 2 floor plan images. To

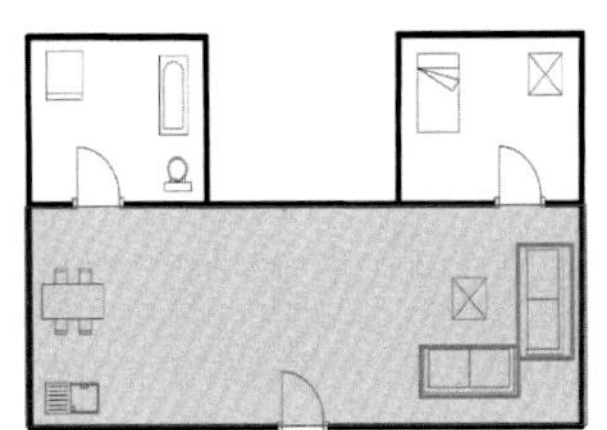

This architectural floor plan has 3 rooms. There is a BATHROOM. It has area of 9.21 sq. ft. Its neighbouring room is BEDROOM. It is located in the north west side of the house. This room has 1 small sofa in the north west side of the room, 1 tub in the north east side of the room, 1 large sink in the south east side of the room. There is a BEDROOM. It has area of 27.45 sq. ft. Its neighbouring rooms are BATHROOM and BEDROOM. It is located in the south side of the house. This room has 1 dining table in the west side of the room, 1 sink in the south west side of the house, 1 coffee table in the east side of the room, 1 bed in the east side of the room, 1 bed in the south east side of the room. There is a BEDROOM. It has area of 9.33 sq. ft. Its neighbouring room is BEDROOM. It is located in the north east side of the house. This room has 1 bed in the west side of the room, 1 coffee table in the north east side of the room.

(a) failure in correct object recognition

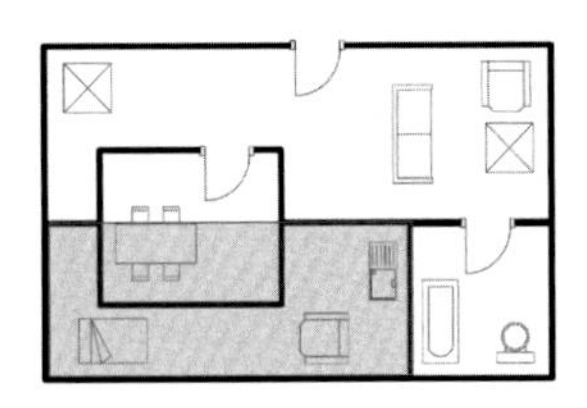

In this architectural floor plan there are 4 rooms. There is a ENTRY. It has area of 29.23 sq. ft. Its neighbouring rooms are HALL and BATHROOM. It is located in the north side of the house. This room has 1 coffee table at the west side of the room, 1 large sofa at the east side of the room, 1 coffee table at the east side of the room, 1 armchair at the north east side of the room. There is a HALL. It has area of 21 sq. ft. Its neighbouring rooms are ENTRY and BATHROOM. It is located in the south side of the house. This room has 1 bed in the west side of the room, 1 dining table in the north side of the room, 1 armchair in the south east side of the room, 1 sink in the west side of the room. There is a BATHROOM. It has area of 11 sq. ft. Its neighbouring rooms are ENTRY and HALL. It is located in the south east side of the house. This room has 1 tub in the west side of the room, 1 large sink in the south east side of the room.

(b) Failure due to asymmetry in the floor plan image

FIGURE 10.12

Two erroneously synthesized descriptions due to failure in correct object recognition.

make the synthesized description easy to understand for the reader, each description is presented in the following manner: (i) the first word of the first sentence about any room is in **bold** face, (ii) each room in a floor plan is highlighted with a color and the same color is used to highlight the room name in the first sentence about the room in the description. Information like name, area, global position in the floor plan, relative position of decors, and neighboring rooms in terms of its accessibility by a door is described for each room in the final output description, along with a room having a door opening to outside of the house is also described. For example, in Fig. 10.10(b), the four rooms are detected and described correctly by our proposed framework.

However, our proposed framework failed to perform on some floor plan sample. Fig. 10.12(a) and (b) depict two failure cases. Fig. 10.12(a) shows an example, where the failure is due to the error in classification of the decors. Also decor present in that room has been classified resulting incorrect classification of the current room. One incorrect object characterization of "large sofa" as "bed" has lead to incorrect room classification. In Fig. 10.12(b), the failure is due to the asymmetry in the floor plan. It can be observed that there is a room in the floor plan which is surrounded by the two rooms. However, in the proposed framework, the inner room is not recognized. As a result, in room segmentation the inner room has not been taken into consideration.

10.5.6.2 Quantitative analysis

We have compared the machine generated description of the floor plan with human written descriptions using 3 metrics, ROUGE [49], BLEU [48] and METEOR [50]. We have textual description of each floor plan by showing the volunteers a floor plan image and then making them write a description in their own words. The textual

Table 10.5 Performance analysis of synopsis synthesis using ROUGE score.

ROUGE	Average recall	Average precision	F score
ROUGE-1	0.5078	0.3397	0.1468
ROUGE-2	0.1820	0.1231	0.1468
ROUGE-3	0.0530	0.0377	0.0440

Table 10.6 SVM classifiers-Results of room label learning by support vector machine.

Variant	Training (R) (%)	Testing (R) (%)	Testing (S) (%)	Testing (R+S) (%)	Testing (R+S) (%)
Linear SVM	89.2	80.00	66.51	89	76.41
Quadratic SVM OVA	89.0	78.67	64.62	90.8	76.92
Cubic SVM OVO	88.6	79.56	55.19	88.7	74.70
Medium Gaussian SVM	88.7	78.44	58.96	89.3	75.21
Quadratic SVM OVO	88.2	80.44	65.57	89.7	75.90

Table 10.7 Performance analysis of synopsis synthesis using METEOR and BLEU score.

METEOR Score				
Average recall	Average precision	F1	F mean	Final score
0.55	0.22	0.33	0.45	0.18
BLEU Score				
0.0107				

description generated by our framework is then compared with the human written descriptions. Table 10.5 depicts the average recall, average precision and F score for ROUGE-1, ROUGE-2, ROUGE-3. As the value of n in n-gram comparison increasing, the ROUGE precision score decreases, which is also clear from Table 10.5. Since ROUGE-1, ROUGE-2, and ROUGE-3 use uni-gram, bi-gram, and tri-gram comparisons respectively, the decreasing nature of average precision is natural. Machine generated descriptions have a fixed pattern for words to be used and the information to be displayed. However, human written descriptions can have any sequence and use of words and phrases. Table 10.7 depicts the METEOR and BLEU score for the description generated, which demonstrates high correlation with human judgments.

10.6 Application to smart homes and buildings

Due to the substantial proliferation in the technology, home devices and appliances are becoming smarter and smarter. These devices not only include assistive appli-

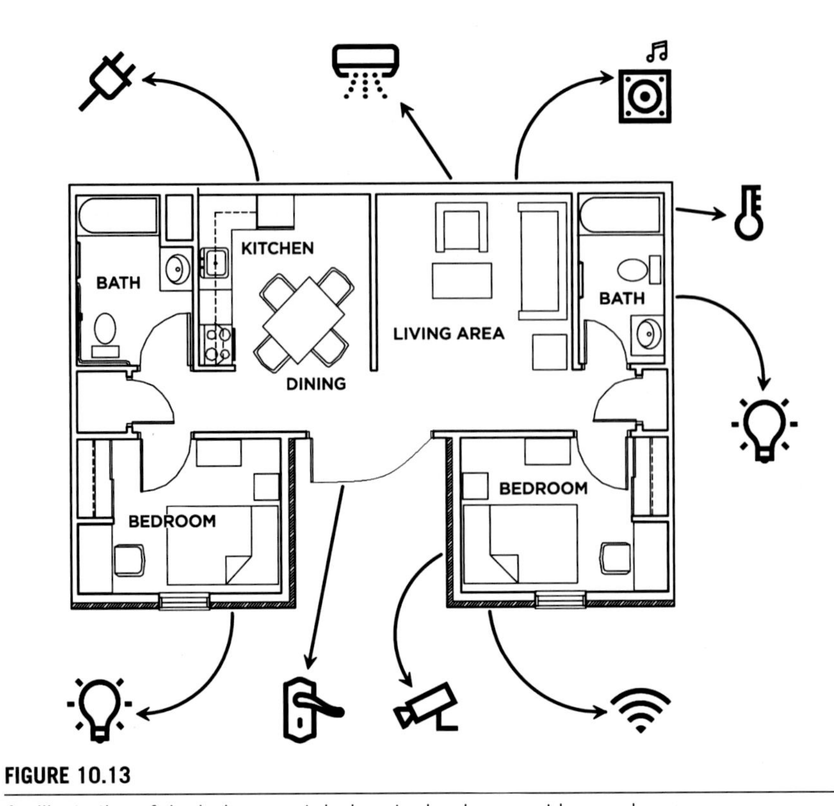

FIGURE 10.13

An illustration of deploying smart devices to develop smart home plan.

ances (such as washers, fridges, and ACs) but also safety and security systems. This scenario has led to an urgent need to automate the current home infrastructure to develop smart homes and thus led to home automatic revolution. The main bottleneck in developing smart homes is the unavailability of synergy among low cost, reduce energy, convenience to the home owner and proper interface to control these smart devices. The integration of the home floor plan plays a remarkable role in developing the smart homes to provide clear view on synergy to an architect and may provide better experience to the user. The benefits of integrating floor plan with the smart home layout can be summarized as follows:

- It helps in displaying all functions on a single floor plan view. This essentially will enable users to switch on/off smart devices in the exact location/room they need.
- Enable users to spot the faults while interacting with home appliances, just looking at the smart home layout integrated on floor plan.
- This will enable the owners/service providers to create a navigation map for iRobots, such as vacuum cleaning robot.
- It will enable developers to develop smart homes with a user friendly interface by avoiding hard to use tree structures of smart homes.

Therefore, it is essential to understand the floor plans to enable it effective usage in smart homes and buildings. An example can be seen in Fig. 10.13, where a single floor plan illustrated all the devices and appliances.

In a nut shell, understanding of building floor plans is not only useful in designing smart home but also to design smart buildings and enabling smart devices in them. Automatic understanding and recognition will enable user to see the home layout, eliminate mistakes when interacting with smart devices. It will also help visually impaired person to understand the layout of the home and interact with smart devices. In order to facilitate smart devices in a home, a floor plan is required to be visualized and understand. Special SDKs can be developed which will help user to configure IoT devices on the plan, and to visualize and interact with the devices on a graphical user interface. With automatic understanding of architectural drawings, these floor plans can be turned into interactive floor plans and user would be able to interact with the smart devices after their installation. With these interactive floor plans, smart security sensors and health sensors can be easily installed and managed by the user.

10.7 Conclusion

In this chapter, a scheme for generating textual narration for floor plan images is discussed. Floor plans are the blue-prints of the interior layout of a building. They capture the details for an indoor scene with all the global relations captured. With the advancement in real-estate business, most of the requirements related to buying and selling are served online. The web-based real estate industry makes every potential buyer and seller look-up their requirement on internet and going door-to-door has become a thing of past. Hence there is a requirement of having a textual narration of all the ad-posting of the houses or any other property. However, describing the indoor scene images without having an intermediate representation lacks the global connection between the scene images, hence, floor plan images are used as representation. In contrast with the state of the art schemes to understand floor plan, advanced machine learning based method is proposed with new feature representation Bag of Decor (BoD). BoD feature representation used with machine learning classifier gives state of the art accuracy in learning room labels for the floor plans. Also an improved decor characterization method is proposed. This chapter presents a semistructured description generation scheme which uses multistaged visual element extraction and generate a holistic description of the house through its floor plan image. In the future work, the framework for generating description from floor plans can be extended to an end-to-end learning pipeline. The generated descriptions can be more flexible in nature and more close to human written form.

References

[1] M. Rusiñol, A. Borràs, J. Lladós, Relational indexing of vectorial primitives for symbol spotting in line-drawing images, Pattern Recognition Letters 31 (3) (2010) 188–201.

[2] A.K. Chhabra, Graphic symbol recognition: an overview, in: IWGR, 1997, pp. 68–79.

[3] M. Rusiñol, J. Lladós, State-of-the-art in symbol spotting, in: SSDL, Springer, 2010, pp. 15–47.

[4] Z. Lu, Detection of text regions from digital engineering drawings, IEEE Transactions on Pattern Analysis and Machine Intelligence 20 (4) (1998) 431–439.

[5] C. Lai, R. Kasturi, Detection of dashed lines in engineering drawings and maps, in: ICDAR, 1991.

[6] O.D. Trier, T. Taxt, A.K. Jain, Data capture from maps based on gray scale topographic analysis, in: ICDAR, vol. 2, 1995, pp. 923–926.

[7] L.A. Fletcher, R. Kasturi, A robust algorithm for text string separation from mixed text/graphics images, IEEE Transactions on Pattern Analysis and Machine Intelligence 10 (6) (1988) 910–918.

[8] J. Almazán, A. Gordo, A. Fornés, E. Valveny, Word spotting and recognition with embedded attributes, IEEE Transactions on Pattern Analysis and Machine Intelligence 36 (12) (2014) 2552–2566.

[9] K. Tombre, S. Tabbone, L. Pélissier, B. Lamiroy, P. Dosch, Text/graphics separation revisited, in: DAS, 2002.

[10] H. Freeman, Computer processing of line-drawing images, ACM Computing Surveys 6 (1) (1974) 57–97.

[11] V. Yadav, N. Ragot, Text extraction in document images: highlight on using corner points, in: DAS, 2016.

[12] X. Hilaire, K. Tombre, Robust and accurate vectorization of line drawings, IEEE Transactions on Pattern Analysis and Machine Intelligence 28 (6) (2006) 890–904.

[13] Y. Aoki, A. Shio, H. Arai, K. Odaka, A prototype system for interpreting hand-sketched floor plans, in: Proceedings of the 13th International Conference on Pattern Recognition, 1996, vol. 3, IEEE, 1996, pp. 747–751.

[14] P. Dosch, K. Tombre, C. Ah-Soon, G. Masini, A complete system for the analysis of architectural drawings, International Journal on Document Analysis and Recognition 3 (2) (2000) 102–116.

[15] C. Ah-Soon, K. Tombre, Variations on the analysis of architectural drawings, in: Proceedings of the Fourth International Conference on Document Analysis and Recognition, 1997, vol. 1, IEEE, 1997, pp. 347–351.

[16] S. Ahmed, M. Liwicki, M. Weber, A. Dengel, Text/graphics segmentation in architectural floor plans, in: ICDAR, 2011.

[17] S. Ahmed, M. Liwicki, M. Weber, A. Dengel, Improved automatic analysis of architectural floor plans, in: ICDAR, 2011.

[18] S. Ahmed, M. Liwicki, M. Weber, A. Dengel, Automatic room detection and room labeling from architectural floor plans, in: 2012 10th IAPR International Workshop on Document Analysis Systems (DAS), IEEE, 2012, pp. 339–343.

[19] C.A. Mello, D.C. Costa, T. dos Santos, Automatic image segmentation of old topographic maps and floor plans, in: SMC, 2012.

[20] L.-P. de las Heras, S. Ahmed, M. Liwicki, E. Valveny, G. Sánchez, Statistical segmentation and structural recognition for floor plan interpretation, International Journal on Document Analysis and Recognition 17 (3) (2014) 221–237.

[21] S. Macé, H. Locteau, E. Valveny, S. Tabbone, A system to detect rooms in architectural floor plan images, in: Proceedings of the 9th IAPR International Workshop on Document Analysis Systems, ACM, 2010, pp. 167–174.

[22] S. Dodge, J. Xu, B. Stenger, Parsing floor plan images, in: 2017 Fifteenth IAPR International Conference on Machine Vision Applications (MVA), IEEE, 2017, pp. 358–361.

[23] L.-P. de las Heras, D. Fernández, E. Valveny, J. Lladós, G. Sánchez, Unsupervised wall detector in architectural floor plans, in: 2013 12th International Conference on Document Analysis and Recognition, IEEE, 2013, pp. 1245–1249.

[24] D. Sharma, N. Gupta, C. Chattopadhyay, S. Mehta, Daniel: a deep architecture for automatic analysis and retrieval of building floor plans, in: 2017 14th IAPR International Conference on Document Analysis and Recognition (ICDAR), vol. 1, IEEE, 2017, pp. 420–425.

[25] H.K. Mewada, A.V. Patel, J. Chaudhari, K. Mahant, A. Vala, Automatic room information retrieval and classification from floor plan using linear regression model, International Journal on Document Analysis and Recognition 23 (4) (2020) 253–266.

[26] I.Y. Surikov, M.A. Nakhatovich, S.Y. Belyaev, D.A. Savchuk, Floor plan recognition and vectorization using combination unet, faster-rcnn, statistical component analysis and Ramer–Douglas–Peucker, in: International Conference on Computing Science, Communication and Security, Springer, 2020, pp. 16–28.

[27] R. Zhu, J. Shen, X. Deng, M. Walldén, F. Ino, Training strategies for cnn-based models to parse complex floor plans, in: Proceedings of the 2020 9th International Conference on Software and Computer Applications, 2020, pp. 11–16.

[28] L.-P. de las Heras, O.R. Terrades, S. Robles, G. Sánchez, CVC-FP and SGT: a new database for structural floor plan analysis and its groundtruthing tool, International Journal on Document Analysis and Recognition 18 (1) (2015) 15–30.

[29] M. Delalandre, E. Valveny, T. Pridmore, D. Karatzas, Generation of synthetic documents for performance evaluation of symbol recognition & spotting systems, International Journal on Document Analysis and Recognition 13 (3) (2010) 187–207.

[30] S. Goyal, V. Mistry, C. Chattopadhyay, G. Bhatnagar, Bridge: building plan repository for image description generation, and evaluation, in: 2019 International Conference on Document Analysis and Recognition (ICDAR), IEEE, 2019, pp. 1071–1076.

[31] A. Kalervo, J. Ylioinas, M. Häikiö, A. Karhu, J. Kannala, CubiCasa5K: a dataset and an improved multi-task model for floorplan image analysis, in: Scandinavian Conference on Image Analysis, Springer, 2019, pp. 28–40.

[32] A. Barducci, S. Marinai, Object recognition in floor plans by graphs of white connected components, in: 2012 21st International Conference on Pattern Recognition (ICPR), IEEE, 2012, pp. 298–301.

[33] M. Delalandre, T. Pridmore, E. Valveny, H. Locteau, E. Trupin, Building synthetic graphical documents for performance evaluation, in: International Workshop on Graphics Recognition, Springer, 2007, pp. 288–298.

[34] A. Dutta, J. Lladós, U. Pal, A symbol spotting approach in graphical documents by hashing serialized graphs, Pattern Recognition 46 (3) (2013) 752–768.

[35] P. Viola, M. Jones, Rapid object detection using a boosted cascade of simple features, in: CVPR, 2001.

[36] P. Le Bodic, P. Héroux, S. Adam, Y. Lecourtier, An integer linear program for substitution-tolerant subgraph isomorphism and its use for symbol spotting in technical drawings, Pattern Recognition 45 (12) (2012) 4214–4224.

[37] N. Nayef, T.M. Breuel, Statistical grouping for segmenting symbols parts from line drawings, with application to symbol spotting, in: 2011 International Conference on Document Analysis and Recognition (ICDAR), IEEE, 2011, pp. 364–368.
[38] A. Fornés, J. Lladós, G. Sánchez, D. Karatzas, Rotation invariant hand-drawn symbol recognition based on a dynamic time warping model, International Journal on Document Analysis and Recognition 13 (3) (2010) 229–241.
[39] C. Ah-Soon, A constraint network for symbol detection in architectural drawings, in: International Workshop on Graphics Recognition, Springer, 1997, pp. 80–90.
[40] E. Valveny, E. Martí, Application of deformable template matching to symbol recognition in handwritten architectural drawings, in: Proceedings of the Fifth International Conference on Document Analysis and Recognition, ICDAR'99 (Cat. No. PR00318), IEEE, 1999, pp. 483–486.
[41] A. Baró, P. Riba, J. Calvo-Zaragoza, A. Fornés, From optical music recognition to handwritten music recognition: a baseline, Pattern Recognition Letters 123 (2019) 1–8.
[42] R.J. Qureshi, J.-Y. Ramel, D. Barret, H. Cardot, Spotting symbols in line drawing images using graph representations, in: International Workshop on Graphics Recognition, Springer, 2007, pp. 91–103.
[43] A. Farhadi, M. Hejrati, M.A. Sadeghi, P. Young, C. Rashtchian, J. Hockenmaier, D. Forsyth, Every picture tells a story: generating sentences from images, in: ECCV, 2010.
[44] G. Kulkarni, V. Premraj, V. Ordonez, S. Dhar, S. Li, Y. Choi, A.C. Berg, T.L. Berg, Babytalk: understanding and generating simple image descriptions, IEEE Transactions on Pattern Analysis and Machine Intelligence 35 (12) (2013) 2891–2903.
[45] S. Li, G. Kulkarni, T.L. Berg, A.C. Berg, Y. Choi, Composing simple image descriptions using web-scale n-grams, in: CoNLL, 2011.
[46] V. Ordonez, G. Kulkarni, T.L. Berg, Im2text: describing images using 1 million captioned photographs, in: NIPS, 2011.
[47] J. Johnson, A. Karpathy, L. Fei-Fei, Densecap: fully convolutional localization networks for dense captioning, in: CVPR, 2016.
[48] K. Papineni, S. Roukos, T. Ward, W.-J. Zhu, Bleu: a method for automatic evaluation of machine translation, in: ACL, 2002.
[49] C.-Y. Lin, Rouge: a package for automatic evaluation of summaries, in: ACL, 2004.
[50] M. Denkowski, A. Lavie, Meteor 1.3: automatic metric for reliable optimization and evaluation of machine translation systems, in: WMT, 2011.
[51] R. Vedantam, C. Lawrence Zitnick, D. Parikh, Cider: consensus-based image description evaluation, in: Proceedings of the IEEE Conference on Computer Vision and Pattern Recognition, 2015, pp. 4566–4575.
[52] P. Anderson, B. Fernando, M. Johnson, S. Gould, Spice: semantic propositional image caption evaluation, in: European Conference on Computer Vision, Springer, 2016, pp. 382–398.
[53] T. Zhang, V. Kishore, F. Wu, K.Q. Weinberger, Y. Artzi, BERTScore: evaluating text generation with BERT, arXiv preprint, arXiv:1904.09675, 2019.
[54] D. Elliott, F. Keller, Comparing automatic evaluation measures for image description, in: ACL, 2014.
[55] D. Sharma, C. Chattopadhyay, G. Harit, A unified framework for semantic matching of architectural floorplans, in: ICPR, 2016.
[56] C. Harris, M. Stephens, A combined corner and edge detector, in: Alvey Vision Conference, vol. 15, Citeseer, 1988, pp. 10–5244.
[57] M.K. Hu, Visual pattern recognition by moment invariants, computer methods in image analysis, IRE Transactions on Information Theory 8 (1962).

[58] T. Ojala, M. Pietikainen, T. Maenpaa, Multiresolution gray-scale and rotation invariant texture classification with local binary patterns, IEEE Transactions on Pattern Analysis and Machine Intelligence 24 (7) (2002) 971–987.
[59] D. Sharma, N. Gupta, C. Chattopadhyay, S. Mehta, DANIEL: a deep architecture for automatic analysis and retrieval of building floor plans, in: ICDAR, 2017.

Index

G

H

I

J

K

L

M

N

O

P

Q

R

Printed in the United States
by Baker & Taylor Publisher Services